Pvt. Ltd.

# Advances in Chilli Research

*Editors*

**Rajesh Kumar** **A.B. Rai** **Mathura Rai**

*Indian Institute of Vegetable Research*
*P.O. Jakhini-Shahanshahpur,*
*Varanasi 221 305*

**H.P. Singh**

*Indian Council of Agricultural Research,*
*Krishi Anusandhan Bhawan II, Pusa,*
*New Delhi 110 012*

**2010**

**Studium Press**

# Advances in Chilli Research



ISBN: 978-93-80012-23-0

*Citation:*
*Kumar Rajesh, Rai AB, Rai Mathura and Singh HP (Eds.) 2010. Advances in Chilli Research, Indian Institute of Vegetable Research-Studium Press (India), pp. 272.*

*Published by:*

**Studium Press (India) Pvt. Ltd.**
*4735/22, 2nd Floor, Prakash Deep Building*
*(Near Delhi Medical Association),*
*Ansari Road, Darya Ganj, New Delhi-110 002*
*Tel.: 23240257, 65150447; Fax: 91-11-23240273;*

*Printed at:*

**Salasar Imaging Systems**
*Delhi-110035 (India)*

# Foreword

Among the vegetables, chilli is an important crop grown in India for domestic consumption and export. Green and dried chillies have great demand in domestic markets, whereas red chillies have immense export potential. Though, chillies originated in South America, India is treated as secondary centre of diversity due to its long history of cultivation, rich genetic variability for fruit size, shape, pungency, colour, and resistance to biotic and abiotic stresses. Besides, non-pungent chillies especially Byadagi type of Karnataka are also commonly grown in several regions of India. The bell/sweet pepper is also grown in different parts of sub-tropical condition during winter season and in hilly region throughout the year for domestic consumption and export.

In fact, useful information on different aspects of research and development in chilli crop has been generated by various research institutions and SAU's in the country, but they are mostly scattered in various journals, proceedings, reports, etc. Considering an importance of the crop and to analyze the current knowledge of research activities to develop strategies for future research, a Brain Storming Meeting was organized at Indian Institute of Vegetable Research, Varanasi.

I am happy that the information presented during brain storming session has been compiled in the form of book entitled **"Advances in Chilli Research"** which covers wide range of issues related to national and international scenario on the research and development of chillies, plant genetic resource management, advances in biotechnological research, breeding for virus resistance, utilization of nuclear and cytoplasmic male sterility systems, current trends in production technology, pest and disease management, post harvest management and value addition. The publication is outcome of sincere and dedicated efforts made by editors who deserve appreciation. I am sure this book will be of great use to all the stake holders including researchers, teachers, students, policy makers and consumers.

Jan. 2010

**(MANGALA RAI)**
Former Secretary. DARE & DG.
ICAR, New Delhi - 110 001

# Foreword

# Preface

The immense horticultural diversity has helped to make chilli globally important as fresh, cooked, spice, and vegetables and processed products, as well. India is recognized as the secondary centre of diversity for the domesticated species of *Capsicum*. The major chilli growing states in India are Andhra Pradesh, Maharashtra, Karnataka, and Tamil Nadu, which together contribute about 75% of the total area. In India, the productivity of green chilli is 9.2 t/ha, however, the potential productivity lies between 30–40 t/ha. Recent advances in chilli research, development of high yielding varieties and better agro techniques, higher economic returns have attracted farmers towards chilli cultivation. Today, India is the largest producer, consumer and exporter of chilli in the world.

The idea of bringing researchers together to deliberate on various issues of important horticultural crops including chilli to formulate future strategies was conceptualized by Dr. H. P. Singh, Deputy Director General (Horticulture), ICAR, New Delhi. It is in this background, a brain storming session on Advances in Chilli Research, second in the series, was organized at Indian Institute of Vegetable Research, Varanasi on 28$^{th}$ March 2008, after the brain storming session on cucurbitaceous crops held at Orissa University of Agriculture & Technology, Bhubaneshwar.

An effort has been made to compile thirteen lead talks delivered and contributed by experts from different parts of the country who have vast experience in different areas of chilli production. In the first chapter, challenges in chilli production have been discussed and subsequently national scenario on the research and development of chilli research has been presented; the third chapter encompasses the international scenario on chilli research. Other chapters include advances in biotechnological research, breeding for virus resistance, utilization of nuclear and cytoplasmic male sterility systems, current trends in production technology, pest management and post-harvest technology. In the end, the overall views and remarks made, and recommendations emerged during technical session have also been included with a view to develop a road map and formulate the future strategies for different aspects of chilli research and development in the country.

The editors are greatly indebted to Dr Mangala Rai, the then Secretary, Department of Agriculture Research and Education (DARE), Government of India and Director General, Indian Council of Agricultural Research

(ICAR), New Delhi for providing necessary guidance and support. We are also thankful to all the contributing authors for their unstinted cooperation and painstaking efforts on various aspects of chilli research. Last but not the least, this task would not have been possible without the support and input of the publisher who have done excellent job in publishing this proceeding in the form of book.

**Editors**

# Contents

# 1

# Challenges in Chilli Production in Twenty First Century

H.P. SINGH*

## 1. INTRODUCTION

Chilli pepper is an important crop in almost every country and is valued for its various uses. Chilli is cultivated in tropical and sub-tropical climates up to 2000 m altitude. In India, *Capsicum annuum* is cultivated for fresh fruits (both sweet and hot types), oleoresin/colour extraction (paprika type) and processing (pickle type). It is consumed in various forms, fresh green chillies as vegetables to dried powders as spices. It is the most widely used universal spice, named as 'wonder spice'. In daily life, chillies are integral and the most important ingredient in many different cuisines around the world as it adds pungency, taste, flavour and colour to the dishes. Indian chilli is considered to be world famous for two important qualities—its colour and pungency level. Some varieties are famous for the red colour because of the pigment capsanthin and others are known for biting pungency attributed to capsaicin. Some accessions of *C. annuum* with dark purple leaf, purple flower petals, and dark purple fruit have higher antioxidant activity, ascorbic acid and phenol content. The active ingredient, capsaicin (and the related capsaicinoids-dihydrocapsaicin, nordihydrocapsaicin and homocapsaicin) is found only in the fruits of the genus *Capsicum*. Extracted carotenoids pigments are important colourants in the processed food industry as well as in cosmetic applications. Chilli is called by several others names like hot pepper, pepper, chili, chilli pepper, chile, paprika, etc.

There are different geographical locations within the country having specialized chilli types. The hottest chilli in the world is 'Nagahari', native to North-eastern region of India, also called as Naga Jolokia or Bhut Jolokia, and having 8,55,000 unit in the Scoville heat unit scale, has replaced Mexican chilli 'Red Savina Habanero' which scales 5,57,000 in Scoville units (Mathur *et al.*, 2000). Mathania is an area near Jodhpur, Rajasthan where

* Deputy Director General (Hort.) ICAR, New Delhi.

large scale chilli cultivation is undertaken which are less pungent, brick red coloured, about 15 cm long having thick flesh with less number of seeds. In North-eastern states, another type, called 'Bird Eye Chilli' is found having very small fruit size with high pungency. In Andhra Pradesh, Karnataka and Jammu, genotypes with less pungency and high colour value are popular and in much demand. Byadagi type chillies having high oleoresin content are popular in southern India, specially Dharwad area of Karnataka. Pickle type chillies are largely grown in some pockets of North India and are utilized for making chilli pickles stuffed with spicy ingredients. Several local land races are popular among the farmers, *e.g.* Faizabadi chilli in Faizabad area and Super in Varanasi area of Uttar Pradesh.

Chilli production in our country is still dominated by the locally available genotypes or open pollinated varieties because of farmers' ignorance, poor extension activities and high cost of improved varieties/ hybrids. A well planned strategy is needed to cover large area under newly developed and identified varieties to boost the chilli production in the country. With the awareness of advantages of cultivation of $F_1$ hybrids, area under chilli production can be extended tremendously.

## 2. PRODUCTION AND EXPORT OF CHILLIES

India is the world leader in growing area devoted to chilli peppers, with China and Indonesia, ranking second and third, respectively. The other major chilli producing countries are Korea, Hungary, Spain, Nigeria, Thailand, Turkey, Kenya, Sudan, Uganda, Japan, Ethiopia, Pakistan, and Mexico. On the global front, according to FAO (2008), the total chilli production was estimated to be around 20.98 lakh tonnes (Table 1.1), down by 2.3 per cent compared to previous year's 21.48 lakh tonnes. The major cause for the decline in production is due to unfavourable weather condition in major chilli producing countries like India, China and Pakistan.

**Table 1.1.** Global chilli area and production

| Years | Area (lakh ha) | Production (lakh tonnes) |
|---|---|---|
| 2000 | 18.81 | 23.87 |
| 2001 | 19.06 | 24.32 |
| 2002 | 18.83 | 22.92 |
| 2003 | 18.32 | 26.41 |
| 2004 | 17.97 | 26.31 |
| 2005 | 17.42 | 26.25 |
| 2006 | 15.82 | 21.48 |
| 2007 | 15.18 | 20.98 |

*Source*: FAO Stat, 2008.

The top ten chilli producing countries, accounting for more than 85 percent of the world production in 2007 were India, China, Ethiopia, Myanmar, Mexico, Vietnam, Peru, Pakistan, Ghana and Bangladesh. The greatest share was taken by India with 37 per cent share in global production, followed by China (11%), Bangladesh (8%), Peru (8%) and Pakistan (6%) (Fig. 1.1).

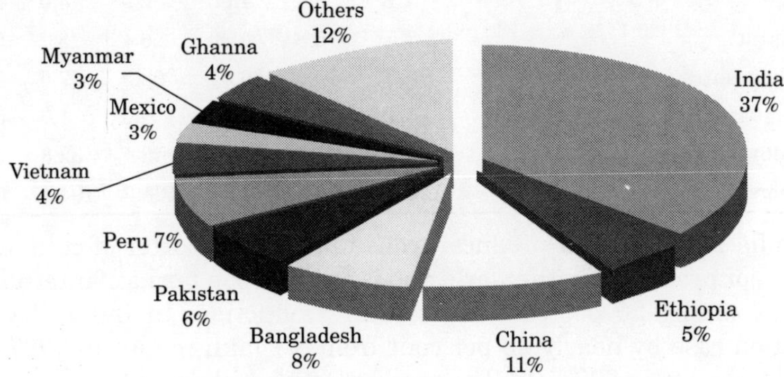


**Fig. 1.1.** Chilli production in global perspective

**Table 1.2.** Area, production and productivity of dry chilli in selected states (Area 000' ha, Production (Prod) 000't and Productivity (Pty) t/ha)

| | 2002–03 | | | 2003–04 | | | 2004–05 | | |
|---|---|---|---|---|---|---|---|---|---|
| | **Area** | **Prod** | **Pty** | **Area** | **Prod** | **Pty** | **Area** | **Prod** | **Pty** |
| 1. Andhra Pradesh | 223.0 | 409.0 | 1.83 | 250.3 | 797.0 | 3.18 | 236.5 | 748.5 | 3.16 |
| 2. Orissa | 75.0 | 62.9 | 0.84 | 75.0 | 63.2 | 0.84 | 76.1 | 63.2 | 0.83 |
| 3. Karnataka | 155.5 | 153.4 | 0.99 | 69.9 | 94.5 | 1.35 | 152.3 | 104.6 | 0.69 |
| 4. West Bengal | 61.7 | 60.5 | 0.98 | 60.5 | 66.3 | 1.10 | 52.2 | 61.4 | 1.18 |
| 5. Bihar | 3.40 | 2.3 | 0.68 | 3.2 | 2.2 | 0.69 | 3.1 | 2.2 | 0.71 |
| 6. Gujarat | 12.1 | 10.5 | 0.87 | 9.1 | 8.4 | 0.92 | 6.4 | 5.7 | 0.90 |
| 7. Maharashtra | 103.0 | 53.0 | 0.52 | 90.0 | 44.0 | 0.49 | 90.0 | 44.0 | 0.49 |
| 8. Chhattisgarh | 6.40 | 5.7 | 0.89 | 6.5 | 4.2 | 0.65 | 5.6 | 3.3 | 0.59 |
| 9. Madhya Pradesh | 44.2 | 31.7 | 0.72 | 49.1 | 38.4 | 0.78 | 45.7 | 39.6 | 0.87 |
| 10. Others | 143.1 | 105.6 | 0.74 | 161.0 | 117.2 | 0.73 | 153.2 | 126.0 | 0.82 |

*Contd...*

| | 2005–06 | | | 2006–07 | | |
|---|---|---|---|---|---|---|
| | **Area** | **Prod** | **Pty** | **Area** | **Prod** | **Pty** |
| 1. Andhra Pradesh | 172.0 | 538.0 | 3.13 | 214.0 | 766.0 | 3.58 |
| 2. Orissa | 75.1 | 63.3 | 0.84 | 76.1 | 61.9 | 0.84 |

**Table 1.2.** *Contd.*

| | 2005–06 | | | 2006–07 | | |
|---|---|---|---|---|---|---|
| | **Area** | **Prod** | **Pty** | **Area** | **Prod** | **Pty** |
| 3. Karnataka | 69.9 | 94.5 | 1.35 | 132.2 | 148.0 | 1.12 |
| 4. West Bengal | 52.0 | 60.7 | 1.17 | 52.2 | 63.6 | 1.22 |
| 5. Bihar | 3.1 | 3.1 | 1.00 | 2.9 | 3.0 | 1.03 |
| 6. Gujarat | 7.5 | 6.8 | 0.90 | 6.4 | 5.7 | 0.90 |
| 7. Maharashtra | 99.0 | 50.0 | 0.51 | 98.0 | 47.0 | 0.48 |
| 8. Chhattisgarh | 5.4 | 1.5 | 0.28 | 5.0 | 1.7 | 0.34 |
| 9. Madhya Pradesh | 43.0 | 35.6 | 0.83 | 42.4 | 32.8 | 0.77 |
| 10. Others | 125.9 | 159.8 | 1.27 | 128.7 | 104.4 | 0.81 |

India is the largest producer, consumer and exporter of chilli among other major producers in the world and is at the top in terms of international trade, exporting 20 per cent of its total production. In India, dry chilli production rose by nearly 43 per cent from 8.7 lakh tonnes in 1997–98 to 12.5 lakh tonnes in 2007–08. The production of chilli in India is dominated by Andhra Pradesh (53% of the total production) followed by Karnataka (9%), Orissa (6%), West Bengal (6%), Maharashtra (5%), Madhya Pradesh (4%) and others (17%) (Table 1.2). The productivity of green chillies in India is 9.2 t/ha as compared to 14.4 t/ha of the world, however, the potential productivity lies between 30–40 t/ha. In 2007–08, the total acreage brought under chilli cultivation was around 7.2 lakh ha, an increase from previous years' 7.0 lakh ha. Rising export demand coupled with higher price realization in the domestic market have motivated farmers to bring more area under chilli cultivation.

## 3. EXPORT OF CHILLIES

India has become world's largest producer and exporter of chilli, exporting to USA, Canada, UK, Saudi Arabia, Singapore, Malaysia, Germany and many countries across the world. It is the leader in export, with 25 per cent share in the world trade, followed by China with 24 per cent share in the global exports. Indian chilli exports are mainly affected by the domestic demand and uneven production which is interrupted by erratic monsoon, drought, and yield factors.

India started chilli export during 1960–61, with 8364 tonnes valued Rs. 176 crores. Since 2001–02, India's export performance has been excellent, with higher international demand. According to the Spices Board, the total export of chillies from India in 2007–08 touched a record high of 2.09 lakh tonnes, valued at Rs. 1097.59 crores, up 41.2 per cent, against 1.48 lakh tonnes valued Rs. 807 crores shipped previous year (Anonymous,

2008) (Fig. 1.2). Currently, India is the main source of red chilli in the international market. It exports in different forms like chilli powder, dried chilli, pickled chillies and chilli oleoresins. Countries like US, UK, Germany and Sweden use chilli for manufacture of oleoresins and extract on a large scale. India's chilli export is showing an increasing trend from the last decade on rising export demand coupled with short supply from other major producers, and the ban by the European Union on import of chillies from Pakistan on account of presence of aflatoxin in its produce. Pakistan's export share in global trade was grabbed by India that resulted in historic high exports from India in the last couple of years. Introduction of sampling and mandatory quality testing of chilli and its products by the Spice Board before shipment for the presence of Sudan I–IV and aflatoxins has boosted the confidence of overseas buyers and helped India's export.

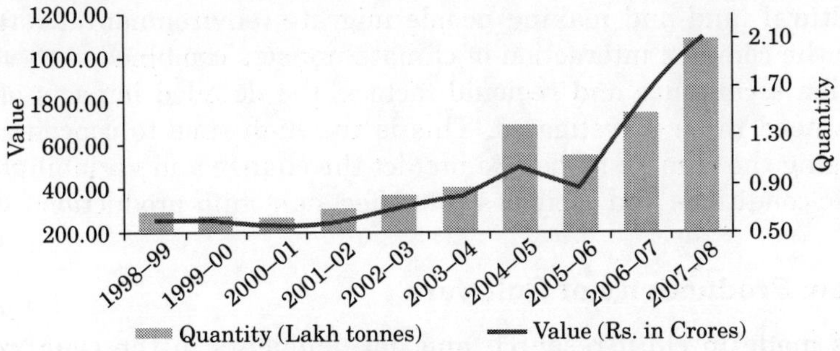


**Fig. 1.2.** Export of chilli from India

Long chilli peppers vary somewhat in size and colour but not in taste. Important fresh fruit quality parameters for processing should be free from blemishes caused by disease or sun bleaching, intense colour (bright or deep red), have good colour stability after processing, and acceptable pungency (pungency level preferences vary according to region). Important dry fruit quality parameters include high dry matter, ease of grinding, good colour retention after drying and grinding, and free from diseases and insects. Short chilli peppers are usually straight, light green or green at the immature stage and bright or dark red at the mature stage. The length ranges from 2–7 cm long and the pungency ranges from medium to very hot. They may be marketed as fresh green fruits, fresh red fruits, dried red fruits, or processed into chilli sauce, chilli powder, etc.

## 4. MAJOR CHALLENGES

### 4.1. Climate Change

Unlike other crops especially cereals, chillies are highly sensitive to abiotic stresses like high temperature, rainfall, humidity, etc. In general, the impact of climate change on agriculture could result in problems with food security and may threaten the livelihood activities upon which much of the population depends. Climate change can affect chilli, as well as the types of chilli that can be grown in certain areas, by impacting inputs such as water for irrigation, amounts of solar radiation that affect plant growth, as well as the prevalence of pests. Erratic change in monsoon pattern causes severe stress on agriculture which is fully dependent on weather. Increase in span of summer increases insect attack on crops. High temperature typically leads to reduction in fruit set and fruit size. The delayed winter hampers the cultivation of winter crops. Increasing humidity leads to incremental phenomenon of vector borne diseases. Climatic alterations may lead to intrusion of saline water into the agricultural land resulting in loss of yield and greater risk to the farmer and causing severe stress on availability of drinking water. Permanent intrusion causes loss of agricultural land and making people migrate (environmental refugees). Due to the complex interaction of climate impacts combined with varying irrigation techniques and regional factors, the detailed impacts of these factors need to be investigated. This is the high time to concentrate on developing the climate models to predict the change and variability of the climatic conditions and their possible effects on chilli production.

### 4.2. Low Productivity of Cultivars

Efforts made in chilli research and developments in the country have resulted in more than 30 improved varieties/hybrids and a number of production and protection technologies. However, productivity of chillies in our country (9.2 t/ha) is comparatively lower than world's average productivity (14.4 t/ha), which can be enhanced to two-three times with the use of improved technologies. The low productivity of cultivars can also be attributed to susceptibility of high yielding varieties to viral and other diseases. Moreover, chilli is predominantly grown in rainfed areas with extremely inadequate irrigation infrastructure. It clearly warrants the attention of all corners to maximize the chilli production by developing better suitable varieties/hybrids and appropriate production technologies. One of the major reasons for low productivity of the crop in India is lack of high yielding and high fruit quality $F_1$ hybrids, although hybrids are gaining popularity in India, China, Indonesia, and some other parts of Asia. It is estimated that only about 2.6 per cent of area of chilli is under hybrids as against 90 per cent in USA and some other developed countries.

### 4.3. Management and Utilization of Genetic Resources

National Bureau of Plant Genetic Resource (NBPGR), New Delhi is the nodal agency for collection, evaluation and maintenance for the germplasm of vegetable crops including chillies. It has more than 2500 collections of different types of genotypes under the genus *Capsicum*. Besides, Indian Institute of Vegetable Research (IIVR), Varanasi is the National Active Germplasm (NAG) site for chillies. At IIVR, around 375 genotypes are being maintained and used for different breeding purposes including hot, sweet- and paprika-types. Major centres/institutes dealing with *Capsicum* in India are IIVR, Varanasi; Indian Institute of Horticultural Research (IIHR), Bangalore; Regional Agricultural Research Station, Lam, Guntur; Indian Institute of Spices Research, Calicut; Spices Board, Cochin; Regional Research Station, Kovilpatti, (TNAU); Chandra Shekhar Azad University of Agriculture & Technology (CSAUA&T), Kanpur; Govind Ballabh Pant University of Agriculture & Technology, (GBPUA&T), Pantnagar; Punjab Agriculture University (PAU), Ludhiana and ICAR Research Complex for NEH region, Shillong. At international level, The World Vegetable Centre (earlier AVRDC), Taiwan; The Chile Pepper Institute, New Mexico State University, USA; Cornell University, Ithaca, USA; The Volcani Center, Israel; INRA, France and Seoul National University, South Korea are major research institutions where chilli pepper is being focused on.

There exist a large number region-specific potential germplasm of chillies suited for different traits. The exploitation of genetic resources collected form area like North-eastern region of India may be made for developing disease resistance varieties along with high level of pungency. Similarly, paprika type chillies can be developed by utilizing the gene repertoire from southern states, especially Karnataka. Consumer preferences for chillies vary from region to region, and thus concerted efforts should be made for the collection and utilization of region specific traits for the development of better varieties and hybrids. Conservation activities of germplasm should involve participation of farmers in addition to the government agencies. Appropriate methods of conservation and emergency strategies, especially for valuable genotypes need to be developed. While technology-intensive methods like *in vitro* conservation and cryopreservation would help conserve genetic diversity, a low cost conservation technology like ultra dry storage is the need of the hour.

### 4.4. Increasing Area under Improved Varieties

Chilli production in our country is largely taken by the locally available genotypes or open pollinated varieties because of poor extension activities, farmers' ignorance, and high cost of improved varieties/hybrids. Well planned strategies should be framed in order to cover large area under newly developed and identified varieties to boost the chilli production in the country. With the help of $F_1$ hybrids having several advantages over pure lines, the area under chilli cultivation can be increased to meet the demand.

### 4.5. Breeding for Resistant/Tolerant Varieties

Breeding resistant varieties is a novel option to avoid the excessive use of chemicals and to reduce the cost of production. Chilli is susceptible to leaf curl viruses, anthracnose, dieback, bacterial wilt, *Fusarium* wilt, bacterial leaf spot, spotted wilt virus, etc. So far work done by various institutions for the development of resistant varieties has yielded several cultivars and breeding lines possessing specific attributes. Punjab Lal (Bal *et al.*, 1995), Pant C1 (Naitam *et al.*, 1990), Perennial, etc. are some of the varieties resistant/tolerant to leaf curl virus under field condition and commercially cultivated on a large scale by the farmers. Genotypes existing in some pockets of the country have immense potential for the resistance to various biotic and abiotic stresses. BS-35, a highly resistant line against pepper leaf curl virus and anthracnose has been characterized (Kumar *et al.*, 2006). Utilization of such lines and other important materials is required for the development of multiple disease resistant varieties. Apart from this, new introduction needs to be made to strengthen the resistant breeding programme in our country.

### 4.6. Availability of Quality Seeds

Lack of availability of quality seeds at the right time continues to be one of the major challenges in increasing chilli production and productivity. Most of the seeds of chilli are produced in open condition except for the sweet peppers in which somewhat controlled conditions are required. Seeds of $F_1$ hybrids of chillies are produced by the private sectors mainly concentrated in southern part of the country. Chilli seeds are low volume high value commodity and demand a high investment particularly for the production of hybrid seeds. Several $F_1$ hybrids have been developed by the public organizations employing hand emasculation and pollination techniques and based on male sterility systems. The initiative taken by the Indian Council of Agricultural Research (ICAR), New Delhi for licensing the parental lines is to be implemented at a faster pace in order to produce seeds of better performing hybrid cultivars by the participating agencies. Under the National Seed Project of Vegetable Crops, a total of 93.6 kg chilli breeder seed was produced upon indent during 2004–05, whereas during 2007–08, 136.36 kg seed of 13 varieties and two hybrids was produced registering an increase of 45.7 per cent.

### 4.7. Development of Varieties and Hybrids

Presently chilli breeding programmes have relied on a relatively narrow genetic base within cultivars of various market types, although huge morphological diversity exists within and between species. This is because of traditional market demand for specific fruits size and shape, the use of pure line or back cross breeding within open pollinated commercial varieties and development of inbred from the commercial hybrid and the utilization of recycled parental lines (Poulos, 1994). The research efforts of various institutions under All India Coordinated Research Project on Vegetable Crops have resulted in development of 20 open pollinated varieties and 11 $F_1$ hybrids of chillies. In addition to it, six hybrids including one pickle type (KCH-3) were identified through Network Project on Development of Hybrids in Vegetable Crops under National Agricultural Technology Project during 1999–2004. A large number of high yielding varieties having better yields (up to 5.0 t/ha of dry chillies), resistance to major pests and diseases, seed content, colour and capsaicin content have been evolved and recommended for different parts of the country. Some of these varieties are dual type (fresh green as well as dry red). There is a need to select the variety suited to the region and the local requirements like chilli fruit type, colour, pungency, resistance to pests and diseases, etc. Although, several hybrids have been developed from public organization (for example, IIVR, Varanasi, IIHR, Bangalore and PAU, Ludhiana) as well as private sectors, there is an urgent need for developing more numbers of $F_1$ hybrids suited to different chilli growing eco-regions and making them available to the growers at affordable seed prices.

### 4.8. Use of Alternative Methods for Hybrid Development

There has been increasing demand of the $F_1$ hybrids by the farmers despite high cost of the seeds of hybrid varieties. This is because, under optimum crop production and protection management, crop raised from the seeds of $F_1$ seeds has several advantages like higher yield, adaptability, uniformity and resistance to biotic and abiotic stresses in comparison to crop raised from the seeds of improved pure lines or population. Chilli express considerable amount of hybrid vigour (20–50%) in the form of $F_1$ hybrids.

#### 4.8.1. *Hybrid Seed Production Using Cytoplasmic Male Sterile (CMS) Lines*

The cytoplasmic male sterility is expressed under the presence of sterile mt-genome (mitochondrial genome) located in cytoplasm (S-cytoplasm) and recessive allele of restorer (maintainer allele; *r*) located in the nuclear genome. Paterson in 1958 isolated for the first time, a male sterile individual in an Indian chilli accession (PI 164835). In Paterson's cytoplasm, pollen fertility has been found to be restored under 25°C and 17°C day and night temperature, respectively, as revealed by Shifriss (1997). Another male sterile cytoplasm was isolated in India during 1990's, which has been transferred in desirable genotypes and utilized for hybrid development at IIHR, Bangalore (Reddy *et al.,* 2002). The Peterson's sterile cytoplasm has been commercially exploited in China, Korea and India, although at limited scale due to the temperature sensitivity of male sterility expression. At IIVR, Varanasi and IIHR, Bangalore, CMS lines have been successfully utilized to develop commercial chilli hybrids. Several reports suggest that restorer allele (*R*) is more frequently distributed in hot pepper lines, while occurrence of maintainer allele (*r*) is more common in sweet and long fruited lines. Since male part (pollen) is sterile, use of CMS line precludes huge cost on manual emasculation. The CMS line (A line) is maintained by making crosses on it using pollen from maintainer line (B line). Through such crossing, 100 per cent male sterile seeds are obtained, unlike in GMS line. Since considerable cross pollination occurs in chilli and sweet pepper, no hand pollination is required for the development of $F_1$ hybrids.

### 4.8.2. *Use of Genic Male Sterility (GMS) in Hybrid Seed Production*

Genic male sterile lines are being increasingly utilized to produce hybrid seeds of chillies in European and other countries. In order to save the labour and time, early identification and removal of male fertile plants in the nursery itself or in hybrid seed production field is very important. Several monogenic recessive male sterile plants have been identified and reported from India (Kumar *et al.*, 2000). In India, MS-12 (*ms-509/ms-10*) line has been utilized to develop commercial male sterile based chilli hybrid, namely, CH-1 and CH-3 and farmers in Punjab are producing seeds of these hybrids. The *ms-10* gene is linked with taller plant height, erect growth and dark purple anthers (Das *et al.,* 2001). Male sterile lines possessing *ms-3* gene are being utilized for hybrid seed production in Hungary.

Using the alternate methods of hybrid seed production, a huge amount of labour, time and cost can be saved. Suitable markers to differentiate the male sterile from male fertile lines in case of genic male sterility needs to be developed. The farmers and entrepreneurs should be given large scale hands-on training for hybrid development by their own. More number of lines carrying either CMS or GMS system has to be developed in various backgrounds for their better utilization in different agro climatic regions.

## 4.9. Molecular Breeding

Modern varieties of crop plants usually represent a small fraction of the variation that exists in their gene pool. Wild *Capsicum* species have been used mainly for the introgression of disease resistance genes, however, these are rarely considered as a potential source for valuable genes for quantitatively inherited traits related to yield because of their inferior performance and the possibility of unfavorable linkages. Molecular breeding has emerged as a potential area in crop improvement research for enhancing the efficiency of selection processes. Quantitative trait loci (QTL) mapping for many traits in diverse species have identified transgressive alleles indicating that alleles with favourable effects on yield related traits exist in the wild parents. The use of wide crosses, combined with molecular mapping to identify the favourable alleles and manage linkage drag will allow deployment of a wider range of alleles in elite *Capsicum* germplasm. Several groups of molecular markers, *e.g.* Simple sequence repeats (SSR), Restriction fragment length polymorphism (RFLP), Amplified fragment length polymorphism (AFLP), and Randomly amplified polymorphic DNA (RAPD) have been employed for the construction of molecular maps of chilli for various traits utilizing recombinant inbred lines (RILs), $F_2$ generation, double haploids and back cross populations.

The marker-assisted selection program documented so far in pepper is for resistance to *Phytophthora capsici*, although allele-specific molecular markers and markers linked to a number of useful traits have been reported. Functional genomics experiments such as determination of expression profiles by microarrays, metabolic profiling and screening mutant populations are currently underway in pepper. These complementary approaches will allow the identification of many new genes and their function in *Capsicum*.

Molecular characterization is very important in the regime of intellectual property rights (IPR), patenting and safeguarding against gene piracy. Molecular marker-aided selection and genetic engineering will have to be employed in future genetic improvement of chillies. Use of molecular methods of transferring male sterility needs to be explored for development of hybrids. In near future, integration of conventional and molecular breeding will have to be resorted to, for genetic improvement of chillies. Special areas, in which genetic enhancement in chilli pepper require attention are selection of superior genotypes with respect to yield, resistance against various diseases and insect-pests using molecular markers, and improvement in processing and nutritional quality.

### 4.10. Host Plant Resistance

Damping off, bacterial wilt, anthracnose, fruit rot, powdery mildew, *Phytophthora* blight, *Cercospora* and *Alternaria* leaf spot, and various geminiviruses, cucumoviruses, potyviruses, and tospoviruses are major threatening diseases. Similarly, mites, thrips, aphid, whitefly, budworm and caterpillar are devastating pests of chillies. Average losses due to disease infestation ranged from 7 per cent in China to 43 per cent in India. To combat insects and diseases, farmers spray tremendous volumes of insecticides and fungicides on their fields. Farmers typically spray highly toxic "cocktails" (mixtures) containing 4–6 different pesticides every other day during the growing season, with often only a one-day waiting period before harvest. Concern about pesticide residues on fresh peppers is growing in several countries.

Improving host plant resistance through varietal selection, induced resistance, and good management practices to strengthen plant tolerance to disease are of economic importance in chilli production. Genetically resistant varieties are the most sustainable strategy for reducing crop losses, while pest control technologies including fungicides, antibiotics, and insecticides can be applied at the time of pest outbreak. Biological control and integrated diseases and pest management strategies can also reduce crop risks, and are often based on the combined use of different natural enemies and minimal use of pesticides. Partially purified toxin and the sorghum based formulation of *Trichoderma harzianum* are very much effective in reducing the anthracnose disease and fruit rot severity in hot pepper under field condition as well as reduce the post-harvest fruit rot and increase the shelf life of fruits by delaying the appearance of disease symptoms.

Leaf curl disease coupled with the infestations of thrips and mites has been threatening chilli production in tropical and sub tropical regions of the world. The satisfactory control of thrips and mites and partial control of viral diseases may be achieved with the application of certain pesticides but complete and environmentally safer protection from the virus through host plant resistance is more preferred and effective option. Leaf curl resistant genotypes under open field conditions have been reported; however, based on the artificial screening, resistant genotypes were reported from IIVR, Varanasi. In an experiment, a total of 307 chilli and sweet pepper genotypes including landraces, improved varieties, inbreds and hybrids representing four cultivated species, *viz. Capsicum annuum, C. frutescens, C. chinense, C. baccatum*, one wild species, *viz. C. chacoense* and six interspecific derivatives were screened under field condition. After the rigorous screening process under field condition, through grafting and thereafter confirmation by polymerase chain reaction, three genotypes, *viz.* GKC-29, BS-35 and EC-497636 were found highly resistant against pepper leaf curl virus disease (Kumar *et al.*, 2006). In another experiment conducted at UAS, Dharwad to assess the effects of four IPM modules on chilli pests, module comprising organics and safer molecule of insecticide was the most effective module against aphids, thrips, mites and *Helicoverpa armigera* (Gundannavar *et al.*, 2007). Higher dry chilli yield (5.13 q/ha) was observed with marigold as trap crop, vermicompost 2.5 t/ha + neem cake 250 kg/ha (without application of recommended dose of fertilizers) superimposed with sprays of neemazal @ 2 ml/l at 5 week after transplanting (WAT), diafenthiuron @ 1g/l (8 WAT), profenofos @ 2 ml/l (11 WAT) and neemazal @ 2 ml/l at 14 weeks after transplanting.

### 4.11. Organic Farming

In their need for higher yields, farmers are using heavy doses of fertilizers, fungicides, pesticides and growth regulators. These results in not only increased cost of production but also drastically change our environment and make the product unsafe to be consumed. In recent years, there has been emphasis on organic farming to obtain pesticide residue-free vegetables, spices and other horticultural commodities. However, scientific technology with sound alternatives is lacking for chillies. Keeping in view the benefits, which this technology offers along with higher returns in international markets for such products, systematic programmes need to be taken up. To ascertain and guarantee the consumer/importer that the produce has been genuinely raised organically, the production has to follow basic organic standards, (Policy paper 13, NAAS, 2001).

### 4.12. Processing and Value Addition

Processed products such as dehydrated chilli, pickle, powder, paste, sauce, etc., can be prepared for higher returns. Chilli is highly perishable in nature and it requires more attention during harvest, storage and transportation. Most of the growers sell their fresh harvest in the local or distant market and it is consumed as green chilli as well as in powder form. Through value addition, its market price, shelf life and the quality can be increased. Utmost care has to be taken in picking, cleaning, sorting, grading, drying, packing, storage and transportation. Hence, farmers must be educated for the processing of chillies. Efforts of scientists, extension agencies and the state government are needed in this area. Dark green fruits are plucked for preparing chilli pickle, whereas for dry powder (pepper spice), picking is done when the fruit is dark red. Pepper spices are the powders that are derived from the pungent, mild pungent or non-pungent fruits. Therefore, the main fruit quality parameters are colour and pungency. Apart from these, colour retention during storage, fruit wall thickness, fruit size, shape and weight are also important quality parameters. Development of aflatoxin in both raw and processed pepper spice is another important quality concern. The aflatoxin level should be checked at less than 5 µg /kg. Fruit drying at low temperature should be preferred because at higher temperatures spice powder will become brown instead of bright red. An optimum moisture content of about 8 per cent is considered to be ideal, as moisture content above 11 per cent allows mould growth and below 4 per cent causes excessive colour loss. Seeds of different cultivars have varying effects on the rate of colour loss, which is most likely due to the presence of varying antioxidant contents in the seeds. For instance, vitamin E, a fat-soluble antioxidant, has an effect on reducing colour loss. Selecting cultivars with seeds having high antioxidant contents therefore, is necessary to produce a colour-stable spice powder. During storage, carotenoids pigments are readily oxidized and the spice powder becomes less intensely coloured. The selection of appropriate cultivars, standardization and adoption of drying and storage methods are the management strategies to reduce the instability of the spice powder colour. For export purpose, picking is done when the fruits are uniform in size and colour. All diseased, deformed and discoloured fruit should be removed before marketing and storage. Cleaning, washing and pre-cooling of the produce enhance shelf life. Dry chilli powder can be stored at room temperature, however green fruit has to be kept in cold storage. Well dried pods after removing the extraneous matters like plant parts, etc. should be packed in clean, dry gunny bags and stored ensuring protection from dampness. Storage is required for off-season consumption and marketing.

## 5. CONCLUSION

The variability existing for important traits in many open pollinated cultivars needs to be integrated into hybrid cultivars by modern chilli breeding programmes. This facilitates protection of proprietary germplasm and allows exploitation of hybrid vigour for yield, earliness and fruit quality. Another area of fruit quality now receiving serious consideration is nutritional value. Peppers are potentially an excellent source of antioxidant compounds and phytochemicals which affect human health. Varieties with uniformly high levels of ascorbic acid, flavonoids and carotenoids needs to be developed and characterized. Breeding for uniform pungency and colour is crucial to the fresh market, warranting immediate need of the researchers in this area.

Identification and exploitation of new resistance genes remains one of the most important goals of pepper breeding programmes in the twenty first century. This includes screening of germplasm to identify resistance to diseases as well insect-pests. Much less focus has been given to identifying genetic mechanisms of insect resistance in peppers. Advances in biotechnology have allowed breeders' access to unique traits not available within the germplasm of various plant species. In addition, new gene mapping and cloning technologies have expedited the process of gene discovery and isolation. Marker-assisted selection and transgene expression are successful if appropriate phenotyping is carried out to confirm stable and effective gene expression. Integration of molecular techniques, such as marker-assisted selection and genetic transformation has become key issues in modern breeding programmes and therefore should be emphatically exercised in order to develop better product(s) of chilli pepper.

## REFERENCES

Bosland PW (1996). Capsicums: Innovative uses of an ancient crop. p. 479–87. *In*: J. Janick (*ed.*), Progress in new crops. ASHS Press, Arlington, VA.

Bosland PW and Votava EJ (2000). Peppers: Vegetable and Spice Capsicums. CABI Publishing, New York.

Gundannavar KP, Giraddi RS, Kulkarni KA and Awaknavar JS (2007). Development of integrated pest management modules for chilli pests. *Karnataka J. Agric. Sci.*, **20**(4): 757–60.

Kumar S, Banerjee MK and Kalloo G (2000). Male sterility: Mechanisms and current status on identification, characterization and utilization in vegetables. *Veg. Sci.* **27**: 1–24.

Kumar S, Kumar S, Singh M, Singh AK and Rai M (2006). Identification of host plant resistance to pepper leaf curl virus in Chilli (*Capsicum* spencs). *Sci. Hortic.* **110**: 359–61.

Mathur Ritesh, Dangi RS, Dass SC and Malhotra RC (2000). The hottest chilli variety in India. *Curr. Sci.* **79**(3): 287–8.

NAAS Policy paper 13 (2001). Hi-Tech Horticulture in India: Policy paper 13, National Academy of agricultural Sciences, New Delhi, October 2001.

Poulos JM (1994). Pepper breeding (*Capsicum* spp.): achievements, challenges and possibilities. *Plant Breed.* Abst. **64**: 143–55.

Reddy MK, Sadashiva AT and Deshpande AA (2002). Cytoplasmic male sterility in chilli (*Capsicum annuum* L.). *Indian J. Genet.* **62**: 363–4.

Shifriss C (1997). Male sterility in pepper (*Capsicum annuum* L.). *Euphytica* **93**: 83–88.

# 2

# Chilli and Paprika Research Scenario in India

Mathura Rai, Rajesh Kumar and A.B. Rai

## 1. INTRODUCTION

The genus *Capsicum* is a dicotyledonous flowering plant and belongs to family Solanaceae. Around 30 species of *Capsicum* have been reported, out of which *Capsicum annuum* is the most widely cultivated species and termed as chilli, pepper, hot pepper, chili, chilli pepper, chile, sweet pepper, bell pepper and paprika. The cultivars of *C. annuum* include both hot pepper (pungent fruits) and sweet pepper (non-pungent fruits). Sweet pepper is often called as bell pepper because majority of sweet pepper cultivars grown worldwide have bell shaped (four-lobbed) fruits. Chilli is cultivated for various market types and exhibits wide range of genetic and morphological diversity in terms of fruit size, shape and their consumption patterns. In India, *C. annuum* is cultivated for fresh fruits (both sweet and hot types), oleoresin/colour extraction (paprika type) and processing (pickle type) (Kumar *et al.*, 2006). Paprika is another type of chilli pepper, which has high natural colour value along with negligible pungency in the fruits or its products. Paprika is the Hungarian word for plants in the genus *Capsicum*. Like chilli, paprika is also a member of the capsicum or chilli family, an annual herb with a wooden stem, white flowers and long fruited peppers. Chilli powder and paprika look alike but they are not the same thing. Paprika is a type of chilli powder but there are specific differences that matter for the both. In India, paprika comes from a pepper called deghi mirch. Paprika oleoresin is a natural food colourant used to obtain a deep red colour in any food that has a liquid/fat phase. The oleoresin is slightly viscous, homogenous red liquid at room temperature. *Capsicum* in a fresh state is very rich in vitamin C (ascorbic acid), as revealed by Dr. Szent Gyorgyi, the Hungarian scientist, who was awarded the Nobel Prize in 1937 for isolating vitamin C from paprika fruits showing that they were one of the richest sources available for this vitamin (Anu and Peter, 2000). History of chilli cultivation in India is very old, however sweet peppers were introduced during British rule.

Indian Institute of Vegetable Research, Varanasi.

The *Capsicum* species includes the vast majority of the cultivated pungent and non-pungent (sweet) capsicum peppers in temperate as well as tropical areas. In the species *C. annuum,* throughout the world there is phenotypic diversity in plant habit and especially in shapes, sizes, colours, pungency, and other qualities of the fruits (Andrews, 1995, 1998; DeWitt and Bosland, 1996; Greenleaf, 1986). This immense horticultural and biological diversity has helped to make *C. annuum* globally important as a fresh and cooked vegetable (*e.g.* salads, warm dishes, pickles) and a source of food ingredients for sauces and powders and as a colourant, which is used in cosmetics as well, (Andrews, 1995; Bosland, 1994; Bosland and Votava, 2000). Moreover, the species is used for medicinal purposes and provides the ingredient for a non-lethal deterrent or repellent to some human and animal behaviour (Krishna De, 2003; Cichewicz and Thorpe, 1996; Reilly *et al.*, 2001). Chilli peppers are also cultivated for ornamental purpose especially for their bright glossy fruits with a wide range of colours. Chilli pepper comprises numerous chemicals including steam-volatile oil, fatty oils, capsaicinoids, carotenoids, vitamins, protein, fiber, and mineral elements (Bosland and Votava, 2000; Krishna De, 2003). Many chilli pepper constituents have importance for nutritional value, flavour, aroma, texture and colour. The ripe fruits are especially rich in vitamin C (Marin *et al.*, 2004). The two chemical groups of greatest interest are the capsaicinoids and the carotenoids. The capsaicinoids are alkaloids that give hot chilli peppers, their characteristic pungency. The rich supply of carotenoids contributes to chilli peppers' nutritional value and colour (Hornero-Mendez *et al.*, 2002; Perez-Galvez *et al.*, 2003).

The major chilli producing countries are India, China, Korea, Hungary, Spain, Nigeria, Thailand, Turkey, Kenya, Sudan, Uganda, Japan, Ethiopia, Indonesia, Pakistan, and Mexico. The world chilli production (green fruits) over the past five years has increased not only due to increase in area under cultivation, but also because of increase in the productivity from 13.2 t/ha in 2000 to 14.4 t/ha during 2004. However, FAO production data of chilli of India is underestimated because of the fact that this data include total production of green fruits only. The average productivity of chillies in India is 9.2 t/ha as compared to 14.4 t/ha of the world, however, the potential productivity lies between 30–40 t/ha. The major chilli growing states in India are Andhra Pradesh, Maharashtra, Karnataka and Tamil Nadu, which together constitute about 75% of the total area. Andhra Pradesh ranks first in dry chilli fruits production followed by Tamil Nadu, Maharashtra, Orissa and Karnataka. Some of the world's hottest chillies are grown in India, *e.g.* Bhut Jolokia or Naga Jolokia in the North eastern region of India. Important reasons for the lower yield of Indian chillies could be unavailability of good quality seeds, attack of pests and diseases and climatic vagaries.

## 2. ORIGIN AND TAXONOMY

The centre of diversity for the genus *Capsicum* is in southern part of South America (Gonzalez and Bosland, 1991). The primary centre of origin for domesticated *C. annuum* is in semi-tropical Mexico (Whitmore and Turner, 2002). India is considered to be the secondary centre of diversity of *Capsicum*. The genus represents a diverse plant group of about 25 wild and five cultivated species. More specifically, the *C. annuum* was originated in highland of Mexico and includes mostly the Mexican chile (syn. chilli) peppers, hot peppers of Asia and Africa and the cultivars of sweet peppers of temperate countries. However, due to the non-adaptability of *C. annuum* in low land tropics of Latin America, its cultivation was replaced by *C. frutescens* and *C. chinense*. The cultivation of *C. baccatum* and *C. pubescens* are mostly restricted to Latin American countries like Peru, Bolivia, Columbia and Brazil. Except *C. pubescens*, wild forms of the remaining four cultivated species are known.

### 2.1. Identification of Domesticated Species

*Capsicum* species are diploid with 24 chromosomes (n = x = 12), with some wild species having 26 chromosomes, n = x = 13 (Pickersgill, 1991; Tong and Bosland, 2003). There are a few species for which the genome is 2n = 2x = 32 (Wang and Bosland, 2006). The cultivated species *C. annuum* has 24 chromosomes with two pairs of acrocentric chromosomes but rest four cultivated species (*C. frutescens, C. chinense, C. baccatum* and *C. pubescens*) have 2n = 24 with one pair of acrocentric chromosomes.

The five domesticated species are differentiated by using morphological characters that rely primarily on colour and morphology of flowers and seeds (Andrews, 1995; DeWitt and Bosland, 1996), as presented in Table 2.1.

**Table 2.1.** Distinguishable morphology of cultivated species

| Cultivated species | | Distinguishable morphology |
|---|---|---|
| *C. annuum* | : | White corolla and white filaments |
| *C. frutescens* | : | Greenish corolla and purple filaments |
| *C. chinense* | : | Presence of annular constriction on pedicel attachment and greenish corolla |
| *C. baccatum* | : | Yellow, brown, or dark green spots on the corolla |
| *C. pubescens* | : | Hairy stems/leaves and black/brown seeds |

All the five cultivated species of *Capsicum* are characterized by genotypes with pungent (hot pepper) and non-pungent (sweet pepper) fruits. Furthermore, fruit size and shape of all the five species have a great variability and occurrence of genotypes with similar fruit morphology across the species is very common, which makes it very difficult to differentiate between the genotypes of cultivated species based on their fruit size, shape and pungency, especially three species within *C. annuum* complex. Nonetheless, certain flower and fruit descriptors can be used to assign a cultivar/genotype to a cultivated species with almost beyond doubt (Table 2.1). It is worth mentioning that occasionally one can find a landrace with over-lapping distinguishable characteristics of *C. annuum, C. frutescens* and/or *C. chinense* because of high amount of natural cross pollination between them, leading to fixation of characteristics of two species in a derivative landrace *e.g.*, Naga Jolokia or Bhut Jolokia in the North eastern part of India.

The genus *Capsicum* consists of a diverse group of plants producing pungent or non-pungent fruits of various sizes, shapes and colours. It consists of about 25 wild and 5 domesticated species (Table 2.2) (Eshbaugh, 1993; Bosland and Votava, 2000).

**Table 2.2.** The species of *Capsicum* and their known natural distributions. The five domesticated species are grouped into the *C. annuum* complex (3 spp.) ($C_A$), the *C. baccatum* complex ($C_B$), and the *C. pubescens* complex ($C_P$)

| Species | | Known or probable natural distribution |
|---|---|---|
| *C. annuum* L. ($C_A$) | : | Southern USA to Colombia |
| *C. baccatum* L. ($C_B$) | : | Peru, Bolivia, Paraguay, Argentina and Brazil |
| *C. buforum* Hunz. | : | Southern Brazil |
| *C. campylopodium* Sendtner; $n$ = 13 | : | Southern Brazil |
| *C. cardenasii* Heiser & P.G. Smith ($C_P$) | : | Northeastern Bolivia |
| *C. chacoense* Hunz. ($C_B$ or $C_A$) | : | Argentina, Paraguay and Bolivia |
| *C. chinense* Jacq. ($C_A$) | : | Northern Amazonian South America |
| *C. coccineum* (Rusby) Hunz. | : | Bolivia and Peru |
| *C. cornutum* (Hiern) Hunz.<br>*C. dimorphum* (Miers) Kuntze | : | Southern Brazil |
| *C. dusenii* Bitter*C. eximium* Hunz. ($C_P$) | : | Colombia |
| *C. flexuosum* Sendtner | : | Southeastern Brazil |
| *C. frutescens* L. ($C_A$) | : | Bolivia and northern Argentina |
| *C. galapagoense* Hunz. ($C_A$) | : | Argentina, Brazil and Paraguay |
| *C. geminifolium* (Dammer) Hunz.<br>*C. hookerianum* (Miers) Kuntze | : | Western Amazon (Colombia to Peru) |
| *C. lanceolatum* (Greenman) Morton & Standley; $n$ = 13 | : | Galapagos Islands (Ecuador) |

**Table 2.2.** *Contd.*

| Species | | Known or probable natural distribution |
|---|---|---|
| *C. leptopodum* (Dunal) Kuntze<br>*C. minutiflorum* (Rusby) Hunz. | : | Colombia and Ecuador |
| *C. mirabile* Mart. ex Sendtner; $n = 13$ | : | Ecuador and northwestern Peru |
| *C. parvifolium* Sendtner | : | Honduras, Guatemala and Mexico |
| *C. praetermissum* Heiser & P.G. Smith ($C_B$) [syn. *C. baccatum* var. *praetermissum* (Heiser and P.G. Smith) Hunz.] | : | Brazil |
| *C. pubescens* Ruiz & Pavon ($C_P$) | : | Argentina, Paraguay and Bolivia |
| *C. rhomboideum* (Dunal) Kuntze [syn. *C. ciliatum* (Kunth) Kuntze]; $n = 13$ | : | Northeastern Brazil, Venezuela and Colombia |
| *C. schottianum* Sendtner; $n = 13$ | : | Southern Brazil |
| *C. scolnikianum* Hunz. | : | Bolivia to Colombia |
| *C. tovarii* Eshbaugh, P.G. Smith & Nickrent ($C_B$) | : | Mexico to Peru |
| *C. villosum* Sendtner | : | Southern Brazil |

*Source*: Tong and Bosland, 1999; Walsh and Hoot, 2001; Jarret and Dang, 2004; Ryzhova and Kochieva, 2004.

## 3. CLIMATE AND SOIL

Chilli crop comes up well in tropical and sub-tropical regions, however it has a wide range of adaptability and can withstand heat and moderate cold to some extent. The crop can be grown over a wide range of altitudes from sea level up to nearly 2100 meters. It is generally a cold weather crop, but can be grown throughout the year under irrigated condition. Black soils which retain moisture for long periods are suitable for rainfed crop, whereas well drained sandy loam soils and deltaic soils are good under irrigated condition.

Chilli peppers are better adapted to warm weather than the sweet pepper, but it does not set fruit well when night temperatures are greater than 24°C and below 13°C. Sexual reproduction in plants is more sensitive to higher temperatures than vegetative processes. High temperatures during flowering have been shown to affect pollen germination, pollen tube growth, fertilization, flower abscission and fruit set in peppers (Wien, 1997; Han *et al.,* 1996; Erickson and Markhart, 2002). Low temperature also results in reduced fruit set in peppers due to inhibition of pollen germination

(Shaked *et al.*, 2004). Erickson and Markhart (2002) reported microspore mother cell meiosis and mature microspores at anthesis as highly temperature sensitive stage of pollen development. The optimum day temperatures for hot pepper growth range from 20–30°C. The optimum soil temperature for seed germination is 25°C and the maximum is 30°C and the optimal temperature for productivity is between 18°C and 30°C. Excessive hot weather produces infertile pollen and poor fruit set. Flower buds will usually abort if night temperatures reach 30°C. Pollen viability is significantly reduced at temperatures above 30°C and below 15°C. The crop can tolerate shade conditions up to 45 per cent of prevailing solar radiation, although shade may delay flowering. Path coefficient analysis of yield and its component traits conducted by Bhalekar *et al.* (2004) under high temperature (40°C) revealed that pollen viability, fruit set and number of primary branches were important yield determiners exerting maximum direct effect on yield. Lower seed yield for plants grown at high temperature conditions are found to be due to decreased pollen viability in chilli peppers (Han *et al.*, 1996; Erickson and Markhart, 2002).

Well-drained loamy soils at pH 5.5–6.8 are most suitable for pepper cultivation. Sandy loam soil is ideal to get early yield, while loam soils are preferred to get high yield. Quality of fruits is better in light soils than in heavy soils. Severe flooding or drought is injurious to almost all cultivars known so far.

### 3.1. Growing Season

The major season for growing chillies in India is kharif season. In northern India, generally two crops, *i.e.* kharif and rabi are taken. In southern India, the three crops, *i.e.* kharif, rabi and summer season crops are grown. Kharif crop is sown/transplanted in the month of June to July, rabi crop in the month of October to November and summer crop in the month of January to February. In the hills, seeds are sown in March to April.

## 4. CROP IMPROVEMENT

### 4.1. Plant Genetic Resources

National Bureau of Plant Genetic Resource (NBPGR), New Delhi holds a large amount of germplasm of chilli and sweet pepper (Table 2.3). Chilli germplasm status at NBPGR, New Delhi was 2506 as on December 2006. Besides, Indian Institute of Vegetable Research (IIVR), Varanasi; Regional Agricultural Research Station, Lam, Guntur; Regional Research Station, Kovilpatti; Chandra Shekhar Azad University of Agriculture & Technology, Kanpur; Indian Institute of Horticultural Research, Bangalore; Punjab

Agricultural University, Ludhiana and other organizations maintains chilli germplasm. At IIVR, all the cultivated lines have been assigned to five cultivated species. At international level, The World Vegetable Center (WVC), Taiwan and Chile Pepper Institute, New Mexico State University, USA have large collections of cultivated and wild species.

**Table 2.3.** Important collection of chilli pepper germplasm at NBPGR, New Delhi

| Species | | Specific traits |
|---|---|---|
| *Capsicum annuum* (EC 582593) | : | Var. Tangerine dream, non-pungent, banana type, heat and drought tolerant |
| *C. baccatum* var. *pendulum* (EC 572240–42) | : | Anthracnose resistant |
| *Capsicum chinense* (EC 589469) | : | Resistant to anthracnose and bacterial wilt |
| *Capsicum annuum* (EC 572236, 39) | : | Resistant to chilli veinal mosaic virus and poty virus Y |
| *C. annuum* (EC 571257–266) | : | Cytoplasmic male sterile (CMS) lines and their maintainers |
| *C. annuum* (EC 572238, 43–45) | : | Resistant to anthracnose |
| *C. annuum* (IC 541402) | : | Non-pungent green chilli |
| *C. chinense* (IC 541444) | : | Highly pungent |

*Source*: Annual Report. All India Coordinated Vegetable Improvement Project (AICRP-VC), 2006. EC—exotic collection, IC—Indigenous collection.

At IIVR, Varanasi, around 375 genotypes of different types of chilli peppers are being maintained. A new set of 37 accessions of chilli introduced through NBPGR, New Delhi during 2005–06 was evaluated and found that most of them were adaptable under Varanasi condition. Several lines have been identified as resistant/tolerant source for more than one biotic stresses. Another set of six pairs of CMS and their maintainers were also introduced from WVC, Taiwan, which are being maintained and utilized in breeding programmes.

### 4.2. Varietal Development

More than twenty horticultural types are predominantly cultivated in one or other parts of world and various grouping have been proposed. Broadly, the breeding objectives for quality traits of pungent as well as non-pungent peppers could be described on the basis of five market types (Table 2.4). Besides developing varieties/hybrids for specific market types, yield and resistance to prevailing biotic stresses (Table 2.5) are also important objectives of pepper breeding.

**Table 2.4.** Important market types of pepper and major fruit traits

| S.No. | Market type | | Important fruit quality traits |
|---|---|---|---|
| I. | Fresh market (green, red, multicoloured whole fruits) | : | Colour, pungency, shape, size, lobe number, flavour, exocarp thickness, endocarp: seed ratio, vitamin A and C |
| II. | Fresh processing (sauce, paste, canning, pickling) | : | Colour, pungency, shape, size, pericarp thickness, endocarp: seed ratio |
| III. | Dried spice (whole fruits and powder) | : | Colour, pungency, shape, size, dry weight, low crude fiber, endocarp: seed ratio |
| IV. | Oleoresin extraction | : | Colour, pungency (essential oils) |
| V. | Ornamental (plants and/or fruits) | : | Colour, pungency, shape, size, dry weight |

*Source*: Poulos JM (1994).

**Table 2.5.** Major diseases and pests of pepper and source of host plant resistance/ tolerance

| Biotic stress | Causal organism | Resistance sources and their utilization |
|---|---|---|
| Anthracnose | *Colletotricum* spp. | *C. chinense* and through back cross breeding pre-breeding lines developed by WVC, Taiwan are available for use. PBC-80 and PBC-81 (*C. baccatum*) are also resistant. |
| Phytophthora blight | *Phytophthora capcisi* | PBC-131, PBC-133 lines of AVRDC, Taiwan |
| Leaf curl | Pepper Leaf Curl Virus (PepLCV) | BS-35, GKC-19 with resistance derived from *C. frutescens* are available at IIVR, Varanasi |
| Bacterial wilt | *Ralstonia solanacearum* | F5-112 |
| Thrips | — | Japani Longi, DC-16, PBC-535, 9852-173, PBC-473 |
| Mites | — | — |

Presently pepper breeding programmes have relied on a relatively narrow genetic base within cultivars of various market types, although huge morphological diversity is apparent within (intraspecific) and between (interspecific) species. This is because of (i) traditional market demand for specific fruits size and shape and (ii) the use of pure line or back cross breeding within open pollinated commercial varieties and development of inbred from the commercial hybrid and the utilization of recycled parental lines. The introduction, mass selection, recurrent selection and heterosis breeding have been utilized to develop varieties with improved yield and resistance to certain diseases. In India, so far 20 open pollinated and 6 $F_1$

hybrids of hot pepper have been identified for cultivation and release under specific agro-climatic zones (Table 2.6). These improved varieties and hybrids are being cultivated for the specific market types, however few varieties/hybrids may be grown for both green and/or red ripe dry fruits.

**Table 2.6.** Varieties and hybrids identified through AICRP-VC.

**Open pollinated variety**

| S.No. | Chilli variety | Developing centre | Recommended zones | Year of identification |
|---|---|---|---|---|
| 1. | G-4 | Lam | — | 1975 |
| 2. | G-5 | Lam | — | 1975 |
| 3. | K-2 | TNAU ( Kovilpatti) | — | 1985 |
| 4. | J-218 | JNKVV | I, IV, V, VI, VII | 1987 |
| 5. | X-235 (LCA-235) | Lam | IV, V, VI, VIII | 1987 |
| 6. | Muslawadi | MPKV | V | 1987 |
| 7. | Sel-1 | IIHR | V, VII, VIII | 1990 |
| 8. | LAC 206-B | Lam | V,VI,VII,VIII | 1990 |
| 9. | AKC-86-39 | Akola | VII | 2001 |
| 10. | BC-14-2 | OUA&T | V, VI | 2001 |
| 12. | RHRC-Cluster Erect | MPKV | VII | 2001 |
| 12. | PMR-57/88-K | IIHR | VII | 2002 |
| 13. | LCA334 | LAM | III, IV, V, VII | 2002 |
| 14. | ASC-2000–02 | GAU, Anand | VII | 2004 |
| 15. | KA-2 | IIVR | IV | 2005 |
| 16. | LCA-353 | RARS, Lam | IV, V, VII, | 2007 |
| 17. | BC-25 | OUA&T, Bhubaneswar | V, VI, VII | 2007 |
| 18. | PC-7 | GBPUA&T, Pantnagar | V | 2009 |
| 19. | HS-HP-154 | SKUA&T, Srinagar | VIII | 2009 |
| 20. | IVPBC-535 (Paprika type) | IIVR, Varanasi | VIII | 2009 |

**Hybrid**

| S. No. | Hybrid | Developing centre | Recommended zone(s) | Year of identification |
|---|---|---|---|---|
| 1. | HOE-888 | Sandoz | IV, VIII | 1997 |
| 2. | ARCH-236 | Ankur Seeds | IV | 1997 |
| 3. | Sungro-86-235 | Sungro | IV, VIII | 2001 |
| 4. | ARCH-228 | Ankur Seeds | IV, V, VI | 2002 |
| 5. | CCH-2 | IIVR | II, IV,V,VI | 2005 |
| 6. | BSS-453 | Bejo Sheetal | II | 2009 |

### 4.3. Male Sterility

Male sterility in *Capsicum* was first documented during 1951 by Martin and Crawford, however first cytoplasmically inherited male sterile plant was isolated from an Indian accession by Peterson (1958).

#### 4.3.1. *Genetic Male Sterility (GMS)*

In *Capsicum*, nearly 20 male sterile mutants have been identified either in natural population or induced through mutagenesis (Wang and Bosland, 2006). A monogenic recessive male sterile line was isolated from the variety 'Big All' and allele was designated as *ms-1*. After that non-allelic gene *ms-2* was isolated from the population at California. Through induced mutagenesis (using *x*- and γ-rays), Daskalov induced six male sterile mutants, *viz., ms-3*, *ms-4*, *ms-5*, *ms-6*, *ms-7* and *ms-8*. Similarly, Pochard in France also induced three male sterile mutants, namely, *mr-3*, *ms-509* and *mc-705*.

Several monogenic recessive male sterile plants have been identified and reported from India (Prakash *et al.*, 1987; Patel *et al.*, 1998). In India, MS-12 (*ms-509*/*ms-10*) line has been utilized to develop commercial male sterile based chilli hybrids, namely, CH-1 and CH-3 and farmers in Punjab are producing seeds of these hybrids. The *ms-10* gene is linked with taller plant height, erect growth and dark purple anthers (Das *et al.*, 2001). Male sterile lines possessing *ms-3* gene are being utilized for hybrid seed production in Hungary.

#### 4.3.2. *Cytoplasmic-Nuclear Male Sterility (CMS)*

The first CMS *Capsicum* plant was identified in an Indian *C. annuum* population (PI-164835) in USA. Subsequently, two more male sterile cytoplasms were isolated, but all the three independently isolated male sterile cytoplasms were found to be genetically identical. Yet another male sterile cytoplasm was located in India during 1990's, which has been transferred in several desirable genotypes and utilized for hybrid development at IIHR, Bangalore. The Peterson's sterile cytoplasm has been commercially exploited in China, Korea and India, although at limited scale due to the temperature sensitivity of male sterility expression (Kumar *et al.*, 2007). At IIVR, Varanasi and IIHR, Bangalore, CMS lines have been successfully utilized to develop commercial chilli hybrids.

##### 4.3.2.1. *Evaluation of CMS derived from C. chacoense*

Male sterile line derived from *C. chacoense* for the first time at New Mexico, USA was introduced at IIVR, Varanasi for evaluation and utilization in breeding programme. However, all the plants raised from the introduced seeds were found to be male fertile with ability to set normal amount of selfed fruits with seeds. Moreover the introduced material is lost at the source. Therefore, it was desired to initiate development of cytoplasmic male sterile plants using *C. chacoense* cytoplasm. Hence, during 2005–06, a cultivated species (*C. annuum*) cv. Kashi Anmol (KA-2) was crossed on *C. chacoense* as seed parent. A total of 100 random primers were screened for polymorphism between the parental lines to confirm hybridity of putative $F_1$ plants, (Kumar *et al.*, 2007). The primer OPS-1 produced unique fragment of 2500 bp, which was male parent specific, *i.e. C. annuum* (Fig. 2.1A). The presence of this fragment in $F_1$ plant confirmed the hybridity and precludes its possible origin from the selfed seeds. The female fertility of this inter-specific cross was normal as $F_1$ plants developed open pollinated cross fruits and the $BC_1F_1$ fruits derived from pollen of *C. annuum* parent were obtained. The $F_1$ plant with rudimentary stamen (Fig. 2.1B) has also been successfully crossed with the recurrent *C. annuum* plant (Kashi Anmol) to create $BC_1F_1$ generation for further backcrossing.

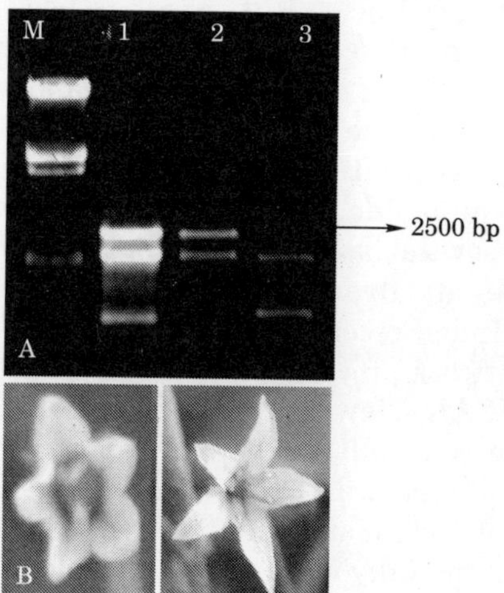


**Fig. 2.1.** A. Confirmation of hybridity: lanes 1: *C annuum*, 2: Hybrid and 3: *C. chacoense.* B. Flowers of hybrid without anther and with anthers

## 4.4. Paprika

Like all capsicums, the paprika varieties are native to South America. Originally a tropical plant, it can now grow in cooler climates. Globally, Hungary and Spain are the two main centres for growing paprika peppers, though these varieties have evolved into much milder forms than their tropical ancestors. Commercial food manufacturers use paprika in cheeses, processed meats, tomato sauces, chilli powders and soups. Its main purpose is to add colour. Its use in cosmetics is also gaining popularity due to its natural colours. Paprika is a fine powder ground from certain varieties of *Capsicum annuum*, which vary in size and shape. They may be small and round or pointed and cone shaped. They are larger and milder than chilli peppers. Paprika is produced from peppers ripened to redness, sometimes called 'pimento' in some countries. The powder can vary in colour from bright red to rusty brown. Paprika is used in seasoning blends for barbeque, snack foods, goulash in the cuisines of several countries including India.

Paprika chilli is being presently grown in a very limited scale in restricted areas in India, although there is a great potential of export of natural colouring agents in the international markets. Globally over 50,000 tons of paprikas are required per annum. Three paprika varieties were introduced at IARI, Regional Station, Katrain, *viz*. KTPL-8, KTPL-18 and KTPL-19 from Spain. At IIHR, Bangalore, Arka Abir was developed from the local variety Byadagi Dabbi (Kumar and Rai, 2005). At IIVR, Varanasi, paprika genotypes have been collected from AVRDC, Taiwan and other sources and are being utilized in different breeding programmes. Recently, a line IVPBC-535 has been identified through AICRP-VC for paprika purpose with negligible pungency and high colour value. Similarly at UAS, Dharwad a number of landraces of paprika lines are available. In Karnataka, the 'Byadagi' type populations have good colour value but pungency is greater than required, while 'Tomato Chilli' populations are less pungent, however colour value is inadequate. In northern Karnataka, several landraces have been identified within Byadagi chilli *e.g*. Byadagi Kaddi, Byadagi Dabbi, Dyavanur Kaddi, Delux and Anthur Benthur. In India, recently the importance of paprika has gained much attention. A network project on the development of paprika hybrids/inbreds funded by ICAR, New Delhi, has been under way with major objectives as—(i) to screen chilli (hot pepper) and sweet pepper germplasm, (ii) to develop and evaluate chilli hybrids and inbred lines with respect to yield, oleoresin content and colour retention, and (iii) to test the adaptability of developed hybrids especially with respect to yield, oleoresin content and quality. Recently, in March 2007, the project was concluded wherein several inbreds and hybrids with good colour value and low pungency were identified (Table 2.7). Pungency below 0.1 per cent and colour value above 100 ASTA units is highly desirable for oleoresin industry.

Padmapriya *et al*. (2004) evaluated different parameters of paprika genotypes for processing purpose and found that the extractable colour was significantly different in all the ten cultivars studied. They found that the genotype KTPL-18 recorded highest value for ascorbic acid (86.85%), TSS (9.96° Brix), capsanthin (176.434 ASTA units) and dry matter (22.70%), while the oleoresin content was highest in Tomato Chilli (28.75%). They concluded that the variety having maximum dry matter, ascorbic acid, increased colour matter, and rich in alcohol soluble matter was suitable for salad as well as for processing in paprika.

**Table 2.7.** Performance of chilli samples for colour and oleoresin content

| S.No. | Genotype | ASTA colour | Oleoresin content (%) | Pungency (%) (capsaicin) |
|---|---|---|---|---|
| 1. | A1 × NIC 268216 | 217.136 | 14.01 | 1.243 |
| 2. | A1 × SP-4C | 213.448 | 14.68 | 0.187 |
| 3. | A1 × Perennial | 223.204 | 18.38 | 1.210 |
| 4. | 9771-16-PT 1 | 143.008 | 14.18 | 0.232 |
| 5. | A1 × California Wonder | 203.606 | 15.64 | 1.055 |
| 6. | A2 × Pusa Jwala | 171.462 | 13.24 | 1.537 |
| 7. | A1 × Kalyanpur Chanchal | 165.394 | 18.25 | 1.814 |
| 8. | A1 × Phuleri | 277.075 | 17.64 | 1.739 |
| 9. | A1 × KDCS-810 | 255.755 | 16.20 | 0.672 |
| 10. | KA-2 | 171.130 | 17.04 | 1.379 |
| 11. | Pant C1 | 158.506 | 11.80 | 1.108 |
| 12. | PBC-535 | 160.392 | 14.48 | 0.006 |
| 13. | PBC-776 | 248.376 | 17.85 | 0.277 |
| 14. | BS-35 | 136.527 | 16.61 | 3.543 |
| 15. | EC-497635 | 186.796 | 15.30 | 0.013 |

### 4.4.1. *Research Needs for Paprika*

Other than general practices adopted for chilli cultivation, the following points should be considered for a better package of paprika production:

1. It is mostly observed that farmers are using excessive doses and in some cases unwanted chemicals leading to the higher residues of pesticides. Thus, it is necessary to have an integrated pesticide management to get paprika acceptable to the international market. Indian standards for pesticide residue should be brought to safe level so that health of the people is protected.

2. The yield per unit area should be increased involving better varieties into package to reduce the cost of cultivation.

3. Indian paprika should have low hotness to be used for fresh, whole and ground products. Capsaicin content should be less than half of the present level.

4. For use as a raw material for oleoresin industry, yield of extracts and total colour should be increased.

5. For use of oleoresin in chicken feed industry the content of *trans* capsanthin and red pigments should be increased.
6. Paprika with a bright red colour will have a great appeal for use as fresh, whole and ground powder.
7. Farmers should be encouraged to dry paprika chilli uninterrupted for use as dry product. This is necessary to control aflatoxins and gives uniform attractive colour.

## 4.5. Biotechnology and Molecular Breeding

In the light of fast acceptance and integration of diverse uses of DNA-based molecular markers by the plant breeders, molecular markers will more likely become an integral component of conventional plant breeding programmes. At IIVR, Varanasi, research on molecular markers has been initiated recently for chilli improvement. So far most of these works have been confined to use of RAPD markers for the tagging of desirable traits of major vegetable crops and genetic purity testing of the hybrid seeds. These accomplishments along with development of mapping populations and works under progress are briefly described in the following text.

### 4.5.1. *Development of Mapping Population*

Mapping population serves for the identification and tagging of desirable genes and quantitative trait loci (QTLs) in crop plants. Recombinant inbred lines (RILs) are the immortal mapping populations and have emerged as one of the best sources for identifying the desirable traits. At IIVR Varanasi, RILs derived from a cross between California Wonder (non-pungent and large fruited) and LCA 235 (pungent and comparatively small fruited) have been morphologically characterized with special reference to fertility restoration locus (Singh *et al*., 2006). Out of the 71, only four RILs *viz*. PT-6, PT-10, PT-11-5 and PT-39a were having the maintainer (*rfrf*) allele and the remaining 67 RILs had restorer (*RfRf*) allele. One of the maintainer RILs (PT-11-5) had pungent and desirable fruit size/shape. Similarly, PT-12-3 was found to be of sweet pepper type. This set of RILs will facilitate tagging/validation of markers for fruit size, shape, pungency, colour and fertility restoration genes/QTLs.

### 4.5.2. *Validation of Markers Linked to Rf Gene and Hybrid Purity Testing*

The distribution of two previously known RAPD markers ($OPW19_{800}$ and $OPP13_{1400}$) associated with fertility restoration (*Rf*) gene in chilli was tested in a series of 47 newly identified restorer and maintainer inbred plants of chilli and sweet pepper. The presence of both these markers often not coincided with *Rf* gene in restorer inbred lines, therefore, it was concluded that both the markers had narrow origin and distribution. However, both these markers were found useful to test genetic purity of a CMS (CCA-4261) based commercial hybrid because both of them were present only in the male parent and are reproducible (Kumar *et al.*, 2006).

### 4.5.3. *Degenerative Primer for Reliable Detection of PepLCV*

In a study using resistant and susceptible chilli lines against Pepper Leaf Curl Virus (PepLCV), the validity of a degenerative primer for the detection of the WFT (white fly-transmitted) geminivirus was tested. The DNA samples were extracted from the leaves of grafted/alternate grafted (susceptible scion/susceptible stock, Pusa Jwala) and field resistant plants of GKC-29, BS-35 and EC-497636 and susceptible plants of KA-2, Pusa Jwala and PBC-534. The DNA samples were subjected to polymerase chain reaction (PCR) using a degenerative primer developed by Wyatt and Brown (Phytopathology, 1996, 86: 1288–1293) to detect presence of WFT geminivirus. In all the three resistant lines *viz.*, GKC-29, BS-35 and EC-497636, PCR reactions were negative, confirming absence of viral genomes in these lines (Kumar *et al.*, 2006). Hence, it may be concluded that the resistance or susceptible reaction against PepLCV of a given chilli genotype can successfully be determined using this degenerative primer. This primer will also facilitate in most reliable and early demarcation of individual susceptible or resistant plants in the segregating generations, which have been developed with the objective to study the genetics of host plant resistance against PepLCV.

## 4.6. Disease Resistance

The most serious diseases are anthracnose/fruit rot/dieback (*Colletotrichum* spp.), powdery mildew (*Levillula taurica*), *Cercospora* leaf spot (*Cercospora capsici*), bacterial leaf spot (*Xanthomonas campestris* pv. *vesietoria*) and wilt (*Pseudomonas solanacearum*). Tobacco Mosaic Virus (TMV), Cucumber Mosaic Virus (CMV), Chilli Mosaic Virus and Pepper Leaf Curl Virus (PepLCV) are the common viral diseases and an early infection of any of these may cause yield losses ranging from 80 to 100%. Resistant sources *e.g. C. baccatum* var. *baccatum* for *Colletotrichum capsici*; Hungarian wax and PI-288982 for *C. gleosporides*; Lorai, Perennial, S-27, S-41-1 for *Phytophthora capsici,* Hungarian Wax for leaf spot, Arka Gaurav tolerant to wilt; *C. baccatum* var. *baccatum*, PI-288982 moderately resistant to powdery mildew; Perennial, S-41-1, IHR-996 resistant to Potato virus Y (PVY) have been identified. None of the sweet pepper lines are resistant to CMV. However, Puri Red, Perennial, IHR 328–9 among chillies are reported as resistant.

A number of multiple disease resistant lines have been developed and among these, Punjab Lal has been most successful under field conditions. It is resistant to CMV, CMMV, wilt and *Spodoptera* leaf eating caterpillar. Besides, other multiple resistant lines are Pant C-1, Pusa Sadabahar, Perennial, BG-1, Lorai, Tiwari, and Indonesian Selection. The multiple diseases resistant lines have many undesirable characteristics like small fruit size, late maturity and low fruit yield. Therefore, there is need to improve undesirable characters by removing the linkage drag.

### 4.7. Insect Resistance

Important insects, which attack chilli and sweet pepper, are thrips (*Scritothrips dorsalis*) mites (*Polyphagotarsonemus latus*), white fly (*Bemisia tabaci*) and aphids (*Aphis gossypii* and *Myzus persicae*). The tolerant sources are Calcapin Red, Chamtkar, NP 46-A, X-1068, X-743, X-1047, BG-4 against thrips; LEC-1, Kalyanpur Red, X-1068, X-204, Goli Kalyanpur, Punjab Lal against mite and LEC-28, LEC-30, LEC-34, Kalyanpur Red and X-1068 against aphid. However, no significant work has been done on breeding resistant varieties to insects in chillies.

### 4.8. Mutation Breeding

The induced mutations are of considerable importance in crop improvement programmes. Chilli pepper has received limited attention for mutation breeding (Venkata Rajam and Subhash, 1986). Kumar *et al*. (2001) obtained sixteen morphological mutants and classified them in four categories in $M_2$ and $M_3$ progenies. These categories are plant height mutants, leaf mutants, maturity mutants and fruit mutants.

## 5. VARIETIES

Description of some popular varieties and hybrids of chilli pepper is given below:

### 5.1. Open Pollinated Varieties

#### 5.1.1. *Kashi Anmol (KA-2)*

This is an improved population derived from a Sri Lankan introduction KA-2, through simple recurrent selection at IIVR, Varanasi. Plants are determinate, bushy, umbrella shaped, bear green attractive pendent fruits and stems are characterized by nodal pigmentation. First picking starts at about 55 days after transplanting, gives an average green fruit yield of 20–25 t/ha in 150–160 days of crop duration. This is also suitable for green fruit production under chilli-wheat cropping system.

### 5.1.2. *Pusa Jwala*

This variety has been developed at Indian Agricultural Research Institute (IARI), New Delhi. Plants are dwarf, bushy and bear pendent fruits. Fruits are light green, 9–10 cm long, highly pungent, tolerant to thrips and mites, suitable for dry fruit production, gives an average green fruit yield up to 15 t/ha.

### 5.1.3. *Pusa Sadabahar*

This variety has been developed at IARI, New Delhi. Plants are erect, 60–80 cm tall with perennial habit and bear upright fruits. Fruits are 6–8 cm long, born in cluster, light green and turn dark red at maturity, highly pungent, reported to be tolerant to CMV and leaf curl complexes. First picking start on 75–80 days after transplanting and gives an average green fruit yield of 14 t/ha.

### 5.1.4. *Bhagyalakshmi (G-4)*

This variety has been developed at Regional Agricultural Research Station (RARS), Lam, Guntur; derived through selection in Thohian chilli from Sri Lanka. Plants bear narrow and dark green leaves. Fruits are olive green, turn dark red on ripening, 7–8 cm long, 3.0–3.5 cm girth with 38–40% seed content, fairly tolerant to diseases and pests and gives an average green fruit yield of 10 t/ha under irrigated and 6 t/ha under rainfed conditions.

### 5.1.5. *Andhra Jyoti (G-5)*

This variety has been developed at RARS, Lam, Guntur, which was derived from the cross G-2 × Bihar collection. Fruits are short, stout with conical shape, 3–4 cm long and 4–4.5 cm in girth and gives an average green fruit yield of 8.5 t/ha.

### 5.1.6. *Arka Lohit (Sel-1)*

Arka Lohit has been developed at IIHR, Bangalore, derived through selection from a local collection (IHR-324). Fruits are dark green, smooth, straight, turn dark red at maturity, highly pungent (capsaicin 0.70%), suitable for cultivation in irrigated and rainfed conditions and gives an average green fruit yield of 15 t/ha in 180 days of crop duration.

### 5.1.7. *Utkal Abha (BC-14-2)*

This variety has been developed at Orissa University of Agriculture & Technology, Bhubaneshwar. Fruits are medium sized, deep red at ripening with high pungency, suitable for both green and dry fruit production, gives an average yield of 11 t/ha as green fruits and 3 t/ha as dry fruits in 125–130 days of crop duration.

#### 5.1.8. *Kashmir Long-1*

This variety has been developed at Sher-e-Kashmir University of Agriculture & Technology, Srinagar. Plants are medium tall, spreading with oval, green leaves and stem having purple pigmentation on nodes. Ribbed fruits are bright red, pendent, long with thick pericarp. The average red ripe fruit yield is 15–17 t/ha.

Some other chilli varieties developed in India are: Sindhur (CA-960), Aparna (CA-1068) CO-1, CO-2, NP 46A, K-1, K-2, MDU-1 (mutant), Pant C-1, Pant C-2, Jawahar-218, Punjab Lal, Musalwadi, etc.

### 5.2. Hybrid Varieties

#### 5.2.1. *CCH-2 (Kashi Surkh)*

This hybrid has been developed at IIVR, Varanasi. This is a $F_1$ hybrid of male sterile line (CCA 4261) × Pusa Jwala. Plants are semi-determinate (1–1.2 m), erect and nodal pigmentation on stem, fruit colour light green, straight, length 11–12 cm, suitable for green as well as red fruit production. The first harvest starts after 55 days of transplanting. Green fruit yield is 24 t/ha, whereas red is 14 t/ha. This is tolerant to thrips, mites and viruses. This variety has been identified for Zone II, IV, V and VI through AICRP-VC.

#### 5.2.2. *CCH-3 (Kashi Early)*

This hybrid has been developed at IIVR, Varanasi. The plant height ranges form 60–90 cm; foliage is light green with anthocyanin pigmentation at nodes. Fruit size is large (10 × 1.2 cm). This has been recommended for green fruit production and it gives green fruit production of 24–25 t/ha and dry fruit yield of 14 t/ha.

#### 5.2.3. *CH-1*

This genetic male sterile line based hybrid has been developed at PAU, Ludhiana. Plants are tall (about 100 cm) and spreading type with light green foliage. Fruits are light green, 6.6 cm long and characterized by 23.05% dry matter, 0.82% capsaicin, 136.5 ASTA unit oleoresin and 98.37 mg vitamin C/100 g. Plants are tolerant to diseases and pests; gives an average yield of 9.5 t/ha (red ripe fruits) and suitable for red ripe fruit production.

### 5.2.4. *CH-3*

This genetic male sterile based hybrid has been developed at PAU, Ludhiana. Plants are medium dwarf (about 70 cm) having pendent fruiting, fruits are dark green, 8.2 cm long and characterized by 22.55% dry matter, capsaicin 0.5%, 145 ASTA oleoresin and 110 mg vitamin C/100 g. This hybrid has advantage over CH-1 in terms of earliness, better fruit colour, mild pungency and longer fruit suitable for processing. Gives an average yield of 9.5 t/ha (red ripe fruits) and is suitable for red ripe fruit production.

### 5.2.5. *Tejaswani*

This hybrid has been developed by private organization and the plants are bushy with 70–80 cm of height and bear pendant fruits. Fruits are dark green (turn dark red at maturity), 9–11 cm long, highly pungent, suitable for both green and red fruit production, first picking starts on 75–77 days after transplanting and gives an average green fruit yield of 34.0 t/ha.

### 5.2.6. *NS-1101*

This hybrid has been developed by private organization and the plants are prolific bearer with dark green foliage providing a good foliage cover. Fruits are lustrous dark green (turn dark red), medium long (10–12 cm), and medium pungent. Suitable for green fruit production, first picking starts 70–75 days after transplanting and gives an average green fruit yield of 30.0 t/ha.

Some other hybrids commonly cultivated are Soldier, ARCH-228, ARCH-236, BSS-453, Indira chilli, INDAM-67, NCH-338, etc.

## 6. INPUT USE EFFICIENCY

In order to get maximum return from any crop with the available resources, the inputs or the resources has to be efficiently and judiciously utilized. The important inputs for growing chillies are seed, manures and fertilizer, water and safe protective measures. Good quality seed is the most critical input in any crop cultivation including chillies. It has been observed that seed germination in chilli can be increased by using NaCl ($10^{-4}$) (Annonymous, 2006). It is always better to treat the seeds while storing or sowing. Seed treatment refers to the application of fungicide, insecticide or a combination of both to seeds, so as to disinfect and disinfest them from seed-borne or soil-borne pathogenic organisms and crop pests both in field and in storage. Seed treatment helps in prevention of spread of plant diseases, protects seed from seed rot and seedling blight, improves germination, and provides protection from insect pests.

Recommended doses of chemical fertilizers along with sufficient amount of organic manures are the important inputs for a good crop. It is universally accepted that the use of chemical fertilizers is an integral part of the package of practices for raising the agricultural production to a higher place. Studies conducted by the Food and Agricultural Organization (FAO) of the United Nations have established beyond doubt that there is a close relationship between the average crop yield and fertilizer consumption level. Moreover, the nutritional requirement of chillies could not be fully met with the use of organic manures like farm yard manure (FYM) and other bulky organic manures like neem cake, castor cake, groundnut cake, etc., for want of their availability in adequate quantities. Hence, there is need for an efficient use of fertilizers as major plant nutrient resource in enhancing the chilli productivity. Other resource of plant nutrients like organic manures, bio-fertilizers, etc., also should be integrated to get the maximum agricultural output from every kilogram of applied nutrient in the form of fertilizers. The recommended dose of manures in chilli crop is 20–25 tonnes of FYM, 120–150 kg nitrogen, 80 kg each of phosphorous and potassium per hectare. In addition to FYM, addition of neem cake @ 4–5 q/ha at the time of last ploughing is beneficial. Whenever an insecticide is sprayed on crop, addition of 1% urea along with protective measures is useful for proper growth and development of plants.

The critical stages of irrigations under chilli crop is just after transplanting, at active flowering and fruit setting stage, failing which results in the non-establishment of plants, flower/bud drop and hence the reduced yield. Under limited water availability conditions, irrigation is provided only at important critical stages of crop so that the yield reduction is not substantial and the water use efficiency is very high, while irrigating chilli crop in heavy soil, care should be exercised that there should not be water stagnation even for few hours. Number of irrigations, interval between two successive irrigations depends on type of soils, age of crop and climatic conditions, such as for light soils: 7–10 days interval, heavy soils: 10–15 days interval and during summer: 4–6 days interval. After every harvesting of chilli, irrigation should be provided (as per the moisture regime of soil). Irrigation should be provided at 40–60 per cent depletion of available soil moisture. Excess irrigation at these stages causes reduction in flowering and yield, higher incidence of fungal diseases and reduction in capture of nutrients, while less irrigation than required causes flower and fruit drop and less branching. After prolonged drought, if irrigation is provided at flowering and fruiting stage, lot of flower and fruit drop will be seen.

All other management practices of crop husbandry will be futile if the crop is not protected against the ravages of pests. Experts' assessments reveal that around 22 per cent of yield losses in major chillies can be attributed to insect pests. In absence of crop protection the yield may be drastically declined. Hence, there is need to reduce if not eliminate these losses by protecting the crops from different diseases and pests through appropriate techniques.

A large number of growth substances have been used to regulate growth and flowering in chilli (Table 2.8). From the findings of several workers on the use of different growth regulators and its effects on chilli, Rajput and Parulekar (1996) made a comprehensive report on these aspects. In addition to the above, timely intercultural operations like weeding, hoeing and earthing up is essential for a good crop return.

**Table 2.8.** Response of growth regulators in chilli

| Growth regulator | Concentration (ppm) | Stage of application | Responses |
|---|---|---|---|
| NAA (Naphthalene acetic acid) | 50 | Full bloom stage | Control of fruit drop and yield increase |
| NAA | 10 | First at flowering and second at 5 weeks later | Increased fruit set |
| NAA | 60 | Bloom stage | Increase in yield |
| MH (Maleic hydrazide) | 3000 | Pre-flowering | Suppression of flowering |
| Ethrel (Ethaphon) | 200 | Pre-flowering | Improved flowering and fruit set |
| GA | 50 | Fruit setting | Decrease in flower shedding and increase fruit set |
| Ethrel (Ethaphon) | 500 | At maturity | Turning colour and increase in yield |
| Tricontanol | 2 | First at 30 days after transplanting and second at bloom stage | Reduction in flower drop and increase fruit yield. |

## 6.1. Quality Planting Material

Quality seed is the most important and crucial factor for a good crop husbandry. Better the quality of the seed, better is the assurance for productivity. Chilli grower use the seeds purchased from the markets (made available either through public organizations or private sectors), exchange from among the growers or their own saved seeds of the last crop harvest. In India, the acreage under improved/notified/identified/released varieties of chilli pepper is lesser which needs utmost attention. The $F_1$ hybrids or the open pollinated (OP) varieties marketed by the private organizations are costlier and sometimes render the farmers to use unimproved local varieties. Seed production of chillies requires sufficient isolation distance (more than 400 m) to maintain the purity. The cross-pollination in chillies ranges up to 90 per cent. Bagging or netting of individual plants or whole plot in field with several genotypes is essential to get the true to type seeds. Public sectors in our country have already developed a large number of open pollinated and hybrid varieties of chilli.

Seeds of open pollinated or hybrid varieties can be obtained from developing centres. For example, seeds of Kashi Anmol can be obtained at commercial scale from IIVR, Varanasi upon prior indenting. Farmers are always advised to purchase seeds from the reliable source(s). However, now the time has come, when farmers should be trained for seed production at least for their own requirement. Therefore, it is advisable that small quantity of seeds of any of the desirable open pollinated varieties should be obtained from developing centres, which can be multiplied by the farmers themselves by maintaining isolation distance of 400 meters from other variety or by bagging/netting of plant(s). However, it is understandable that all the farmers may not be able to maintain 400 m isolation distance. Therefore, they should be advised to use nylon cages to produce genetically pure seeds. Cages of small size may be used for individual plants or bigger cages for the plot as a whole.

## 7. BIOTIC AND ABIOTIC STRESSES

A number of diseases and insect-pests attack chilli crop. Apart from this, often the climatic vagaries also hinder proper growth and development of the crop. Many of them are common throughout the world, while few of them are specific to certain regions. A brief description on symptoms and their control measures of economically important diseases and insect pest have been given in the following paragraph.

### 7.1. Biotic Stresses

The diseases caused by pathogens and viruses transmitted by insect-pests are described here under:

#### 7.1.1. *Anthracnose*

This fungal disease is also called as 'dieback' disease which occurs worldwide and is caused by *Colletotrichum gloeosporioides, C. capsici, C. acutatum,* and *C. coccodes*. The fungus affects all growth stages including post harvest stage. On ripe fruit small, sunken and circular depressions up to 30 mm in diameter is seen. The center of the lesions becomes tan in color while the tissue beneath the lesion is light colored and dotted with many dark-colored fruiting bodies of the fungus that form concentric rings in the lesion. The twigs are water soaked, brown and die back. Green fruit may also be infected but symptoms will not appear until the fruit ripens at harvest time. Such an infection is called latent infection (*AVRDC Fact sheet*). In India, among the *Colletotrichum* spp. that affect chilli pepper, *C. capsici* and *C. gloeosporioides* have the widest host range among solanaceous crops.

These pathogens are seed-borne and may also persist on alternate hosts such as other solanaceous crops (tomato, potato and eggplant), cucumber, and many other cultivated crops and weeds. The optimum temperature for fruit infection is 20–24°C with fruit surface wetness, although infection may occur from 10–30°C.

***Management***: Seeds should be collected from anthracnose-free fruit and/or treat seeds with a fungicide. Use of healthy transplants is also beneficial. Crop rotation with potato, soybean, tomato, and eggplant should be avoided. Harvest fruit as soon as it ripens since anthracnose develops more readily as the fruit matures. Weeding should be done regularly and fruit injury should be avoided. Control by fungicide application will depend upon proper dosage and good coverage of fruit. Seed treatment is done with Bavistin @ 0.25 per cent. Seedlings spray with either copper oxychloride @ 0.2 per cent or carbendazim 0.1 per cent are also beneficial.

### 7.1.2. *Damping Off*

Damping Off disease is caused by *Pythium* spp., *Phytophthora* spp., *Fusarium* spp., *Rhizoctonia solani* and other fungi. It is prevalent in every chilli growing areas of world. Seedlings may be infected before or after emergence. If infected before emergence, the germinating seeds/plants become soft, brown and decay. If infected after emergence, water-soaked lesions above or near the soil line is formed. The stem softens and cannot support the seedling, which collapses and dies. Affected plants usually occur in patches in nursery beds or lower parts of fields. Symptoms of the infections may vary with age and stage of development of the plant. When infection occurs late after emergence, the infections are not lethal but plant growth and yield may be affected with foliar yellowing, reduced vigor and stunting. Plants severely affected in the root region may wilt in warm or windy weather. Symptoms of nutrient deficiency may occur in the foliage because nutrients are prevented from moving up the plant by extensive root rotting.

Seedlings during the three weeks after sowing are particularly susceptible. Several factors favour the disease development such as sowing seeds in disease-infested soils, over-watering, poor drainage, inadequate light, overcrowding, poor ventilation and applying excess levels of nitrogen to soil. The presence of nematodes may exacerbate disease development (*AVRDC Fact sheet*).

***Management***: Well-drained seed beds are chosen for nursery locations. To reduce soil moisture, sowing seeds on raised beds, and appropriate drainage in the field to avoid waterlogged conditions should be ensured. Treatment of seed with hot water, followed by application of a fungicidal seed protectant, and sowing into sterilized soils free from fungi is recommended. It is important to determine which fungus is responsible for disease development. Fumigate nursery beds or apply a fungicide as soil drench, if the disease appears. Soil treatment with carbendazim @ 0.3 per cent and seed treatment before sowing followed by drenching with copper oxychloride with 0.3 per cent is found beneficial.

### 7.1.3. *Bacterial Leaf Spot*

Bacterial leaf spot of pepper is caused by *Xanthomonas campestris* pv. *vesicatoria*. This bacterium survives in crop debris in the soil and on seed. Disease develops only when there is plenty of moisture and warm temperatures. During the summer monsoon sufficient moisture may be present for infection when the bacteria are splashed or blown on to leaves. Disease severity also depends on the level of tolerance of pepper varieties to disease. Symptoms often appear first in the lower foliage as water-soaked lesions. The center of the lesions may die out leaving spots that have a purple-brown border and a lighter brown center. Disease is less severe in well fertilized and irrigated plants.

***Management***: Application of streptocyclin @ 200 ppm is effective for disease control. Disease free seed should always be used for planting. Contaminated seed should be treated with a 40 min soak of 20 per cent bleach solution using one gallon of solution per pound of seed (AVRDC Fact sheet). Shake the seed continuously while soaking, and dry promptly afterward.

### 7.1.4. *Root-Knot Nematode*

The causal organisms of root knot disease are *Meloidogyne incognita, M. javanica, and M. hapla,* in order of importance. The disease is predominant particularly in warm climates. Symptoms consist of stunting, yellowing and a general unhealthy appearance of plants, wilting and death may occur in dry hot weather. The plant will show reduced fruit and leaf size with consequent low yield. Below the ground, the primary and secondary roots will have obvious galls or knot-like swellings in the root tissue. These swellings prevent movement of water and nutrients in the plant resulting in stunting. Soil-borne fungi and bacteria more easily infect plants affected by nematodes. This secondary infection may lead to severe discoloration of internal stem and root tissue, and rapid plant death.

***Management***: Rotation of pepper crops with crops that tolerate nematodes, such as sorghum, Brassicas or onion. Soil solarization for 4 to 8 weeks in small area is also possible. It will be most effective when conducted during the hottest season of the year. Most plant parasitic nematodes are killed between 44 and 48°C. The depth of penetration with solarization is about 5 to 10 cm. Application of Phorate and Nemacur has been found effective in the disease control. Steam sterilization and soil fumigation in greenhouses are found to be beneficial.

### 7.1.5. *Whitefly-Transmitted Geminiviruses*

Whitefly-transmitted geminiviruses affect several crops including chilli peppers. These viruses have almost similar symptoms but are biologically and genetically distinct. Symptoms depend upon the geminivirus and the pepper variety involved. Common symptoms are stunting, curling, or twisting of the leaves, bright yellow mosaic, distortion of leaves and fruit, and reduced yields. One of the most destructive viruses, PepLCV (Pepper leaf curl virus) has very common occurrence in all over the Indian sub-continent. The typical symptoms are: leaves curl towards midrib, plants remain stunted, flower buds abscise and pollen development is hampered, no fruit set or development of tiny fruits with no market value. Losses up to 80 per cent have been reported in northern India, however, incidences up to 100 per cent at the farmers' field are not uncommon (Prakash and Singh, 2006).

***Management***: It is very difficult to control geminiviruses once a crop becomes infected. Few cultural practices can be used to prevent infestations, *e.g.* growing seedlings in an insect-proof net house or seedbed to prevent from insect (vector) infestation. Chemical control methods include the use of systemic insecticides as soil drenches or overhead sprays during the seedling stage. Applications in the field may be needed to control adults that infest after transplanting. Rotation of insecticides is recommended to prevent insects developing resistance to the chemicals. Chemical control may not be effective in areas where disease incidence is high. Oil sprays may also be effective in reducing levels of infestation. Neem seed extracts control young nymphs, inhibit the growth and development of adults, and reduce egg lying by adults.

### 7.1.6. *Aphid-Transmitted Viruses*

This is the largest genus of plant viruses consisting of several species. Some of the common viruses affecting chilli production are Potato virus Y (PVY; Jeyarajan and Ramkrishnan, 1969), pepper veinal mottle virus (PVMV; Prasad Rao, 1976), tobacco etch virus (TEV; Prasad Rao, 1976), pepper vein banding virus (PVBV; Prasad Rao, 1976) and chilli veinal mottle virus

(CVMV; Prakash *et al*., 2002). Most of these viruses have wide host range, transmitted by mechanical sap inoculation and by several species of aphids (*i.e. Aphis gossypii, A. craccivora, A. fabae,* and *Myzus persicae*) in a non-persistent manner (Prakash *et al.*, 2002). The symptoms of PVY includes mosaic, mottle, dark green vein banding and vein clearing of leaves, whereas PVMV exhibits vein clearing and mosaic mottling on just emerging leaves as an initial symptom which later on elicit distortion and extremely reduced leaf lamina leading to stunting of plants with short internodes. TEV produces dark green mottling, mild chlorosis, fain yellow flecking, vein clearing severe stunting and wilting of plants.

Reddy *et al*. (2004) collected six isolates and identified all of them as chilli veinal mottle virus (CVMV) based on the host range, serological relationship, electron microscopy and phylogenetic analysis of coat protein sequences. They inoculated 25 chilli pepper genotypes with CVMV isolates and found three genotypes as immune, 10 as highly resistant, two as resistant and ten genotypes as susceptible after confirming through symptoms and ELISA test.

***Management***: Use of insecticides during the growing season is mostly ineffective, however control of aphids early in the season prior to seeding or planting may be useful. Spray weeds bordering the field with an aphicide prior to seeding or planting the field to prevent the aphids moving from weeds to main crop. Destroy all annual weeds in the field, including those in hedge or fence rows. Aphid populations should be monitored early in the season and mineral oil or other insecticide treatments are applied when needed. The mineral oil sprays will reduce the frequency of transmission of the virus by the vector and thereby delay development of the disease in the pepper crop. Use of a finer mesh netting to exclude aphids from transplants before they are set into the field saves from the infection. Planting chillies close to established tomato, tobacco and pepper fields should be avoided since these fields may harbour aphids.

### 7.1.7. *Aflatoxins*

Aflatoxins are toxins produced by moulds which occur throughout the world. Aflatoxin may be produced, both before and after harvest. Beside chillies and paprikas, groundnuts, pistachios and other edible nuts and cereals are identified as highly susceptible to aflatoxin contamination. Aflatoxins are the most potent carcinogens of mycotoxins and are the most commonly found toxin in chilli and paprika spice. Aflatoxins are produced by some strains of *Aspergillus flavus* and *Aspergillus parasiticus* moulds if they encounter appropriate environment. The most potent aflatoxins are B1, B2, G1 and G2, all of which have been found in chilli and paprika spice. In addition occasionally ochratoxin-A produced by *Aspergillus ochraceus* or

*Aspergillus carbonarius* may occur in chillies. Aflatoxins are chemically very stable. Therefore, if contamination has occurred, it is very difficult to destroy them by processing or cooking. Recently ICAR has given due attention on the aflatoxin research work in India under the network project 'Prevention and management of mycotoxin contamination in commercially important agricultural commodities'.

Drought stress favours pre-harvest contamination, while post-harvest contamination mostly occurs during the rainy season. Diseased fruits that are not sorted after harvest bear a risk of mould contamination that could lead to aflatoxin production. Moisture content of the product above 11 percent during storage allows fungal growth and above 14 per cent allows aflatoxin accumulation.

In a survey, the preliminary observations reveal that out of 166 samples tested, 70 samples yielded *A. flavus* and 10 samples yielded *A. ochraceous* on blotter paper test. The moisture gradient is more determining than inoculum's dose as far as infection of *A. flavus* on chilli is concerned. Relative humidity of 90 to 95 per cent is the best for getting maximum infection by *A. flavus* on chilli fruits. The spore inoculums load of $1 \times 10^4$ was optimum threshold to create proper infection on red chilli fruits. Geographically different chilli samples collected from various parts of the country including 9 commercial chilli powder and 25 chilli fruit samples were tested for aflatoxin contamination through ELISA. The maximum (782.3 µg/kg) aflatoxin $B_1$ was recorded in Tirupati sample-4, while minimum toxin (15.3 µg/kg) was obtained in Kanpur and Delhi samples. In commercial chilli powder samples, the maximum toxin was recorded in Nagpur sample (551.3 µg/kg) followed by Siliguri-21 (85.6 µg/kg). The minimum toxin level of 15.3 µg/kg was recorded in Pune and Agartalla chilli samples. Among the 70 isolates of *A. flavus*, 21 isolates were observed as toxicogenic and rest of the isolates was non-toxicogenic. DNA extraction from 57 isolates of *A flavus* and 9 isolates of *A. ochraceous* were done by CTAB method. However, only 36 *A. flavus* and five *A. ochraceous* samples gave detectable DNA. Total 34 random and four universal primers were screened in which only one primer, 5′-CAGGCGCACA-3′ was best for maximum amplification and polymorphism of different *A. flavus* isolates. The radial growth of *A. flavus* was inhibited by different *Trichoderma* species on 5th day at maximum level and remains constant up to 10 days after inoculation (DAI). It indicated that pathogen was completely restricted by antagonist and none of the *A. flavus* isolate has capability to over come the antagonist even after 10 DAI. Forty varieties artificially screened against *A. flavus* by pin-pricked and dip method and PBC-535 (0%), Tabasco (0%), CCH-1 (2.0%), LCA-424 (2.50%) and Capsicum chilli (2.5%) were found resistant against *A. flavus*.

***Management***: Varieties of chilli and paprika should be selected having a thin pulp with as high in dry solid content as possible. This makes drying easier and reduces heat level of the fruits. Plants should be stable and have a fruit setting, which ensures that fruits do not get in contact with the soil. In case of severe drought before harvesting, crops should be irrigated to avoid stress. Contact of irrigation water with the fruits should be avoided as this increases risk of microbial contamination. Time between harvesting and drying should be as short as possible. This includes transport from the field to post-harvest facilities; sorting and preparation for rapid drying. After harvesting, diseased and injured fruits must be removed. Cutting of fruits into small pieces (2.5 × 2.5 cm for paprika) to achieve rapid drying is recommended. This method reduces drying time by 50 to 80 per cent. But cutting also gives fungi access to fruit tissues. Thus, cutting must be done quickly and should be done by an automatic shredder. Time between cutting and drying must be minimized. It is important to reach quickly 'safe' moisture content. That means moisture content of 8 per cent, which is attained by drying within 48 hours. If the chillies are at 'unsafe' moisture content for longer than 48 hours, mould can grow and mycotoxins be produced. Drying methods which needs more than 48 hours to reach this level must be improved. Solar drying is a very economical method, but utilizes high temperatures and does not quickly dry fruits to low moisture content. High temperatures do result in brown discoloration of the spice. If the weather conditions are not favourable due to rain, cloud, etc., drying temperatures are reduced and drying times are extended, potentially allowing aflatoxin accumulation.

### 7.2. Insects-Pests

The two most important insects affecting chilli crops are thrips (*Scirtothrips dorsalis*) and mite (*Polyphagotarsonamus latus*). Besides, white fly and aphids also harm the crop in various ways. These sucking insects-pests create cosmopolitan problems resulting in average loss of over 35 per cent (Ahmed *et al*., 1987) through various types of chilli leaf curl diseases. Satpathy *et al*. (2006) studied the pattern of thrips infestation in chilli and found that the treatment of methomyl @ 300 g ai/ha was the most effective followed by dimethoate @ 300 g ai/ha.

### 7.3. Abiotic Stresses

The abiotic factors affecting the chilli crop are described as below:

#### 7.3.1. *Temperature*

The principal abiotic stresses affecting chilli pepper cultivation are temperature and drought. Loss of flower buds and flowers primarily occurs due to these factors. Depending on when the stress occurs, the disorder can result in the delay of anthesis, prolongation of the flowering period without fruit set or early termination of fruit setting. High night temperature is more detrimental to fruit setting than the high day temperature (Aloni *et al.*, 1991). When high temperature is coupled with moisture stress, abscission is further increased. Low temperature generally leads to production of small seedless fruits.

***Management***: Irrigation at flowering and fruit set stage help in reducing flower/fruit drops. Foliar application of 50 ppm NAA at full bloom stage effectively control the fruit drops and Triacontanol (Vipul 1 ml/2 liter water) also reduces flower and fruit drops.

### 7.3.2. *Sun scald*

Affected areas become straw-colored or white, soft, sunken and wrinkled. These dead areas form only on the side exposed to the sun, in contrast to blossom end rot where the symptom appears on unexposed areas, as well. The dead areas eventually become papery in texture, and may become dark-colored if infection by secondary fungi occur. Fruit affected by sun scald is unmarketable. This occurs on fruit suddenly exposed to intense sunlight at high temperatures and high humidity. Sudden exposure may be due to loss of foliage cover, defoliation, or prolonged wilting. Also, breakage of branches due to rough handling during harvest or from heavy rains may expose fruit to sun scald.

***Management***: Sufficient nitrogen should be provided for healthy plant growth. Keeping foliage healthy by controlling diseases and insect pests helps control the disorder. Prolonged dry spell is avoided. If feasible, provide support for pepper plants by use of stakes or use of string running along the rows, or wire running horizontally along the beds.

## 8. POST HARVEST AND VALUE ADDITION

The post harvest operation starts immediately after the harvest of the crop. The different operation performed in chilli harvesting and post harvest care is discussed below:

### 8.1. Harvesting

Pepper fruits can be harvested either at the green immature or red mature stage. For vegetable use, green fruits are harvested. For pickles, picking is done at ripe or green stage. For the preparation of dry chilli, the chilli fruits should be in fully red condition or more than 80 per cent red stage. Under well managed situation 10–12 pickings can be harvested as green chilli or 7–9 pickings as red chilli stage. Harvested red mature fruits should be kept at room temperature for two days for the development of red colour of partially red fruits before exposure to sunlight for drying. Laying material like mats or gunny bags should be used to dry the pods under sunlight. Good quality dry chilli can be produced 5–6 days after drying under sunlight.

### 8.2. Post Harvest Management

Fruits are stored in a cool, shaded, dry place until they are sold. At typical tropical ambient temperature and humidity (28°C and 60% relative humidity), fruits may last unspoiled for 1–2 weeks. Anthracnose is the major cause of fresh fruit spoilage and for dry fruits, colour retention ability of mature fruits is important. Drying of fruits in the sun is a common practice, but this tends to bleach the fruits, and rainfall and dew promote fruit rot. Dry chillies can be kept for longer period in dry places. The moisture content of dry pods is to be kept at 8–10 per cent. Well dried pods after removing the extraneous matters should be packed in clean, dry gunny bags and stored ensuring protection from dampness.

### 8.3. Value Addition

Capsaicin, chilli oil, powder and oleoresin are used to impart pungency. Nayaki (2004) has reported that extracts of chilli are used in the production of ginger beer and other beverages. Capsaicin has many medicinal properties, especially as an anti-cancerous agent and instant pain reliever. It has been observed that capsicum pigment is incorporated in poultry feed. In Mexico and some other countries, pigments are concentrated and blended in feed mix for chickens (Prasad and Saini, 2004). This gives a reddish tinge to the chicken meat, which is more valued. It is believed that yolks of eggs of such chickens are also more coloured and healthy looking. Chilli is highly perishable in nature and it requires more attention during harvest, storage and transportation. Most of the growers sell their fresh crop in the market and it is consumed as green chilli. Through value addition, its market price, shelf life and the quality can be increased. To obtain quality product, care has to be taken during picking, cleaning, sorting, grading, drying, packing, storage and transportation. Processed products such as dehydrated chilli, pickle, powder, paste, sauce, etc. can be prepared for higher returns. Almost all chilli growers sell it directly. Real return can come only from processed products (Prasad and Saini, 2004). Efforts of scientists, extension agencies and the state governments are needed to educate the farmers regarding the processing of chilli.

## 9. CONCLUSION

Chilli is valued for its diverse commercial uses. India is a major producer, exporter and consumer of chilli. Bangladesh, Bahrain, Israel, Japan,

Malaysia, USA and UAE are among the leading importers of chilli from India. It is gaining importance in the global market because of its byproducts like powder, oleoresin, capsanthin, chilli paste and chilli oil. Presently India has become world's largest producer and exporter of chilli contributing over 20 per cent of world's total production. Efforts are needed for the better management practices with respect to cultivation, harvesting and most importantly on the post-harvest management practices.

## REFERENCES

Aloni B, Pashkar T and Karni L (1991). Partitioning of $^{14}C$ sucrose and acid invertase activity in reproductive organs of pepper plants in relation to their abscission under heat stress. *Ann. Bot.* **67**: 371–7.

Andrews J (1995). Peppers: The Domesticated Capsicums, University of Texas Press, Austin. pp. 186.

Andrews J (1998). The Pepper, Lady's Pocket Pepper Primer. University of Texas Press, Austin. pp. 190.

Anu A and Peter KV (2000). The chemistry of paprika. *Capsicum and Eggplant Newslett.*, **19**: 19–22.

AVRDC Fact sheet (2009). The World Vegetable Center, Taiwan, www.avrdc.org.

Bhalekar MN, Dutonde SN, Gupta NS and Warade SD (2004). Evaluation of chilli (*Capsicum annuum* L.) genotypes for variability, path analysis and genetic divergence under high temperature. *In*: Proc. of the XII EUCARPIA Meeting on Genetics and Breeding of Capsicum and Eggplant, May 17–19, 2004, The Netherlands, p. 84.

Bosland PW (1994). Chiles: History, cultivation, and uses. *In*: G. Charalambous, (*ed.*), Spices, Herbs and Edible Fungi. Developments in Food Science Vol. 34. Elsevier, Amsterdam, pp. 347–66.

Bosland PW (1996). Capsicums: Innovative uses of an ancient crop. *In*: J. Janick (*ed.*), Progress in new crops. ASHS Press, Arlington, VA.

Bosland PW and Votava EJ (2000). Peppers: Vegetable and Spice Capsicums. Crop Production Science in Horticulture. CAB International Publishing, Wallingford, England, UK. pp. 204.

Cichewicz RH and Thorpe PA (1996). The antimicrobial properties of chile peppers (*Capsicum* species) and their uses in Mayan medicine. *J. Ethnopharmacology* **52**: 61–70.

Das SS, Kumar S and Singh JN (2001). Cytomorphological characterization of a nuclear male sterile line of chilli pepper (*Capsicum annuum* L.). *Cytologia* **66**: 365–71.

DeWitt D and Bosland PW (1996). Peppers of the world: An identification guide. Ten Speed Press, Berkeley, California. pp. 219.

Erickson AN and Markhart AH (2002). Flower development stage and organ sensitivity of bell pepper (*Capsicum annuum* L.) to elevated temperature. *Plant Cell Environ.* **25**: 123–30.

Eshbaugh WH (1993). Peppers: History and exploitation of a serendipitous new crop discovery. pp. 132–139. *In*: J. Janick and J.E. Simon, (*eds.*), New Crops. John Wiley & Sons, New York.

FAO (2003). Food and Agriculture Organization of the United Nations Production Yearbook 2001, Vol. 55, Statistics Series No. 170. FAO, Rome. pp. 333.

Gonzalez M and Bosland PW (1991). Strategies for stemming genetic erosion of *Capsicum* germplasm in the Americas. *Diversity* **7**(1&2): 52–53.

Greenleaf WH (1986). Pepper breeding. p. 67–134. *In*: Mark J. Bassett (*ed.*), Breeding vegetable crops. AVI, Westport, CT.

Han XB, Li RQ, Wang JB and Miao C (1996). Effect of heat stress on pollen development and pollen viability of pepper. *Acta Hort.* **23**: 359–64.

Hornero-Mendez D, Costa-Garcia J and Mínguez-Mosquera MI (2002). Characterization of carotenoid high-producing *Capsicum annuum* cultivars selected for paprika production. *J. Agric. Food Chem.* **50**: 5711–6.

Jarret RL and Dang P (2004). Revisiting the *waxy* locus and the *Capsicum annuum* L. complex. *Georgia J. Science* **62**: 117–33.

Jeyarajan R and Ramakrishnan K (1969). Potato virus 'Y' on chilli (*Capsicum annuum* L.) in Tamil Nadu, India. *Madras Agric. J.* **56**: 761–6.

Krishna Reddy M, Madhavi Reddy K, Lakshminarayn Reddy CN, Smitha R and Jalai S (2004). Molecular characterization and genetic variability of chilli veinal mottle virus and its reaction on chilli pepper genotypes. *In*: Proc. of the XII EUCARPIA Meeting on Genetics and Breeding of *Capsicum* and Eggplant, May 17–19, 2004, The Netherlands, pp. 171–7.

Krishna De A (2003). Capsicum: The genus *Capsicum*. Medicinal and aromatic plants—industrial profiles Vol. 33. Taylor & Francis, London and New York. pp. 275.

Kumar OA, Anitha V, Roseline Subhashini K and Raja Rao KG (2001). Induced morphlogical mutations in *Capsicum annuum* L. *Capsicum and Eggplant Newslett.* **20**: 72–75.

Kumar S, Kumar R and Singh J (2006). Cayenne/American pepper (*Capsicum* species). *In*: Peter KV (*ed.*), Handbook of Herbs and Spices, Vol 3, Woodhead Publishing, Cambridge, UK, pp. 299–312.

Kumar S, Singh M, Kumar R, Prasanna HC and Rai M (2006). Molecular markers: progress at Indian Institute of Vegetable Research (IIVR). *In*: Souvenir, National Symp. Molecular Breeding in Crop Plants, March 20–21, 2006, IIVR, Varanasi, pp. 92–94.

Kumar S, Kumar S, Singh M, Singh AK and Rai M (2006). Identification of host plant resistance to pepper leaf curl virus in chilli (*Capsicum* species). *Sci. Hortic.* **110**: 359–61.

Kumar S, Singh V, Singh M, Rai SK, Kumar S, Rai M and Kalloo G (2007). Genetics and distribution of fertility restoration associated RAPD markers in pepper (*Capsicum annuum* L.). *Sci. Hortic.* **111:** 197–202.

Kumar S, Kumar S, Rai A, Kumar R, Singh M, Yadav DS and Rai M (2007). Male sterile interspecific hybrid between *Capsicum chacoense* and *Capsicum annuum*. *In*: Proc. XIII EUCARPIA meeting on Genetics and Breeding of Capsicum and Eggplant, September 5–7, 2007, Warsaw, Poland.

Marin A, Ferreres F, Tomas Barberan FA and Gil M (2004). Characterization and quantitation of antioxidant constituents of sweet pepper (*Capsicum annuum* L.). *J. Agric. Food Chem.* **52**: 3861–9.

Padmapriya S, Chezhiyan N and Subramanian S (2004). Evaluation of Paprika (*Capsicum annuum* var. *longum*) cultivars for processing. *In*: Proc. of the XII EUCARPIA Meeting on Genetics and Breeding of Capsicum and Eggplant, May 17–19, 2004, The Netherlands, p. 124.

Patel JA, Shukla MR, Doshi KM, Patel SA, Patel BR, Patel SB and Patel AO (1998). Identification and development of male sterile lines in chilli (*Capsicum annuum* L.). *Veg. Sci.* **25**: 145–8.

Perez-Galvez A, Martin HD, Sies H and Stahl W (2003). Incorporation of carotenoids from paprika oleoresin into human chylomicrons. *British J. Nutr.* **89**: 787–93.

Peterson PA (1958). Cytoplasmically inherited male sterility in *Capsicum*. *Amer. Nat.* **92**: 111–9.

Pickersgill B (1991). Cytogenetics and evolution of *Capsicum*. *In*: T. Tsuchiya and PK Gupta (eds.), Chromosome Engineering in Plants: Genetics, Breeding, Evolution, Part B. Elsevier, Amsterdam, pp. 139–60.

Poulos JM. 1994. Pepper Breeding (*Capsicum* spp.): Achievements, challenges and possibilities. *Plant Breed.* Abst. **64**: 143–55.

Prakash NS, Lakshimi N and Harini I (1987). Male sterility and manipulation of yield in *Capsicum*. *Capsicum and Eggplant Newslett*. **6**: 35.

Prakash S and Singh SJ (2006). Insect transmitted viruses of pepper: a review. *Veg. Sci*. **33**(2): 109–16.

Prakash Satya, Singh SJ, Singh RK and Upadhyaya PP (2002). Distribution, incidence and detection of a potyvirus on chilli from eastern Uttar Pradesh. *Indian Phytopathol*. **55**: 294–8.

Prasad Rao RDVJ (1976). Characterization and identification of some chilli mosaic viruses. Ph.D. thesis, UAS, Hebbal, Bangalore, Karnataka, India.

Prasad SP and Saini GS (2004). Look for value addition to chilli. Agriculture Tribune, The Tribune, May 24, 2004, online edition, Chandigarh, India.

Rajput JC and Parulekar YR (1996). *Capsicum*. *In*: Handbook of Vegetable Science and Technology, Marcel Dekker, Inc., New York.

Reilly CA, Crouch DJ and Yost GS (2001). Quantitative analysis of capsaicinoids in fresh peppers, oleoresin capsicum and pepper spray products. *J. Forensic Sci.* **46**: 502–9.

Ryzhova NN and Kochieva EZ (2004). Analysis of microsatellite loci of the chloroplast genome in the genus *Capsicum* (pepper). *Russ. J. Genet*. **40**: 892–6.

Satpathy S, Kumar A, Swamy TMS, Rai AB and Rai M (2006). Field efficacy of methomyl against thrips, *Scirtothrips dorsalis* (Hood) in chilli. *Veg. Sci*. **33**(2): 164–7.

Shaked R, Rosenfeld K and Pressman E (2004). The effect of night temperature on carbohydrates metabolism in developing pollen grains of pepper in relation to their number and functioning. *Sci. Hortic.* **102**: 29–36.

Singh TK, Kumar S, Singh M, Kumar S, Singh G and Rai Mathura (2006). Characterization of recombinant inbred lines (RILs) of pepper (*Capsicum annuum* L.). *Veg. Sci*. **33**(2): 201–2.

Tong N and Bosland PW (1999). *Capsicum tovarii*, a new member of the *Capsicum baccatum* complex. *Euphytica* **109**: 71–77.

Tong N and Bosland PW (2003). Observations on interspecific compatibility and meiotic chromosome behavior of *Capsicum buforum* and *C. lanceolatum*. *Genetic Resources and Crop Evolution*. **50**: 193–9.

Venkata Rajam M and Subhash K (1986). A cytogenetic study of lanciolate narrow leaf, sterile mutants induced by X-irradiation and EMS in chilli (*Capsicum annuum* L.) *J. Cytol. Genet*. **21**: 1–5.

Walsh BM and Hoot SB (2001). Phylogenetic relationships of *Capsicum* (Solanaceae) using DNA sequences from two non-coding regions: The chloroplast *atpB-rbcL* spacer region and nuclear *waxy* introns. *International J. Plant Sci*. **162**: 1409–18.

Wang Deyuan and Bosland PW (2006). The Genes of Capsicum. *HortScience* **41**(5): 1169–87.

Whitmore TM and Turner BL (2002). Cultivated Landscapes of Middle America on the Eve of Conquest. Oxford University Press, New York. pp. 338.

Wien HC (1997). Peppers. *In*: Wien HC (*ed.*), The Physiology of Vegetable Crops. CAB International, Wallingford, pp. 259–93.

# 3

# International Scenario on Research and Development in Chillies

M.L. Chadha[1] and Paul A. Gniffke[2]

## 1. INTRODUCTION

Chilli peppers (*Capsicum* spp.) of the family Solanaceae are native to the New World, but have spread around the globe to become one of the most important tropical and subtropical crops on the basis of its high consumption, nutritional contribution, and cash value to farmers and consumers, both in developed and developing countries. They are available in different form, shape, size and colours. The *Capsicum annuum* is divided into the non-pungent groups such as sweet (bell) pepper, and the pungent group called hot chilli, while the *C. frutescens* and *C. chinense* are generally pungent and used in fresh or dry form (Ali, 2006). The size, color and pungency of chilli vary according to its type and use. Currently, the most pungent pepper recognized is a variety of *C. frutescens* grown in Assam, India ('Naga Jolokia' or 'Bhut Jolokia'), which has as much as 6.25 per cent (dry weight basis) capsaicinoids.

Chilli peppers are consumed in various forms such as fresh green and red fruit in raw salads, dried pods, flakes and powder as seasonings; processed condiments such as pastes, sauces and pickles; tender leaves as a green vegetable; even various non-food applications such as medicine, self-defense sprays and ornamental decorations. They enhance palatability in otherwise bland diets and are rich in ascorbic acid and phenolic compounds, important antioxidants with numerous benefits for human health.

## 2. CHILLI SPECIES

Cultivated chillies belong to five main species: *C. annuum, C. frutescens, C. chinense*, *C. baccatum* and *C. pubescens*; the first three being most

1 The World Vegetable Center-Regional Center for South Asia, Patancheru, (AP), India.
2 The World Vegetable Center, P.O. Box 42, Shanhua, Tainan, Taiwan.

commonly grown globally. *C. baccatum* is cultivated mainly in South America, while *C. pubescens* cultivars are grown under cool upland tropical conditions in South and Central America; its production has been recently introduced in Indonesia. Several additional wild and semi-domesticated species have been identified, but are not well established in commerce. Each of the cultivated species has somewhat distinct flavour and aroma profiles, and local preferences are often very pronounced.

## 3. CHILLI PRODUCTION

Chilli pepper is cultivated on 1.98 million ha worldwide, with the production of 2.76 million tonnes in dry weight output. Asia is the largest chilli pepper producer in the world and it covers about 70 per cent of the area and production. Africa and Latin America/Caribbean regions stand second and third in the area and production, respectively (Table 3.1). In terms of dry production, India, China, Bangladesh, Peru and Ethiopia are the major chilli pepper producing countries in the world (Table 3.2). Comparing to the area of production, India, Ethiopia, Bangladesh, Myanmar and Vietnam are the major chilli pepper growing countries, while the maximum productivity of chilli pepper is in Reunion, Cape Verde, Morocco, Hungary and Senegal with the productivity of 14.545, 13.750, 10.306, 8.741 and 8.011 tonnes per hectare. India's rank is 36th in productivity (1.249 t/ha). Sweet pepper has become the third most important crop for greenhouse cultivation due to increasing demand of the coloured fruits in the markets of big cities (Singh and Kumar, 2006).

**Table 3.1.** Area, production and productivity of dry chilli pepper by continent

| Continent | Area ('000 ha) | Production ('000 t) | Productivity (t/ha) |
|---|---|---|---|
| Africa | 463.1 | 467.9 | 1.0 |
| Asia | 1,416.6 | 1,944.3 | 1.4 |
| Europe | 49.3 | 119.6 | 2.4 |
| North and Central America | 33.8 | 59.6 | 1.8 |
| Oceania | — | 0.3 | — |
| Latin America & Caribbean | 60.6 | 237.8 | 3.9 |
| World Total | 1,989.6 | 2,769.6 | 1.4 |

*Source*: FAO STAT, 2006.

During 2003, half of the chilli area planted in Asia was with hybrid varieties, 15 per cent with open pollinated varieties and one-third area was devoted to the local landraces. The annual per capita availability of chilli increased both in Asia and the world at an annual rate of 4.1 and

3.1 per cent, respectively, in 1991–2003. During 2003 per capita availability of chilli stood at 5.75 and 4.47 kg in Asia and the world, respectively (Ali, 2006).

**Table 3.2:** Area and productivity of ten major dry chilli pepper producing countries in the world

| Country | Production ('000 t) | Area ('000 ha) | Productivity (t/ha) |
|---|---|---|---|
| India | 1193.0 | 954.7 | 1.2 |
| China | 245.0 | 38.0 | 6.4 |
| Bangladesh | 185.6 | 154.8 | 1.2 |
| Peru | 161.7 | 21.2 | 7.6 |
| Ethiopia | 118.3 | 288.4 | 0.4 |
| Viet Nam | 81.0 | 50.8 | 1.6 |
| Ghana | 78.0 | 12.0 | 6.5 |
| Myanmar | 65.7 | 101.4 | 0.6 |
| Pakistan | 61.9 | 42.6 | 1.5 |
| Mexico | 59.6 | 33.8 | 1.8 |
| World Total | 1,989.6 | 2,769.6 | 1.4 |

*Source*: FAO STAT, 2006.

## 4. CHILLI FOR NUTRITION

Hot pepper fruits contain more dry matter (13.8% on average) than sweet pepper fruit (10.2% on average). The red or orange peppers are very rich in vitamin A, and even the green peppers are also rich in vitamin A and precursors, hidden by the green chlorophyll. Chillies have the highest level of vitamin C per gram of tissue of any plant, including citrus fruits and are rich in minerals (Table 3.3). Hot chillies may have over 100 milligrams of vitamin C and 16,000 units of vitamin A per ounce, three to four times the daily dietary requirements. The maximum accumulation of vitamin C occurs at the onset of pigmentation of the fruit (76 days) and later decreases (Cruz *et al.,* 2007). Vitamin C is an indispensable nutrient for humans, especially in clinical implications such as the prevention of cancer and other diseases. High genetic variability for carotenoids and antioxidants is found in *Capsicum* accessions, and nutritional contribution of chilli peppers may be improved through varietal selection (Hanson *et al.,* 2004). Slightly more vitamin C (214.5 mg on average) and total sugars (187.6 mg or 4.14% fresh matter on average) are present in hot pepper than in sweet pepper fruits (Buczkowska and Najda, 2002). Some accessions of *C. annuum* with dark purple leaf, purple flower petals and dark purple fruit have been noted to have high antioxidant activity (AOA), ascorbic acid (AsA) (226 mg/100 g

FW), and T-Phenol content (1,158 mg chlorogenic acid equivalent/100 g FW). Young shoots and leaves of *Capsicum* pepper were observed to generally have high T-Phenol content (817 mg chlorogenic acid equivalent/ 100 g FW) which appeared to be promising leafy vegetables. The leaves, which are usually soft, thin narrow shaped and glabrous form a very popular vegetable in gardens in Southeast Asia (AVRDC, 2003).

**Table 3.3.** Nutritional value (per 100 g) from green and red chilli pepper

| Nutrient | Units | Green | Red |
|---|---|---|---|
| **Proximate** | | | |
| Energy | kcal | 40 | 40 |
| Protein | g | 2.00 | 1.87 |
| Total lipid (fat) | g | 0.20 | 0.44 |
| Carbohydrate, by difference | g | 9.46 | 8.81 |
| Fiber, total dietary | g | 1.5 | 1.5 |
| Sugars, total | g | 5.10 | 5.30 |
| **Minerals** | | | |
| Calcium, Ca | mg | 18 | 14 |
| Magnesium, Mg | mg | 25 | 23 |
| Phosphorus, P | mg | 46 | 43 |
| Potassium, K | mg | 340 | 322 |
| Sodium, Na | mg | 7 | 9 |
| **Vitamins** | | | |
| Vitamin C, total ascorbic acid | mg | 242.5 | 143.7 |
| Thiamin | mg | 0.090 | 0.072 |
| Riboflavin | mg | 0.090 | 0.086 |
| Niacin | mg | 0.950 | 1.244 |
| Vitamin B-6 | mg | 0.278 | 0.506 |
| Folate, food | mcg | 23 | 23 |
| Choline, total | mg | 11.1 | 10.9 |
| Vitamin A | IU | 1179 | 952 |
| Vitamin A, RAE | mcg-RAE | 59 | 48 |
| Vitamin K (phylloquinone) | mcg | 14.3 | 14.0 |
| Amino acid Lutein + zeaxanthin | mcg | 725 | 709 |

*Source*: USDA National Nutrient Database for Standard Reference, Release 20 (2007).

The active ingredient, capsaicin (and the related capsaicinoids dihydrocapsaicin, nordihydrocapsaicin and homocapsaicin) is a phenolic similar in chemical structure to vanillin, and is found only in the fruit of

*Capsicum* spp. It can be detected at concentrations of less than 1 part per million (ppm). It activates nerve endings related to pain sensation, and causes piquancy, or hotness. Extracted carotenoid pigments are important colourants in the processed food industry as well as in cosmetics applications. The concentration of capsaicinoids is reduced in all varieties after 58 to 96 days of development.

## 5. CROP MANAGEMENT

Chilli pepper, especially sweet pepper, is a relatively delicate vegetable crop, and needs special crop management to get high yield and quality. Chilli pepper crops are mostly grown in field, but may also be produced in polyhouse, net-house and greenhouse systems, either in soil, artificial media in pots, or in hydroponics systems. Selection of the appropriate variety and assurance of seed quality are important steps in achieving high yields and market value. Several seed treatments and dressing technologies are being employed in the world, including application of fungicides, insecticides, growth regulators, and various coatings can improve seedling viability and vigour. Seed priming with high osmotic potential fluids (*e.g.* Polyethylene glycol) or solids ('Micro-Cel G') to accelerate germination may also be applied to increase the vigour, yield and fruit quality of the hot pepper (Da-browska *et al.,* 2002). Seeds soaked in solutions of –1.2 MPa and –1.5 MPa of potassium nitrate ($KNO_3$) improves the germination performance (Kikuti *et al.,* 2005). Korkmaz (2005) has reported that priming seeds in 0.1 mM of acetyl salicylic acid or 3 μM and methyl jasmonate incorporated into the $KNO_3$ solution can be used as an effective method to improve low temperature performance of sweet pepper seeds and that these seeds can be stored for 1 month at 4°C and still exhibit improved germination performance at 15°C.

Likewise, seedlings are the next main factor for production and yield of any transplanted crop. Thus substrate for seedling growth plays an important role in seedling production. The mixture of 60 per cent decomposed rice hull, 10 per cent zeolite, and 30 per cent vermiculite, coir or *Sphagnum* peat moss improves root growth and lower top: root ratio. In these substrates, nitrogen (N) content of seedlings is less and potassium (K) content are more than *Sphagnum* peat moss-based substrate (Lee *et al.,* 2000a) and the water content of such type substrate is 10–15 per cent higher than that of fresh expanded rice hull under 1–60 kPa moisture tension. Growth of seedlings is best in the substrate with a particle size <0.85 mm as pH is more stable and electrical conductivity (EC) is higher in the substrate with smaller particle sizes than that with larger particle sizes (Lee *et al.*, 2000b). Moreover, Li *et al.* (2007) observed that compound substrata of mushroom mixed with perlite or vermiculite residue could increase sweet pepper plant height, stem thickness, leaf area, root volume and sound seedling index than the seedlings on the control substrate in pot cultivation.

Vegetative propagation in some sweet pepper cultivars has also been reported to be successful by using phenol foaming resin, vermiculite and granular moulding as rooting media and dipping in 0.5 per cent IBA to induce roots on pepper cuttings. Medium-sized shoots with a terminal internode shorter than 2 cm and second internode longer than 4 cm, cut at the bottom of the second internode, rooted and grew vigorously (Shirai and Hagimori, 2004). Plant spacing significantly affected all the yield parameters with the exception of number of branches per plant in sweet pepper. The best treatment combinations for plant stand and yield is a double row 60 cm apart (on a 120 cm wide bed) with 20 cm in-row spacing and 0.3 cm sowing depth. Low tunnel cultivation also increases plant stand rate and yield of hot pepper (Shin *et al.,* 2006). Jan *et al.* (2006) observed the highest number of leaves (106.6), fruits (17.91), fruit weight (66.93 g) and yield per plant (1161.89 g) at 75 × 60 cm plant spacing. However, Buczkowska (2004) has reported the highest sweet pepper average marketable yield (4.06 kg/m$^2$) at a density of 5.88 plants/m$^2$.

The use of white plastic mulch for chillies results in doubling of yield (53.0 t/ha) as compared to traditional methods (Ibarra-Jimenez *et al.,* 2004) and the plastic mulching, without cover, register the highest early and total yields (Ibarra-Jimenez and Rosa-Ibarra, 2004). The use of dark colored mulches is considered the safest solution in the northern Hungary because even in the case of high air temperature and solar radiation the soil does not warm to a harmful degree (Locher *et al.,* 2005). Compared to the uncovered level ground treatment, the covered raised bed with black polythene mulching treatment increases chilli yields by 14–19 per cent (Locher *et al.,* 2003). It's also reported that microclimate of polythene greenhouse cover for one year improves the precocity and increases the yield of hot pepper plants compared to two year covering (Denden *et al.,* 2002). Maximum and minimum air temperature, relative humidity, soil temperature, number of branches, number of leaves, internode diameter, total fresh and dry weight, total carbohydrates in leaves, vitamin C, percentage of the total soluble solids, total acidity and total carbohydrates contents in fruit, early and total fruit weight, under red greenhouse cover are higher than the other treatments (Saleh *et al.,* 2002). Training of extra shoot in sweet pepper increases fruit yield and also the yield of late fruits remarkably (An ChulGeon *et al.,* 2002). With respect to plant productivity, the system of pruning side shoots above the fourth leaf is most efficient, resulting in the highest early yield of 3.31 kg/m$^2$ and total yield of 11.63 kg/m$^2$ and second fruit set without a significant decrease in the commercial value of sweet pepper fruits (Cebula and Kalisz, 2001). Upright training is suitable under low plastic film greenhouse cultivation on sweet pepper in the highlands, which has the highest fruit setting ratio (41%) and marketable yield 43 per cent higher than that of inclination training. The number of fruits per plant of the upright training can be about 5.2 per cent more than that at inclination training (Lee *et al.,* 2007). As a general conclusion, the use of a high wire training system with three main stems can also be good to obtain fruits of better quality, and higher early and total yield (Salas *et al.,* 2003). Gibberellic acid (30 ppm) spray at 7-day intervals and 15 per cent flower thinning increases the normal fruit size of sweet pepper and consequently, its weight (Gaafer, 2006).

Polyhouse/tunnel-greenhouse/plastic shelters or protected cultivation of the chilli pepper are becoming popular and beneficial technologies. The greatest early and final commercial yield is being obtained in a tunnel-greenhouse covered with the white co-extruded 3 layer UV-absorbing white thermic polyethylene (PE) film which also shows lowest population of thrips and whiteflies, the most destructive pest of the chilli-pepper (Gonzalez *et al.,* 2001). Moreover, the plastic shelters are showing their effectiveness in overcoming soil-related and biotic constraints to vegetable production where rainy-season tomatoes and sweet peppers are almost totally destroyed by geminiviruses. Plastic shelters are reported to increase sweet pepper yield by 169 per cent without any use of pesticides and plants maintain greenness and production well into the fourth month after transplanting. The number of whiteflies visiting the plastic-shelter plants is less, only about 28 per cent of that in the open-field (Arya *et al.,* 2000). Cultivation in the greenhouse induced plants to have numerous but small fruits, whereas cultivation in the field results in smaller plants with large and well-developed fruits (Ramirez-Luna *et al.,* 2005).

## 6. CROP DISEASE MANAGEMENT

Damping off, bacterial wilt, anthracnose, fruit rot, powdery mildew, phytophthora blight, Cercospora and alternaria leaf spot, and various cucumo-, poty-, gemini- and tospo-viruses are major threatening diseases, while mites, thrips, aphid, whitefly, bud worm and caterpillar are most devastating insects of chilli pepper. Average losses due to disease infestation ranged from 7 per cent in China to 43 per cent in India from 1998–2002 (Ali, 2006).

Improving host plant resistance through varietal selection, induced resistance, and good management practices to strengthen plant tolerance to disease are of economic importance in agriculture. Genetically resistant varieties are the most sustainable strategy for reducing crop losses, while pest control technologies including fungicides, antibiotics and insecticides can be applied as disease threats arise. Biological control and integrated disease and pest management strategies can reduce crop risks, and are often based on the combined use of different natural enemies and minimal use of pesticides. Success of these methods is quite variable, and often difficult to quantify. Moreover, pathogens frequently generate new races that display increased tolerance/resistance to the pesticides. For example, the silverleaf whitefly (*Bemisia argentifolii*), which transmits extremely damaging plant viruses and induces a systemic, irregular ripening disorder, have developed the highest resistance level for endosulfan ($LC_{50}$ value of 1034.6 µg/ml) (Servin-Villegas *et. al.,* 2006).

Chlorothalonil, bavistin, dithane M 45, mencozeb, benlate, blitox 50 are reported to provide lowest incidence of *Alternaria, Cercospora, Stemphylium,* anthracnose, and karathane, wettable sulphur and aureofungin for powdery mildew. Only cuprous oxide and cuprous oxide + mancozeb reported best to reduce the bacterial population levels (Aguiar *et al.,* 2003). Cyproconazole (7 g *a.i.*/100 liter $H_2O$) and sodium bicarbonate (125 g *a.i.*/100 liter $H_2O$) are also efficient chemical treatments for good control of *Odium taurica* and was not phytotoxic at this concentration (Souza and CafeFilho, 2003). However, application of strains BSCBE4 and PA23 (BSCBE4 and *Pseudomonas chlororaphis* strain PA23 (=*P. aureofaciens*) at the rate of 20 g/kg of seed significantly increases the growth of hot pepper seedlings, induces development of plant defense-related enzymes and suppresses the incidence of damping off (Nakkeeran *et al.,* 2006) and strain Tx-1 strongly antagonize *Pythium aphanidermatum* in the pepper roots and maintain productivity of peppers (Khan *et al.,* 2003; Ramamoorthy *et al.,* 2002). *P. putida* YLFP14 selected from foliar fluorescent pseudomonas is also reported to control bacterial leaf spot of sweet pepper caused by *Xanthomonas axonopodis* pv. *vesicatoria in vitro* (Tsai *et al.,* 2004). Sid *et al.* (2004) have observed about 4 (identified as *Bacillus* spp.) out of 500 bacterial species as good biocontrol agent of diseases caused by *Phytophthora capsici* and *A. alternata* in sweet pepper. Some of these beneficial bacteria can be co-cultured *in vitro* with plant explants (micropropagation) and colonize the emerging plantlets. The bacterized plantlets appear more vigorous and withstand transplant stress better than non-inoculated controls (Nowak *et al.,* 2007).

Sharma *et al.* (2005) have observed that the partially purified toxin and the sorghum based formulation of *Trichoderma harzianum* are very much effective in reducing the anthracnose disease and fruit rot severity in hot pepper under field condition (61.1 and 63.7%) as well as reduce the post-harvest fruit rot (89.8%) and increase the shelf life of fruits by delaying the appearance of disease symptoms (12.2 days after harvesting). Implementation of either 'friendly' (*Ampelomyces quisqualis* AQ10/*T. harzianum*T39/sulfur/neem seed extract) or chemical (sulfur/pyrifenox/ polyoxin AL/myclobutanil/azoxystrobin) spray regimes successfully reduce powdery mildew disease severity in net house (Elad *et al.,* 2007). Heliosulfur, trichodex, neemgard and AQ10 in the warm greenhouse climate are also reported to be effective disease control agents, similar to the standard alternation and mixture of chemical agents (Dissevelt and Ravensberg, 2002; Brand, 2002). The botanical fungicide *Tithonia diversifolia* shows a good control against fruit rot anthracnose and cercospora leaf spots, equal to the application of mancozeb 0.2 per cent (Duriat, 1999).

Application of salicylic acid (SA) at 1 to 5 days interval between SA-treatment and challenge inoculation with *Phytophthora capsici* was reported to be effective to induce disease resistance in pepper. The resistance could persist more than 20 days after treatment with SA (Mao AiJun *et al.,* 2005). Phosphite has also been recommended to enhance plant resistance against *Phytophthora*. Treatment of the cotyledons with 50 µM and 100 µM of a jasmonic acid (JA) solution causes the plant to acquire strong oviposition deterrence against the leaf-miner, *Liriomyza trifolii*. The induced compound identified to be caffeoyl putrescine result in a significant decrease in the number of punctures made by *L. trifolii*, indicating that the JA treatment enhances the deterrence against the leaf-miner by inducing CP accumulation (Tebayashi *et al.,* 2007). The plant growth regulators increase the fruit and flower setting in the crop and results in higher yield (Ramirez-Luna *et al.,* 2005).

## 7. CROP PEST MANAGEMENT

Several predators are found including *Coccinella* spp. on aphids; *Amblyseius cucumeris* [*Neoseiulus cucumeris*] on thrips and mites; *Aphidius* spp., a parasite of nymph aphids; and *Telenomus spodopterae,* parasite of army worm egg. A weak cucumber mosaic virus (CMV) constructed from viral RNA and satellite RNA (CARNA-5) provide good plant protection against a virulent CMV. Methyl eugenol and *Melaleuca bracteata* oil are good fruit fly attractants and can be used as chemical trap. Silver coloured plastic mulch results in more healthy plants and higher fruit yield in North East of Central Java (Duriat, 1999). *Orius strigicollis* reduces the densities of thrips (*Frankliniella occidentalis*) on green and sweet pepper by 14.3–99.5 and 21.6–98.3 per cent, respectively (Kim *et al.,* 2006). Release of green lacewing (*Mallada basalis*) larvae on sweet pepper effectively control *Aphis gossypii*, *Bemisia argentifolii* and *Plenacoccus* sp. and can be a good biological control agent against these pests (Lu and Wang, 2006). The *Neoseiulus californicus* and a synthetic sexual aggregation pheromone have been reported for controlling spider mites at low humidity (Weintraub *et al.,* 2006; Gomez *et al.,* 2006). A highly significant increase in whitefly mortality was observed in *Verticillum lecanii*–treated plants compared to mortality in untreated plots and imidacloprid treatment which indicated the high level of acquired resistance among the whiteflies against the available chemical insecticides in horticultural areas in southern Spain (Blom *et al.*, 2005). *Beauveria bassiana* is effective for the control of tarnished plant bug (TPB) and western flower thrips (WFT). The bumblebees have been proven to be effective to disseminate conidia of *B. bassiana* from hive-mounted dispensers to greenhouse sweet peppers in greenhouse trials using large screened enclosures (Al-mazraawi *et al.*, 2006). Methanolic extracts of some accessions of *C. baccatum* and *C. annuum* may have a great potential for repelling spider mites and need to be field-tested on a large-scale to assess their value in managing populations of two-spotted spider mite, *Tetranychus urticae*, which could reduce reliance on synthetic acaricides (Antonious *et al.*, 2006). A one minute soil layer (0–16 cm) treatment with steam at 60 m/h, 65°C, obtained by using plastic mulching reduced *R. solani* and *P. capsici* root rot on sweet pepper (cv. Quadrato d'Asti) (Gilardi *et al.*, 2007).

For the control of pests, *i.e.* mite (*Polyphagotarsonemus latus*), white fly (*B. tabaci*), aphids (*Myzus persicae* and *A. gossypii*), and fruit borer *Spodoptera* and *Helicoverpa armigera,* non-chemical methods and with judicious use of insecticides on different cultivars throughout the growing season are effective alone or in different combinations: need-based 3 applications of lime+sulfur at 10+10 g/l for the mite; 9 sprays on pest appearance at weekly intervals of Biocatch (*Verticillium lecanii*)+ Nimbecidine 0.03% AzEC (*Azadirachta indica*) at 10 g+1 ml/l and need-based 2 applications of malathion 50 EC at 4 ml/l for the aphids; 9 sprays from March-June at weekly intervals of Helivax (nuclear polyhedrosis virus (NPV))+ Spodovax (NPV)+Nimbecidine 0.03% AzEC at 2+2+10 ml, need-based 3 sprays of Thiodan 35 EC (endosulfan) at 0.075 per cent and 3 sprays in mid-April, mid-May and mid-June of Neem Nutrene (neem cake) at 1.5 t/ha as soil application for *Spodoptera* and *H. armigera* on different cultivars throughout the growing season (Singh *et al.*, 2004). Similarly, the botanical insecticide NeemAzal-T/S (azadirachtin A at 1%) alone at 0.5 per cent can be effective against the larvae and adults of the 2-spotted mite (*Tetranychus urticae),* white fly (*Trialeurodes vaporariorum),* green peach aphid (*Myzus persicae)* and onion thrip (*Thrips tabaci)* on sweet pepper where the control efficacy for white flies ranged 73.2–98.2 per cent; aphids 71.05–96.9 per cent; and 2-spotted mites and thrips, over 90 per cent. Releasing the parasitoid *Aphididus colemani*, a few days before the predator *Aphidoletes aphidimyza* is reported as an efficient treatment against the aphid *M. persicae* (Wiethoff *et al.,* 2002).

Pepper crops can be severely damaged from soil diseases and flooding during the off-season. The grafting of pepper seedlings onto disease resistant rootstocks can lead to significantly higher yields. Possibility of grafting high quality chilli pepper on resistant rootstocks has been proved for getting good yield. Santos and Goto (2004) have reported grafting as an alternative to control of *Phytophthora* blight infecting greenhouse grown sweet pepper. Attia *et al.* (2003) have observed highest yield, fruit length, diameter and weight in soil infested with *F. oxysporum* var. *redolens* by grafting cv. Gedeon onto California Wonder. Moreover, grafting onto Chilli Putih Kelanten gave the tallest plants and the largest number of flowers. Use of net screens could reduce the need for application of insecticides. Sweet pepper grown under a 50 × 50 holes/cm mesh net provides the lowest leaf curl virus (LCV) incidence transmitted by *Bemisia tabaci* (16.8%), followed by the 40 × 40 holes/cm (22.7%) and the 30 × 30 holes/cm (55.2%) mesh nets (Singh *et al.,* 2006). White-blue and white traps are being more effective in controlling infestation of thrips (*F. occidentalis*) than that in yellow trap (Larrain *et al.,* 2006). Similarly, Roditakis *et al.* (2001) observed that the shade of blue was the most attractive on 70, 150 and 200 cm above the ground in a greenhouse cropping system and the fuchsia traps attract more females than the other coloured traps. Teulon *et al.* (1999) reported that the yellow water traps with anisaldehyde controls between 11 and 15 times more female and 3 and 20 times more male adults thrips than yellow traps without anisaldehyde.

*Meloidogyne incognita* and *M. javanica* are the major nematode pathogens of chilli pepper and several races have been reported for *M. incognita* (races 1, 2, 3, 4) (Peixoto *et al.,* 1999). Biofumigation in combination with soil solarization is reported suitable for controlling *M. incognita*, increase the highest marketable yield, similar to methyl bromide soil treatment (Guerrero *et al.,* 2006). Moreover, combination of grafting and biofumigation (using sheep and chicken manure) plus solarization can be a feasible alternative to methyl bromide in sweet pepper greenhouses in controlling *M. incognita* (Ros *et al.,* 2005). Grafted plants in soil treated with 1, 3-dichloropropene+chloropicrin results in levels of pest control similar to methyl bromide (Ros *et al.,* 2006).

## 8. WATER MANAGEMENT

Pepper yield increases with the increasing amount of irrigation water applied for all spacing treatments. Yields also increase with increasing plant populations. Water use efficiency (WUE) reduces as the applied irrigation water increases because yield increases at decreasing increments compared to the amount of irrigation water applied. Increasing in-row plant spacing significantly reduces WUE. The highest yield and WUE can be obtained with 30 cm in-row plant spacing (Al-Gharibi and Abu-Awwad, 2005). Different irrigation methods are reported for chilli pepper cultivation, the highest mean potential dry yield with 1.58 t/ha can be obtained using drip method followed by sprinkler (1.36 t/ha), basin (1.13 t/ha), and closed-end furrow (0.81 t/ha) methods. In terms of plant mortality, crops having basin irrigation has the highest (93.9%), followed by lessen amount to the sprinkler (3.2%), closed-end furrow (3.1%), and drip methods (1.7%) because the basin irrigation method stimulated the development of late blight disease. WUE for drip method is slightly higher than sprinkler method, but considerably higher than basin and closed-end furrow methods (Gencoglan *et al.,* 2005).

Dorji *et al.* (2005) observed that deficit irrigation and partial root zone drying save 170 and 164 liter of water, respectively compared to common irrigation method and they could be feasible irrigation strategies for hot pepper production where the benefit from saving water outweighs the decrease in total fresh mass of fruit. While, Kang *et al.* (2001) reported that alternate drip irrigation on partial roots gave high yield up to 40 per cent more in hot pepper, better WUE and reduction in irrigation compared to fixed drip irrigation on partial roots and drip irrigation on whole roots. Drip irrigation treatments with polyethylene mulching have been found to be an efficient technique to reduce irrigation number and water volume applied in the sweet pepper (Klar and Jadoski, 2004). Mahajan *et al.* (2007) obtained the highest red hot pepper yield (21.61 t/ha) when 50 per cent of the recommended amount of N was supplied through drip at 0.5 × Epan. WUE and nitrogen use efficiency (NUE) increased by 232.1 and 38.7 per cent over check basin method of irrigation, thus this method was beneficial for red hot pepper in term of yield, better morphological characters *viz*., plant height, number of branches, root length, size and weight of fruits along with 58.6 per cent saving of irrigation water.

## 9. FERTILIZER MANAGEMENT

As compared to inorganic fertilizer, mixed organic fertilizer and organic-mineral complex fertilizer greatly promoted the growth and development of sweet pepper. The increase in yield by 43.3–109.8 per cent, vitamin C by 39.1–59.5 per cent, and decrease in $NO_3$ contents by 17.9–35.7 per cent under organic fertigation has been reported. Such fertilizers optimized soil microorganisms increased the activity of enzymes in the soil and improved soil fertility. The fresh weight of plant aerial part, number of fruits per plant and plant yield increase with the application of increasing N rates. The highest yield (16.5 t/ha) was obtained with the application of 450 kg N/ha (Chaves *et al.,* 2006). The greatest plant height (29.13 cm), number of fruits per plant (15.36), fruit weight (67.05 g), yield per plant (1040 g) and yield/ ha (43.19 t) in sweet pepper were recorded with the application of 125:90:70 kg N:$P_2O_5$:$K_2O$/ha (Jan *et al.,* 2006). The nitrate forms of nitrogen fertilizer give taller plants with higher girths, higher yields, look greener and have more fruits than the other sources. Ammonium nitrate gives 72–101 per cent greater yields than ammonium sulfate, the commonest form of nitrogen used in Ghana (Owusu *et al.,* 2000). Potassium fertilizer application increases yield, yield components and marketable yield as well as fruit dry weight in sweet pepper, while paclobutrazol foliar application increases yield and marketable yield but did not affect yield components and fruit dry weight (El-Masry, 2000). Calcium application stimulates fruit maturation, and number of fruits with better quality, and carotene and mass of ripe fruits of hot pepper cultivars increased by about 17–27 per cent (Perucka and Materska, 2004). Increase in combined salinity and boron levels disrupts plant nutrient balance and water use, and induced production of secondary toxic substances leading to an increased plant tissue concentration of $H_2O_2$, and suppression of growth in hot pepper (Supanjani and Lee, 2006).

The application of 25 or 50 t/ha compost increase yield of hot pepper production in arid lands and the application of higher doses also improve physical soil conditions in the short term (Nieto-Garibay *et al.*, 2002). Use of pigeon manure at a rate of 45 kg/540 $m^2$ during irrigation or pigeon manure at 22.5 kg + chicken manure at 22.5 kg/540 $m^2$ during irrigation to produce high yield and quality of sweet pepper under unheated plastic houses (Salama and Zake, 2000). The package consisting soil:FYM:coir pith (2:1:1) as growing medium, irrigation at 20 kPa, 50 kg each of NPK/ha as basal using straight fertilizer and 150 kg each of NPK/ha as water-soluble fertilizer, and mulching found to be most effective in improving the growth, yield and quality of sweet pepper for commercial cultivation under polyhouse conditions (Sasikala *et al.*, 2004; Sasikala *et al.*, 2007).

An application of 7.2 N: 6.2 P: 6 K kg/ha starter solution could substitute for 30–50 per cent of inorganic fertilizer and 50 per cent of organic fertilizer use during the cropping season. It also reduces residual mineral N in soil, which ultimately minimizes environmental pollution after cultivation. Maximum yields of chilli pepper can also be obtained using a basal application of chicken manure compost (1 and 2 times 14 t/ha for sweet pepper; 10.4 t/ha for chilli pepper), an application of a starter solution at transplanting, and then followed by various side dressings of supplemental fertilizers, depending on crop and season (Ma and Kalb, 2006). Classical soluble fertilizers some times may create environmental and economic problems. As an alternative, the effects of direct applications of rock phosphate and potassium rock along with phosphate- and potassium-solubilizing bacteria on the growth and yield of hot pepper increased P availability from 12 to 21 per cent and K availability from 13 to 15 per cent in the soil compared to the control. The treatment subsequently improves the N, P and K uptake of the plant as well as increases plant photosynthesis by 16 percent and leaf area by 35 per cent as compared to control plants. Biomass harvest and fruit yield of the treated plants increases by 23 and 30 percent higher, respectively as compared to the control (Supanjani *et al.*, 2006). The application of hot pepper with the arbuscular mycorrhizal (AM) fungus, *Glomus intraradices* as cold-stored mixed inoculum is reported to increase plant dry weight and chlorophyll concentration significantly (Kim *et al.,* 2002).

## 10. WEED MANAGEMENT

Several weeds are reported to grow in chilli pepper fields depending on the region and the locations where it grows. Generally weeding is done by hand or by weeding equipments and using chemicals. Biological alternatives are also being utilized to effectively manage the weeds to get high yield and to reduce the insect-pest inoculums that grow and multiply on weeds.

Perennial ryegrass (*Lolium pratense*) and white clover (*Trifolium repens*) intercropping are reported to provide the best weed control and the conventional tillage gives greatest yields of hot pepper (Biazzo and Masiunas, 2000). The maximum weed (*Anagallis arvensis, Argemone mexicana, Chenopodium album, Convolvulus arvensis, Melilotus indica, Vicia sativa, Avena ludoviciana* [*A. sterilis* subsp. *ludoviciana*] and *Phalaris minor*) control efficiency (77.2%) and the lowest weed index (5.84), maximum plant height (83.1 cm), weight (63.5 g) and yield (337.2 q/ha) of fruits in sweet pepper were recorded with the application of fluchloralin at 2.0 kg/ha+hand weeding 60 days after transplanting (DAT, Singh, 2005). However, Gutierrez *et al.* (2002) have observed weed control higher than 90 percent with 2 hand-weeding and application of herbicide halosulfuron methyl [halosulfuron] 15 days after application (DAA). The halosulfuron methyl mixed with acetochlor also provides better weed control than halosulfuron methyl alone. Weed growth under the plastic shelters is also reported to be negligible, and plants maintained greenness and production well into the fourth month after transplanting (Arya *et al.,* 2000).

## 11. MALE STERILITY IN CHILLI PEPPERS

Chillies are hermaphrodites in nature and have both male and female parts in same flower. For the need of hybrid production as well as crossing the lines for important traits, male sterility has its great importance. The male sterile lines have been reported and several studies on genetics of fertility restoration in *C. annuum* have been carried out to find out presence or absence of restorer/maintainer genes. Cytoplasmic genic male sterility (CGMS) has been found in pepper and most researchers agree that a single nuclear gene, designated *rf1*, interacts with S (sterile) cytoplasm to produce sterility, and the restorer allele *Rf1* restores fertility. Plants with N (normal) cytoplasm are fertile regardless of whether they have the *Rf1* or *rf1* allele. Kumar *et al.* (2003) have reported that fertility restoration is inherited as a monogenic trait in 'Pant C 1' studied with cytoplasmic male sterile line 'CCA 4261'. Further they have reported that *Rf* gene associated two markers *viz.*, $OPW19_{800}$ and $OPP13_{1400}$ were not frequently distributed in the restorer inbred lines because presence of marker bands often does not coincide with the presence of *Rf* gene identified in many restorer inbreds. Good restoration has been found in pungent chilli pepper accessions, but restoration of fertility in CMS sterile sweet peppers is relatively poor when utilizing sweet pepper lines with the *Rf* restorer genotype, compared to restoration using a hot chilli pepper restorer line. Consequently, CMS strategies are used commonly in hot pepper hybrid varieties, but rarely in sweet pepper varieties. Wang ShuBin *et al.* (2006) have observed a significant difference in parthenocarpy between the CMS lines. The male sterile cytoplasmic genes has some positive effects on different agronomic traits such as plant size, fruit setting number, fruit size and yield of the $F_1$ generation, but have no effect on seed propagation. Maintaining CMS lines and producing hybrid seeds by artificial supplementary pollination or natural pollination in a plastic tunnel is more efficient than in an open isolated plot or plastic tunnel.

## 12. MOLECULAR RESEARCH

Molecular methods have been employed successfully to differentiate various plant genetic materials, their breeding characteristics, population of pathogens from many hosts for the resistance against these pathogens as well as the nutritional and quality characteristics. Using plastid-lipid-associated protein-simple sequence repeats (PAP-SSR) in *Capsicum,* through fragment analysis and nucleotide sequences, many alleles have been located. It was found that these markers could be effective in marker-assisted selection (MAS) programmes for sweet pepper breeding purposes (Sugita *et al.,* 2005). To estimate the levels of heterosis exhibited by the $F_1$ or to estimate the genetic distances among elite inbred lines, specially focusing on the breeding objectives of long-fruited hot pepper variety type, two marker systems, *i.e.* restriction site amplified polymorphism (RSAP) and simple sequence repeat (SSR), are found useful (Du *et al.,* 2007). Also the randomly amplified polymorphic DNA (RAPD) marker specific to the female and male parent have been useful in hot pepper hybrid production and seed purity identification (Wang *et al.,* 2002). Locus-specific co-dominant polymerase chain reaction (PCR)-based markers that allow semi-automated, high-throughput investigation technologies, microsatellites are ideal tools for genotype identification (Juhasz *et al.,* 2006). RAPD-PCR for identification of noteworthy populations, suitable for processing, (Masi *et al.,* 2007) and PCR-based RAPD primers has been used for quantitative trait loci (QTL) analysis (Millawithanachchi *et al.,* 2006). Sequence characterized amplified region (SCAR) markers have been observed to be effective in MAS programmes for the introduction of the $L^3$ gene derived from PI 159236 (*C. chinense*) into sweet pepper for breeding objectives (Sugita *et al.,* 2004).

Tobacco stress-induced gene 1 (*Tsi1*) may play an important role in regulating stress responsive genes and pathogenesis-related (PR) genes. The transgenic hot pepper plants expressing *Tsi1* exhibited resistance to pepper mild mottle virus, cucumber mosaic virus, *Xanthomonas campestris* pv. *vesicatoria* (*Xcv*) and to *Phytophthora capsici* (Shin *et al.,* 2002). Similarly, Yoo *et al.* (2004) have isolated a gene encoding putative *Ornithine decarboxylase* (ODC) by differential screening of a cDNA library from the resistant hot pepper inoculated with a virulent tobacco mosaic virus (TMV) pathotype $P_0$. Transcripts of the *CaODC1* (*C. annuum ODC1*) gene started to accumulate at 24 h post-inoculation of *TMV-*$P_0$, the signal spreads systemically and the transcript level increased rapidly but not in a susceptible hot pepper after inoculation which suggests the possible role(s) for *CaODC1* in plant defense against a broad range of pathogens including viruses and bacteria. Moreover, Kim *et al.* (2005) reported that *CaLEA6* induced protein plays a potentially protective role when water deficit is induced by dehydration and high salinity, but not low temperature.

Lee *et al.* (2004) developed a cDNA clone, from a pathogen-inoculated hot pepper, encoding a putative transcription factor that contains a single *ERF/AP2* DNA binding domain identified as *pCaERFLP1* which results in partially improved tolerance in transgenic tobacco plants against *Pseudomonas syringae* and salt stress (100 mM NaCl). Similarly, Yi *et al.* (2004) have isolated *CaPF1* gene from *C. annuum* and the transgenic *Arabidopsis thaliana* as well as tobacco plants expressing *CaPF1* displayed tolerance against freezing temperatures and enhanced resistance to *P. syringae* pv. tomato DC3000. The results also indicated that *CaPF1* is an *ERF/AP2* transcription factor in hot pepper may play dual roles in response to biotic and abiotic stress in plants. Hot pepper plants exhibiting a hypersensitive response (HR) upon infection by Tobacco mosaic virus (TMV) pathotype $P_0$ is due to the presence of *CaAlaAT1* gene. The expression of *CaAlaAT1* gene triggers by salicylic acid (SA) and ethylene and the *CaAlaAT1*, a protein known to be involved in metabolic reactions, might be one of the components in the plant's defense signal pathway against pathogens (Kim *et al.,* 2006).

Transgenic bacterial wilt resistant pepper lines with *CP* genes of TMV-B and CMV-P1 showed resistance against both bacterial wilt, TMV and/or CMV (AVRDC, 2001). Transgenic sweet pepper (virus resistance) has been released for field production in China. More genes of agronomic significance were cloned, and molecular markers identified may be very useful to plant breeding (Xu, 2006). Transgenic pepper plants co-expressing coat proteins (CPs) of cucumber mosaic virus (CMV-Kor) and tomato mosaic virus (ToMV) showed resistance against CMV-Kor and pepper mild mottle virus (PMMV) (Shin Ryoung *et al.,* 2002). Transformed sweet pepper plants with a coat protein (*CP*) gene of CMV cloned from a Chinese CMV (with the *CaMV* promoter and *NOS* terminator) confer resistance to CMV (Chen ZhangLiang *et al.,* 2003). This shows possibility that transgenic pepper plants might be a useful marker system for the transgenic screening and useful for classical breeding programmes of developing virus resistant hot pepper genotypes. Moreover, the use of an anther culture system coupled with genetic transformation in breeding programmes may greatly facilitate the genetic improvement of pepper plants. The progenies derived from transgenic anthers may have desired traits, grow normally and show similar phenotypes to the selected-type (Kim *et al.,* 2007).

## 13. POST HARVEST MANAGEMENT

Fruit maturity, harvest conditions, and post-harvest storage and handling methods can influence pepper fruit quality. As pepper fruits mature, fruit length, diameter, volume, weight, and total soluble solid, ascorbic acid and carotene contents increase, while moisture content and chlorophyll decrease. Surface colour change and firmness can be used to determine fruit maturity, 40 DAA (days after anthesis) is the appropriate harvesting stage for green sweet pepper when total soluble solid and ascorbic acid contents and acidity are greatest (Behera *et al.,* 2004). Although harvest at the green fruit stage can increase sweet pepper yield because of the higher number of fruits per plant, but harvesting at the yellow or red ripening stage increases mean fruit weight, length, basal diameter and mesocarp thickness (Miccolis *et al.,* 1999). The highest capsaicinoid content has been observed in fruits harvested at the initial colouring stage, followed by those in fruits of reddening and greening stages, but the content is not influenced by post harvest storage period or storage temperature. Decay developed rapidly in fruits of the greening stage compared to those in fruits of initial colouring and reddening stages. Decay rates are higher at high temperature than at low temperature (Park *et al.,* 2001). Foliar application with 0.3 and 0.5 per cent $CaCl_2$ before harvest is observed to be effective in improving storage quality of red sweet pepper (Park SungMin *et al.,* 2001). Calcium chloride at 5 g + hydrated lime at 15 g, followed by calcium chloride at 5 g + hydrated lime at 30 g are also reported for reducing decay and increasing storage life and quality for pepper (El-Sheikh *et al.,* 1997).

Longer pre-cooling time decreases decaying of red hot-pepper fruits, while increases in red sweet pepper. Decaying can be decreased with lower pre-cooling temperature (Park *et al.,* 2001). Treatment of 1-methyl-cyclopropene (1-MCP) at 250 nmol/liter improves storage quality of hot pepper fruits at room temperature. It slower down the rates of chlorophyll degradation and softening, release less ethylene and have a higher marketable percentage than untreated fruits (Huang *et al.,* 2003). Post-harvest dip in 100 ppm benzyladenine with 4 per cent Vapor Gard can be the most effective treatment to extend the shelf-life up to 18 days and to reduce the post-harvest losses without any adverse effects on the quality of sweet pepper fruits (Yadav and Dashora, 2003). Fruit storage at 6.5°C in perforated polythene packaging produced the best results in terms of skin green colour and weight retention and lower incidences of chilling injury and diseases (Nyanjage *et al.*, 2005).

## 14. ATTRIBUTES FOR PEPPER BREEDING

Desirable breeding attributes on chilli peppers include fruit size, shape, pungency, colour and oleoresin contents. Many different types of hot and sweet pepper have arisen to meet market needs and consumer preference. Therefore, local consumer preferences, and specialized market requirements for fruit types and quality traits *i.e.* color, pungency, flavor, and aroma, must be included in a pepper breeding strategy. Genetic resistance to several diseases and insect-pests are as important in new chilli pepper varieties as high yield potential. The genetic and cytoplasmic male sterility may be needed to make seed of high yielding hybrid varieties affordable.

## 15. PEPPER IMPROVEMENT PROGRAMMES AT THE WORLD VEGETABLE CENTER

The World Vegetable Center (WVC) had initiated chilli pepper programmes in 1986. Like other vegetable crops, the chilli pepper has numerous pest, disease and physiological constraints to its cultivation. Changing pathogen and pest races are the emerging problems, which limit chilli pepper production in the world. Moreover, consumer preference, demands and specialized market demands require different fruit types and quality traits *i.e.* color, pungency, flavor and aroma. The primary biotic constraints are complex of viral, fungal and bacterial diseases *i.e.* ChiVMV, CMV, PVY, geminiviruses, anthracnose, phytophthora blight and bacterial leaf spot, and insect-pests such as aphids, thrips, mites and fruit worms. The abiotic constraints like flooding, high temperature, late maturity and long crop duration, stress induces flowers and fruit abscission (sweet pepper), reduced fruit size throughout the growing season, poor balance in nutrient availability and uptake. Knowledge is limited in approaches to cost-effective sustainable production, improved post harvest and marketing practices, and consumption recommendations. Production of low cost, high quality seed is needed to translate genetic improvements into improved farmer practices and results.

In this regard, a comprehensive survey was carried out and the International Symposium for Tomato and Pepper Production in the Tropics was held in 1988 at The World Vegetable Center HQs. Major production constraints and priority areas for these crops identified in this symposium became the basis for active disciplinary research at the initial phase of the Pepper breeding programmes, which began in 1986. The World Vegetable Center is particularly well situated to conduct pepper breeding for global impact because of the extensive germplasm collection that it maintains (including more than 7500 accessions of *Capsicum* spp.), and its location close to the Tropic of Cancer, which produces distinct seasonal conditions throughout the year. Winter field trials can simulate the temperature conditions encountered in tropical highland areas, while the typhoon-prone summer season mirrors the conditions encountered in the hot humid tropical lowlands. Consequently, succeeding generations of pepper breeding materials can be subjected to 'shuttle breeding' conditions, while monitored and reselected with high quality pathology and crop management professionals. The following R&D mandates for pepper programmes were adopted: (1) to conduct strategic and applied research aimed at improving the genetic and environmental components of pepper productivity in the hot humid tropics; (2) to enhance yield and crop quality in existing areas of productivity; and (3) to adapt peppers to new regions of production in the hot humid tropics for the benefit of commercial growers and home gardeners.

## 15.1. Screening of Germplasm and Resistant Breeding

In cooperation with WVC's staff of pathologists, the pepper breeding programmes has emphasized the incorporation of host plant resistance into a diverse array of chilli pepper backgrounds. Principal diseases addressed have been Cucumber Mosaic Virus (CMV), Chilli Veinal Mottle Virus (CVMV), Pepper Virus Y (PVY), Phytophthora blight (*Phytophthora capsici*), Anthracnose (*Colletotrichum spp.*), and bacterial wilt (*Ralstonia spp.*). Secondary attention has also addressed Tobacco Mosaic Virus (TMV), Tomato Mosaic Virus (ToMV), bacterial spot (*Xanthomonas* spp.), and Pepper Leaf Curl Virus (geminivirus). Early bulk population improvement strategies have given way to conventional pedigree breeding strategies, in which cross combinations are carefully designed, and segregating populations are systematically screened for resistance to diseases under carefully controlled conditions. Progenies are tracked and reselected to the $F_6$ or $F_7$ generation, and then tested under replicated yield trials under two or more distinct seasonal conditions. The most promising lines identified are then shared with cooperating researchers and National Agriculture Research Systems (NARS), for potential release to farmers in countries around the world. Private sector seed companies are also given access to the new lines, which can be used in further variety development, generation of hybrid varieties, or even direct commercialization. All seed lines distributed by WVC are considered to be Global Public Goods, and consequently not subject to restrictive intellectual property rights applications.

### 15.2. International Pepper Nursery Networks

The principal method used in making improved chilli pepper varieties available has been WVC's International Chilli Pepper Nursery (ICPN) system. Begun in 1986 as the International Hot Pepper Trial (INTHOPE) network, this programme includes pepper researchers in government, academia and private industries around the world. The revised International Chilli Pepper Nursery (ICPN) was begun in 1990, and a parallel International Sweet Pepper Nursery (ISPN) was initiated in 1999. Typically, a set of entries chosen at WVC from among the most promising advance lines and germplasm accessions is distributed to cooperators, who evaluate them in replicated variety trials under their unique production conditions. Data is collected on the trial locations and plant performance, and returned to WVC for multi-location analysis. Over the seventeen years of activity, the number of entries has been reduced from 30 to currently 10 new lines each year, while at the same time, the inclusion of advanced lines bred entirely at WVC has increased from 0 per cent to 100 per cent. The number of entries displaying resistance to one or more diseases has also increased strongly. The International Chilli Pepper Nursery (ICPN) and International Sweet Pepper Nursery (ISPN) were started in 1990 and 1999, respectively to evaluate the pepper lines with the NARS partners for screening their quality, performance and adoption. In 2007, 88 sets of ICPN materials, and 67 sets of ISPN materials were distributed to cooperators in more than 30 countries. Most of the recipients were scientists in NARS stations, while approximately 35 per cent of the cooperators were private sector seed companies.

The lag time between distribution of candidate lines through the ICPN/ ISPN system, and public release by the NARS cooperators can be substantial. For instance, two varieties released in Ghana in 2005, were originally shared with cooperators in 1998 or earlier. It is the nature of local variety development systems that results may not be evident for many years, and may await the needed agricultural, economic, and political conditions for the potential of a variety to be recognized and acted upon. Thus, WVC seeks to distribute candidate lines to as many sincere cooperators, so that the seed will be accessible when conditions become ripe. The WVC has recently hired an "International Variety Development Coordinator", who will give more attention to nurturing the opportunities for WVC varieties to find their appropriate place in the farming systems of our partners in developing countries.

While the WVC pepper breeding programmes releases a dozen new lines each year for evaluation by cooperators, it also continues to make available lines that were released in previous years' ICPN and ISPN nurseries. Characterization of each line has been compiled into a tabular display of morphological traits (plant height and width, fruit shape, size, and colour), performance (days to flowering and maturity, heat tolerance) and disease resistances (Tables 3.4 and 3.5). Currently, 57 hot cultivars, and 35 sweet cultivars are listed. Interested cooperators are invited to review the listing in the light of local constraints and preferences, and request seed samples of appropriate candidate lines.

**Table 3.4.** Disease resistances in entries in the first five international sweet pepper nurseries (ISPN)

| | ISPN1 | ISPN2 | ISPN3 | ISPN4 | ISPN5 | Overall |
|---|---|---|---|---|---|---|
| **Recipients** | 46 | 51 | 38 | 45 | 25 | |
| **New entries** | 10 | 7 | 8 | 8 | 8 | 41 |
| **AVRDC crosses** | 5 | 3 | 3 | 4 | 6 | 21 |
| **PVY res** | 6 | 3 | 1 | 4 | 1 | 18 |
| **CVMV res.** | 5 | 2 | 0 | 0 | 0 | 5 <60% res |
| **ToMV res.** | 0 | 2 | 0 | 1 | 0 | 3 < 60% res |
| **BW res.** | 0 | 1 | 1 | 0 | — | 2 <50% res |
| **BS res.** | — | 0 | 2 | 1 | 0 | 2 |
| **PC1 res** | 6 | 0 | 0 | 1 | 0 | 7 <90% res |

Resistances: PVY: Potato Virus Y; CVMV: Chilli Veinal Mottle Virus; ToMV: Tomato Mosaic Virus; BW: Bacterial wilt; BS: Bacterial Spot; PC1: *Phytophthora capsici*, mild strain, res.: resistance.

**Table 3.5.** Disease resistances in entries in six recent international chilli pepper nurseries (ICPN)

| | ICPN10 | ICPN11 | ICPN12 | ICPN13 | ICPN14 | ICPN15 | Overall |
|---|---|---|---|---|---|---|---|
| **Recipients** | 36 | 30 | 33 | 31 | 77 | 51 | |
| **New entries** | 20 | 8 | 5 | 7 | 5 | 6 | 51 |
| **Crosses** | 12 | 7 | 9 | 8 | 9 | 8 | |
| **PVY res.** | 9 | 6 | 2 | 5 | 0 | ? | 22 > 90%res |
| **CVMV res.** | 2 | 1 | 1 | 1 | 1 | 0 | 6 > 67%res |
| **BW res.** | 0 | 4 | 3 | 6 | 4 | 4 | 21 > 67%res |
| **PC3 res.** | 0 | 0 | 0 | 0 | 2 | 1 | 3 > 67%res |

Resistances: PVY: Potato Virus Y; CVMV: Chilli Veinal Mottle Virus; ToMV: Tomato Mosaic Virus; BW: Bacterial wilt; BS: Bacterial spot; PC3: *Phytophthora capsici*, severe strain.

The WVC has targeted resistance to anthracnose as a key goal since early work by staff mycologists in the late 1980's. Wide-ranging surveys of germplasm were successful in identifying promising sources of resistance in accessions of *Capsicum chinense* and *C. baccatum*. This resistance was introgressed into more commercial backgrounds of chilli pepper through programmes of interspecific crossing and backcrossing, combined with innovations in laboratory assay technologies. Several lines have been released through the ICPN programmes, as well as in the course of direct collaboration with researchers in several nations. Awareness of the disease, and progress in reducing its damage led to the organization in conjunction with the Korean Rural Development Administration and Seoul National University, of the First International Symposium on Anthracnose in Pepper in September, 2007. This symposium brought almost 80 researchers together to present 31 papers on the topic, including status reports from seven countries.

### 15.3. International Networking among Pepper Researchers

In conjunction with the pepper breeding programmes's general research and development at WVC, it is active in several externally funded international projects. This mix is a significant benefit to both the target countries, and to the general WVC programmes. One programme, funded by the German government, seeks to develop pepper varieties for four cooperating countries (including India, Indonesia, Thailand, and China), carrying host plant resistance to four important pepper diseases (CMV, ChiVMV, anthracnose, and bacterial wilt). After six years of activity, involving cooperative efforts of pathologists, economists, horticulturalists, and breeders of WVC and the target countries, several new lines have been developed and identified which have demonstrated resistance to all pathogens, and work is underway to expand that array of defenses to include phytophthora blight, and eventually, geminiviruses. The candidate lines will be evaluated under farmer field conditions, and released to commerce as quickly as possible. This project was instrumental in developing molecular markers associated with resistance to anthracnose, which are now being utilized in improving conventional breeding progress. Another externally funded project was recently initiated to establish vegetable breeding stations in four countries in Africa, with the goal of generating new varieties that are specifically adapted to the challenges, opportunities, and preferences of African farmers and consumers in several distinct agro-ecological zones. Initial target crops will include tomatoes, onions, sweet peppers, and chilli peppers.

Private seed companies have recognized the significant value offered by the WVC pepper breeding unit. As noted above, dozens of countries routinely receive and review new lines developed at WVC. Some incorporate the most appropriate materials into their in-house breeding programmes, while others look to directly commercialize the lines. Several lines have been reported to provide good combining ability in hybrid variety crosses, and hybrids utilizing one or more WVC parental lines are currently in the market in India. A recent survey of 29 Asian seed companies found that 72 per cent of the companies use improved lines from WVC. The survey indicated that 16 per cent of chilli pepper cultivars to be released in the near future are developed from WVC germplasm. The commercial seed sector makes use of an array of chilli pepper lines developed at WVC which carry cytoplasmic male sterility (CMS), and which can thus serve as a foundation for rapid hybrid variety development and low-cost seed production. Ten or more sets of these CMS lines (both sterile A-line and maintainer B-line) are requested annually.

### 15.4. Germplasm Enhancement and Evaluation

The WVC began with 590 accessions in 1972, and currently has more than 56,000 accessions of diverse vegetables including 7,855 pepper accessions, making it the world's largest collection of vegetable germplasm. WVC not only characterizes, evaluates and standardizes vegetable germplasm resources, but also shares useful germplasm material with NARS, public and private sectors world wide. For example in the year 2004, the WVC distributed 14,560 vegetable accessions including 1,158 pepper germplasm accessions to Asian countries.

The WVC pepper breeding programmes seeks to combine superior traits into broadly useful backgrounds of chilli peppers for use in production regions around the world. Adaptation to warm, humid climates, and resistance to several important diseases are WVC's main breeding goals, but it continues to select for acceptable fruit shapes, sizes, pungencies and flavors as well as high yield potential.

### 15.5. Technology Development and Promotion

The WVC has developed a number of technologies for germplasm screenings and crop production management. *In vitro, in vivo* and *in situ* screening methods for economic important diseases and insects' resistance were standardized and several source of resistance were identified. Capsicum dry matter content, fiber, crude fat and colour can be routinely screened by the use of near-infrared spectrophotometery which is a rapid method and can provide results of 1000 samples a day. The methods for quantitative analysis of capsaicin and oleoresin content in pepper lines were standardized and being utilized. The WVC has developed both phenotypic and molecular criteria in order to overcome the difficulty of using morphological characteristics alone in species identification. Molecular characterization using specific-PCR primers has been considered a reliable and efficient tool (AVRDC, 2000).

Crop production and management technologies like raised bed, shelter, drip irrigation, grafting, and starter solution technology (SST) for high and better fruit quality production have been developed by WVC and are being utilized in many countries. SST include a starter solution (ST) applied at a rate of 240 N: 206.4 P: 199.2 K mg in 50 ml water per plant (equivalent to 7.1 N: 6.1 P: 5.9 K kg/ha) for one application after transplanting which increase yield 24–26 per cent higher than the standard inorganic fertilizer and are less costly to growers (AVRDC, 2003). The technology of growing under shade was developed where 25 per cent shading (white nylon net) gave the highest pollen viability (84%), pollen germination (61%), fruit set (62%), seed yield and seed quality over the 40 per cent (blue nylon net) and 60 per cent (black saran) shading (AVRDC, 1997). Plastic film rain shelter at 2.5 m height technology has proved to be the best for getting high marketable yield higher than no cover and net cover (AVRDC, 2002).

Grafting technology developed for tomato production during the hot-wet season that is being adopted by a growing number of farmers in Southeast Asia. Similar technology is being developed for sweet pepper production to enhance yield, improve quality and flood tolerance of up to 48 per cent. Moreover the grafting and shelters technology provided resistance to bacterial wilt, and led to high yield during hot-wet season (AVRDC, 2002). In summary, promising lines of chilli rootstock for production of grafted sweet pepper during the hot-wet and hot-dry seasons performed equally well with WVC recommended rootstocks (AVRDC, 2004).

Combination of drip irrigation and single-bed shelter has been proved to give highest marketable fruit yield (43 t/ha) and number of marketable fruits compared to the open field and double-bed shelter. Drip irrigation water use is 39.3 l against 100 l/per plant in furrow and shows 60.7 per cent improved WUE. Although nitrate N uptake was high under shelters, plant uptake in general was lower in drip than in furrow irrigation. Moreover, in the sweet pepper vines lower branches pruning and supporting other branches on a suspended horizontal 12 × 12 cm mesh net trellis and rain shelter with net covering technology provided highest yield (AVRDC, 2003).

### 15.6. Improving Nutritional Value and Storage Quality

Due to malnutrition problems mainly among women and children living in developing countries, WVC aims to evaluate nutrition contents of germplasm and breed for improved nutritional quality and long shelf-life of locally adapted varieties and germplasm. Dry matter and total capsaicin contents, antioxidant activity (AOA), ascorbic acid (AsA), and total phenol (T-phenol) content are continuously being accessed in the germplasm accessions. Several accessions have been identified with high level of antioxidant contents (Hanson *et al.*, 2004). Accession of *C. annuum* with dark purple leaf, purple flower petals, and dark purple fruit are identified to have high AOA, AsA (226 mg/100 g FW), and T-Phenol content (1,158 mg chlorogenic acid equivalent/100 g FW) and appeared to be promising leafy vegetables. Among the four species (*C. annuum*, *C. frutescens*, *C. chinense* and *C. baccatum*), it has been observed that young shoots and leaves of *C. annuum* are more suitable as leafy vegetable compared to *C. frutescens* and *C. chinense*, which have tougher leaves. Growing conditions in Southeast Asia are not favorable to *C. baccatum*. In addition to the selected accession with dark purple leaves, selection for green leaved accessions will also be important in the future (AVRDC, 2003).

## 15.7. Develop and Integrate Biotechnology into Traditional Breeding

Biotechnology has contributed a great deal in improving field crops, *e.g.*, soybean, cotton, and maize. The WVC has developed and integrated molecular marker and screening techniques into traditional breeding process to facilitate inbred line development and germplasm enhancement. WVC has also contributed the technique of application of phenotypic and molecular criteria for the identification of resistant lines as well as the pathogen species/strains.

DNA fingerprinting of *C. capsici* and germplasm screening using microsatellite markers and gene tagging have been started for the characterization and screening of resistant lines. RAPD analysis of $F_2$ progenies of inter-specific pepper hybridization to introduce pest and disease resistance was found effective to reduce the time needed in backcrossing to reach homozygosity (AVRDC, 2000) and to allow selection for resistance to disease at the seedling stage, rather waiting for immature fruit to be assayed under tedious, expensive and error-prone conditions. The WVC has molecularly characterized, cloned and sequenced *Ralstonia solanacearum* strains, *Colletotrichum* species and geminivirus, ChiVMV for finding out the strains homozygosity as well as study the genetic diversity for their virulence for further use in resistant breeding and suitable IPM technology development. Sequencing of the internal transcribed spacer (ITS) region (ITS1-5.8S-ITS2) in phenotypic-characterized *Colletotrichum* isolates was accomplished for analyzing nucleotide divergence among four *Colletotrichum* species (Zong-ming Sheu *et al.*, 2007).

Surveys were conducted in Asia (Bangladesh, China, Cambodia, India, Indonesia, Laos, Malaysia, Myanmar, Nepal, Pakistan, Philippines, Sri Lanka, Taiwan, Tanzania, Thailand, Vietnam) on pepper (11), tomato (14), eggplant (6) and weeds (10) for the presence of nine viruses: cucumber mosaic virus (CMV), chilli veinal mottle virus (ChiVMV), potato virus Y (PVY), tomato mosaic virus (ToMV), pepper mild mottle virus (PMMV), broad bean wilt virus (BBWV), watermelon silver mottle virus (WSMV), pepper veinal mottle virus (PVMV) and whitefly-transmitted geminiviruses. The geminiviruses infecting tomato, chilli pepper, and weeds in selected countries in Asia were characterized to know the sequence homology of the viruses collected from different parts of the region (AVRDC, 2001). Genetic diversity of whitefly-transmitted geminiviruses affecting solanaceous crops was studied for the implementation of effective IPM programmes.

## 15.8. Capacity Building

Training has been one of WVC's big contributions to the agriculture development in developing countries. The WVC provides training in research techniques, breeding methods, production practices for all its collaborating as well as other interested institutes. WVC has online cultivation and production technologies as well publishing its results regularly in standard journals for public awareness and their use in pepper improvement, production and crop management. Moreover, WVC has published Fact Sheets and guide on pepper cultivation, disorders on its website. These guides included detailed information on identifying and controlling the major diseases and the production technologies of pepper in developing countries. Training on seed treatment, pepper breeding, mulching, grafting etc. are continuously provided by WVC to the government, non-governmental organizations and private sector scientists, extension specialist, students and the farmers. Staff researchers participate in training workshops at headquarters in Taiwan, as well as at regional centers in Tanzania and Thailand.

## 15.9. Crop Priority and Future Work Plan of the WVC

The WVC is focused on increasing vegetable production and quality and to fulfill its mission "to alleviate poverty and malnutrition in the developing world through the increased production and consumption of safe vegetables". AVRDC has designated the following research priorities for pepper and aims to improve specific traits, which limit production in target regions:

- Collect, characterize, evaluate and conserve pepper germplasm
- Screen germplasm for disease resistance, especially anthracnose, bacterial wilt, ChiVMV, CMV, and phytophthora blight, and for tolerance of flood and heat stresses
- Characterize disease causing agents and their vectors for further development of suitable and effective disease management technologies

- Develop multiple disease resistant, heat tolerant, high yielding lines for hot and wet conditions
- Develop integrated crop and pest management (ICPM) technologies and recommendations
- Develop CMS lines (maintainer/restorer) in hot and sweet pepper for hybrid variety seed production
- Identify and promote high antioxidant and nutritious pepper lines
- Conduct fingerprinting and fine mapping of QTLs, and develop molecular markers for marker assisted selection (MAS)
- Conduct multi-location trials, seminar/trainings and symposia.

## 16. CONCLUSION

Chilli pepper is an important world vegetable, especially in Asia, in terms of total production, cultivation area, people engaged in its production, farm and retail value, and processing and marketing activities. A total of 4.2 million farmers families were engaged in chilli production in Asia during 2003 out of which 40 per cent were in China (Ali, 2006). Chilli peppers provided full time yearly jobs to 3.8 million people at the farm level and a similar number was engaged in its marketing and processing activities. International trade in chillies (import and export) has reached US$5.6 billion. Emerging problems of new diseases and new races of pathogens with resistance to common fungicides threaten chilli pepper production in developing countries. There is a need to develop lines with high nutritional value and durable resistance to multiple diseases. Moreover, emphasis should also be given on year-round and safe chilli pepper production to resolve nutritional as well as seasonal vegetable availability problems. Strong networking among chilli pepper researchers will provide better understanding of chilli production constraints, and will develop and share scientific discoveries and technical innovations leading to improved chilli pepper production and consumption around the world.

## ACKNOWLEDGEMENTS

Thanks are due to Dr. Satish Kumar Sain and Dr. Roohani Pal for their contribution in compiling this paper.

## REFERENCES

Aguiar LA de, Kimura O, Castilho AMC, Castilho KSC, Ribeiro R de LD, Akiba F and Carmo MGF Do (2003). Effect of copper formulations on resident *Xanthomonas campestris* pv. *vesicatoria* populations on sweet pepper leaf surfaces. *Horticultura Brasileira*. **21**: 44–50.

Al-Gharibi ER and Abu-Awwad AM (2005). Effect of irrigation water quantities and planting spacing on hot pepper yield and efficient water use. *Dirasat. Agril. Sci.* **32**: 27–37.

Ali M (2006). Chilli (*Capsicum* spp.): Food chain analysis: setting research technical priorities in Asia. Shanhua, Taiwan: AVRDC—The World Vegetable Center, Technical Bulletin No. **38**, AVRDC Publication 06–678.253 pp.

Al-Mazraawi MS, Shipp L, Broadbent B and Kevan P (2006). Biological control of *Lygus lineolaris* (Hemiptera: Miridae) and *Frankliniella occidentalis* (Thysanoptera: Thripidae) by *Bombus impatiens* (Hymenoptera: Apidae) vectored *Beauveria bassiana* in greenhouse sweet pepper. *Biological Control*. **37**: 89–97.

Antonious GF, Meyer JE and Snyder JC (2006). Toxicity and repellency of hot pepper extracts to spider mite, *Tetranychus urticae* Koch. *J. Environ. Sci. Health*. Part B, Pesticides, food contaminants, and agricultural wastes. **41**: 1383–91.

An ChulGeon, Kang DalSoon, Rho ChiWoong, Kang HoSung and Jeong ByoungRyong (2002). Effect of training an extra shoot on growth and yield of sweet pepper (*Capsicum annuum* 'Jubilee' and 'Fiesta'. Kor. J. *Hort. Sci. Tech*. 20: 217–20.

Arya LM, Pulver EL and Genuchten MT van (2000). Economic, environmental, and natural resource benefits of plastic shelters in vegetable production in a humid tropical environment. *J. Sustainable Agric*. **17**: 123–43.

Attia MF, Arafa AM, Moustafa MA and Mohamed MA (2003). Pepper grafting, a method of controlling soilborne diseases and enhancement of fruit yield: 1. Improvement of pepper resistance to Fusarium wilt. *Egyptian J. Phytopath*.. **31**: 151–65.

AVRDC. (1997). AVRDC—The World Vegetable Center Annual report.

AVRDC. (2000). AVRDC—The World Vegetable Center Annual report.

AVRDC. (2001). AVRDC—The World Vegetable Center Annual report.

AVRDC. (2002). AVRDC—The World Vegetable Center Annual report.

AVRDC. (2003). AVRDC—The World Vegetable Center Annual report.

AVRDC. (2004). AVRDC—The World Vegetable Center Annual report.

Behera TK, Pal RK, Sen Nita and Singh Manoj (2004). Effect of maturity at harvest on physicochemical attributes of sweet pepper (*Capsicum annuum* var. *grossum*) varieties. *Indian J. Agric. Sci*. **74**: 251–3.

Biazzo J and Masiunas JB (2000). The use of living mulches for weed management in hot pepper and okra. J. *Sustainable Agric*. **16**: 59–79.

Blom J van der, Lafuente M, Galeano M, Perez E, Urbaneja A, Pas R van der and Ravensberg W (2005). The efficacy of *Verticillium lecanii* against whitefly in sweet pepper and tomato in Spain. *In*: Papierok, B. (*ed*.), Bulletin OILB/SROP. **28**: 77–80.

Brand M, Mesika Y, Elad Y, Sztejnberg A, Rav-David D and Nitzani Y (2002). Effect of greenhouse climate on biocontrol of powdery mildew (*Leveillula taurica*) in sweet pepper and prospects for integrated disease management. *In*: Elad, Y. and Köhl, J. and Shtienberg, D. (*eds*.), Bulletin OILB/SROP. **25**: 69–72.

Buczkowska H (2004). Effect of plant density on growth, yielding of sweet pepper cv. 'Mino'. Folia Universitatis Agriculturae Stetinensis, *Agricultura*. **95**: 27–32.

Buczkowska H and Najda A (2002). A comparison of some chemical compounds in the fruit of sweet and hot pepper (*Capsicum annuum* L.). *Folia Horti*. **14**: 59–67.

Cebula S and Kalisz A (2001). The effect of side shoots pruning on the growth and fruiting of sweet pepper plants trained to one main shoot in greenhouse production. *Veg. Crops Res. Bulletin*. **54**: 91–98.

Chaves SWP, Azevedo BM de, Aquino BF de, Viana TV de A and Morais NB de (2006). Hot pepper yield in function of different levels of nitrogen. *Revista Ciencia Agronomica*. **1**: 19–24.

Chen ZhangLiang, Gu HongYa, Li Yi, Su YiLan, Wu Ping, Jiang ZhiCheng, Ming XiaoTian, Tian JinHua, Pan NaiSui and Qu LiJia (2003). Safety assessment for genetically modified sweet pepper and tomato. *Toxicology*. **188**: 297–307.

Da-browska B, Capecka E, Suchorska-Tropilo K and Senatorska-WA (2002). The effect of presowing seed conditioning and protecting with fungicide on the vigour of seeds and seedlings and the yield of two cultivars of hot pepper (*Capsicum annuum* L.). *Folia Hort*. **14**: 105–17.

Da-browska B, Suchorska K and Szalacha E (2000). The value of matriconditioned seeds of hot pepper (*Capsicum annuum* L.) after one-year storage. Part I. Speed and capability of emergence and vigour of seedlings. Annales Universitatis Mariae Curie-Sklodowska. Sectio EEE, *Horticultura*. **8**: 363–8.

Denden M, Bouslama M, Morjene H, Mathlouthi M, Bouaouina T and Cheour F (2002). Comparison of effects polyethylene greenhouse cover age on hot pepper (*Capsicum annuum* L.) growth and development. *Tropicultura*. **20**: 4–9.

Dissevelt M and Ravensberg WJ (2002). The effect of cultural and environmental conditions on the performance of *Trichoderma harzianum* strain T-22. *In*: Elad, Y., Köhl, J. and Shtienberg, D. (*eds*.), Bulletin OILB/SROP. **25**: 49–52.

Dorji K, Behboudian MH and Zegbe-Domínguez JA (2005). Water relations, growth, yield, and fruit quality of hot pepper under deficit irrigation and partial root zone drying. *Sci. Hortic*. **104**: 137–49.

Du XiaoHua, Wang DeYuan and Gong ZhenHui (2007). Estimation and comparison of genetic distances among elite inbred lines in hot pepper (*Capsicum annuum* L.) by RSAP and SSR. *J. Northwest A & F University*-Natural Science Edition. **35**: 97–102.

Duriat AS (1999). Non-chemical control of pests and diseases of hot pepper. *Indonesian Agric. Res. Dev. J*. **21**: 21–2687.

Elad Y, Messika Y, Brand M, David DR and Sztejnberg A (2007). Effect of colored shade nets on pepper powdery mildew (*Leveillula taurica*). *Phytoparasitica*. **35**: 285–99.

El-Masry TA (2000). Growth, yield and fruit quality response in sweet pepper to varying rates of potassium fertilization and different concentrations of paclobutrazol foliar application. *Ann. Agric. Sci. Moshtohor*. **38**: 1147–57.

El-Sheikh TM, El-Rahman SZA and Hassanen SM (1997). Effect of calcium chloride and hydrated lime on keeping quality of sweet pepper and cucumber fruits. *Ann. Agric. Sci. Moshtohor*. **35**: 2371–89.

Gaafer SA (2006). Effect of gibberellic acid and flower thinning on sweet pepper (*Capsicum annuum* L.) fruit yield and quality under unheated plastichouse. *Ann. Agri. Sci. Moshtohor*. **44**: 615–24.

Gencoglan C, Gencoglan S and Ucan K (2005). Effect of different irrigation methods on yield of red hot pepper and plant mortality caused by *Phytophthora capsici* Leon. *J. Environl. Biol*. **26**: 741–6.

Gilardi G, Kejjii S, Minuto A, Gullino, ML and Garibaldi A (2007). Steam soil disinfestation with an innovative equipment: observations on the efficacy against some soilborne pathogens. *Informatore Fitopatologico*. **57**: 63–70.

Gomez M, García F, GreatRex R, Lorca M and Serna A (2006). Preliminary field trials with the synthetic sexual aggregation pheromone of *Frankliniella occidentalis* on protected pepper and tomato crops in South-east Spain. Castane, C. and Sanchez, J.A. (*Eds.*), Bulletin OILB/SROP. **29**: 153–8.

Gonzalez Benavente-Garcia A, Rodriguez R, Banon S, Franco Jose A and Fernandez Hernandez JA (2001). The influence of photoselective plastic films as greenhouse cover on sweet pepper yield and on insect pest levels. (*eds.* Fernandez JA, Martinez PF and Castilla N). No. **559**: 233–8.

Guerrero MM, Ros C, Martínez MA, Martinez MC, Bello A and Lacasa A (2006). Biofumigation *vs.* biofumigation plus solarization to control *Meloidogyne incognita* in sweet pepper. *In*: Castane, C. and Sanchez, J.A. (*eds.*), Bulletin OILB/SROP. **29**: 313–8.

Gutierrez W, Medrano C, Baez J L, Pinto H, Villalobos Y and Medina B (2002). Efficacy evaluation of the herbicide halosulfuron methyl alone and mixed with acetochloro in weed control in sweet pepper *Capsicum annum* L. in the Maracaibo Plain, Zulia state, Venezuela. *Revista de la Facultad de Agronomia, Universidad del Zulia*. **2**: 87–94.

Hanson PM, Yang Ray-yu, Lin Susan, CS Tsou Samson, Lee Tung-Ching, Wu Jane, Shieh Jin, Gniffke Paul and Ledesma Dolores (2004). Variation for antioxidant activity and antioxidants in a subset of AVRDC—the World Vegetable Center Capsicum core collection. Plant Genetic Resources, **2**: 153–166 Cambridge University Press.

Huang XueMei, Zhang Zhao Qi and Duan XueWu (2003). Effects of 1-methylcyclopropene on storage quality of hot pepper (*Capsicum frutescens*) at room temperature. *China Vegetables*. **1**: 9–11.

Ibarra-Jimenez L and Rosa-Ibarra M de la (2004). Comparison between microtunnels with polyethylene and polypropylene in cucumber and sweet pepper with plastic mulching. *Revista Chapingo. Serie Horticultura*. **10**: 133–9.

Ibarra-Jimenez L, Flores J, Quezada MR and Zermeño A (2004). Mulching, irrigation and microtunnels in tomato, Anaheim chilli and sweet pepper. *Revista Chapingo. Serie Horticultura*. **10**: 179–87.

Jan NE, Khan IA, Sher Ahmed, Shafiullah and Rash Khan (2006). Evaluation of optimum dose of fertilizer and plant spacing for sweet peppers cultivation in Northern Areas of Pakistan. *Sarhad J. Agri*. **22**: 601–6.

Juhasz AG, Stagel A, Acs S, Zatyko L and Nagy I (2006). Microsatellite markers and automated fragment analysis techniques for efficient and precise hybrid identification and genetic purity testing in pepper (*Capsicum annuum* L.). *In*: Paldi, E. (*ed.*), *Acta Agron. Hungarica*. **54**: 141–6.

Kang ShaoZhong, Zhang Lu, Hu XiaoTao, Li ZhiJun and Jerie P (2001). An improved water use efficiency for hot pepper grown under controlled alternate drip irrigation on partial roots. *Sci. Hortic*. **89**: 257–67.

Khan A, Sutton JC and Grodzinski B (2003). Effects of *Pseudomonas chlororaphis* on *Pythium aphanidermatum* and root rot in peppers grown in small-scale hydroponic troughs. *Biocontrol Sci. Tech*. **13**: 615–30.

Kikuti ALP, Kikuti H and Minami K (2005). Physiological conditioning in sweet pepper seeds. *Revista Ciencia Agronomica*. **36**: 243–8.

Kim HyungSae, Lee JeeHyun, Kim JaeJoon, Kim ChangHoon, Jun SungSoo and Hong YoungNam (2005). Molecular and functional characterization of *CaLEA6*, the gene for a hydrophobic LEA protein from *Capsicum annuum*. *Gene*. **344**: 115–23.

Kim JeongHwan, Byeon YoungWoong, Kim YongHeon and Park ChangGyu (2006). Biological control of thrips with *Orius strigicollis* (Poppius) (Hemiptera: Anthocoridae) and *Amblyseius cucumeris* (Oudemans) (Acari: Phytoseiidae) on greenhouse green pepper, sweet pepper and cucumber. *Kor. J. Appl. Entomol.* **45**: 1–7.

Kim KiJeong, Park ChangJin, Ham ByungKook, Choi SooBok, Lee BooJa and Paek KyungHee (2006). Induction of a *cytosolic pyruvate kinase 1* gene during the resistance response to Tobacco mosaic virus in *Capsicum annuum*. *Plant Cell Rep.* **25**: 359–64.

Kim YoungSoon, Kuk YongIn and Kim KyungMoon (2007). Inheritance and expression of transgenes through anther culture of transgenic hot pepper. *Zeitschrift für Naturforschung*. Section C, *Biosciences*. **62**: 743–6.

Kim KY, Cho YS, Sohn BK, Park RD, Shim JH, Jung SJ, Kim YW and Seong KY (2002). Cold-storage of mixed inoculum of *Glomus intraradices* enhances root colonization, phosphorus status and growth of hot pepper. *Plant Soil.* **238**: 267–72.

Klar AE and Jadoski SO (2004). Irrigation and mulching management for sweet pepper crop in protected environment. *IRRIGA*. **9**: 217–24.

Korkmaz A (2005). Inclusion of acetyl salicylic acid and methyl jasmonate into the priming solution improves low-temperature germination and emergence of sweet pepper. *Hort. Sci.* **40**: 197–200.

Kumar S, Rai SK, Singh Major and Kalloo G (2003). Genetics of fertility restoration and identification of restorers and maintainers in pepper (*Capsicum annuum* L.). *Cap.& Eggplant Newslett.* **22**: 79–82.

Larrain SP, Varela UF, Quiroz EC and Grana SF (2006). Effect of trap colour on catches of *Frankliniella occidentalis* (Pergande) in sweet peppers (*Capsicum annuum* L.) Agricultura Tecnica, **66**: 306-11.

Lee JaeHoon, Hong JongPil, Oh SangKeun, Lee SanghYeob, Choi Doil and Kim WooTaek (2004). The ethylene-responsive factor like protein 1 (*CaERFLP1*) of hot pepper (*Capsicum annuum* L.) interacts *in vitro* with both GCC and DRE/CRT sequences with different binding affinities: possible biological roles of *CaERFLP1* in response to pathogen infection and high salinity conditions in transgenic tobacco plants. *Plant Mol. Biol.* **55**: 61–81.

Lee JiWon, Lee ByoungYil, Son JungEek, Kim KiSun and Lee YongBeom (2000a). Water retentivity and several vegetable seedling growth in decomposed expanded rice hull substrates with different particle sizes. J. *Korean Society for Hort. Sci.* **41**: 245–8.

Lee JiWon, Lee ByoungYil., Kim KwangYong and Son JungEek. 2000b. Growth of vegetable seedlings in decomposed expanded rice hull-based substrates. *J. Kor. Soc. Hort. Sci.* **41**: 249–53.

Lee JongNam, Lee EungHo, Im JuSung, Ryu SeungYeol and Yong YeoungRok (2007). Suitable training method under low plastic film greenhouse cultivation on sweet pepper (*Capsicum annuum* 'Special') in highland. *Kor. J. Hort. Sci. Tech.* **25**: 97–102.

Li XiaoQiang, Guo ShiRong, Bu ChongXing and Wang Xu (2007). Effects of compound substrata of mushroom residue on growth of sweet pepper seedlings. *Acta Agriculturae Shanghai*. **23**: 48–51.

Locher J, Ombodi A, Kassai T and Dimeny J (2005). Influence of coloured mulches on soil temperature and yield of sweet pepper. *Europ. J. Hort. Sci.* **70**: 135–41.

Locher J, Ombodi A, Kassai T, Tornyai T and Diményi J (2003). Effects of black plastic mulch and raised bed on soil temperature and yield of sweet pepper. *International J. Hort. Sci.* **9**: 107–10.

Lu ChiuTung and Wang ChinLing (2006). Control effect of *Mallada basalis* on insect pests of nethouse sweet peppers. *J. Taiwan Agril. Res.* **55**: 111–20.

Ma CH and Kalb T (2006). Development of starter solution technology as a balanced fertilization practice in vegetable production. *In*: Tei F, Benincasa P and Guiducci M (*eds.*), *Acta Hort*. No. **700**: 167–72.

Mahajan G, Singh KG, Rakesh Sharda and Mukesh Siag (2007). Response of red hot pepper (*Capsicum annum* L.) to water and nitrogen under drip and check basin method of irrigation. *Asian J. Pl. Sci*. **6**: 815–20.

Mao AiJun, Wang YongJian, Feng LanXiang, Geng SanSheng and Xu Yong (2005). Study on the resistance induced by salicylic acid against *Phytophthora capsici* in pepper (*Capsicum annuum*). *Agril. Sci. in China* **4**: 192–8.

Masi L de, Siviero P, Castaldo D, Cautela D, Esposito C and Laratta B (2007). Agronomic, chemical and genetic profiles of hot peppers (*Capsicum annuum* ssp.). *Mol. Nutr. Food Res*. **51**: 1053–62.

Miccolis V, Candido V and Marano V (1999). Influence of harvest time on yield of some sweet pepper cultivars grown in greenhouses. *In*: Tuzel Y, Burrage SW, Bailey BJ, Gul A, Smith AR and Tuncay O (*eds.*), *Acta Hortic*. No. **491**: 205–8.

Millawithanachchi MC, Perera ALT, Peiris BCN and Fonseka HM (2006). Development of new high yielding chilli hybrids (*Capsicum annuum* L.) based on heterobeltiosis and characterization of parental germplasm for DNA polymorphisms. *Trop. Agric. Res*. **18**: 209–16.

Nakkeeran S, Kavitha K, Chandrasekar G, Renukadevi P and Fernando WGD (2006). Induction of plant defense compounds by *Pseudomonas chlororaphis* PA23 and *Bacillus subtilis* BSCBE4 in controlling damping off of hot pepper caused by *Pythium aphanidermatum*. *Biocontrol Sci. Tech*. **16**: 403–16.

Nieto-Garibay A, Murillo-Amador B, Troyo-Diéguez E, Larrinaga-Mayoral JA and García-Hernandez JL (2002). The use of compost as an ecological alternative for sustainable production of hot pepper (*Capsicum annuum* L.) in arid zones. *Interciencia*. **27**: 417–21.

Nowak J, Veilleux RE, Nowak J and Turgeon S (2007). Priming for transplant stress resistance in *in vitro* propagation via plantlet bacterization. *In*: Santamaria JM and Desjardins Y (*eds.*), *Acta Hortic*. No. **748**: 65–75.

Nyanjage MO, Nyalala SPO, Illa AO, Mugo BW, Limbe AE and Vulimu EM (2005). Extending post-harvest life of sweet pepper (*Capsicum annuum* L. 'California Wonder') with modified atmosphere packaging and storage temperature. *Agricultura Tropica et Subtropica*. **38**: 28–32.

Owusu EO, Nkansah GO and Dennis EA (2000). Effect of sources of nitrogen on growth and yield of hot pepper (*Capsicum frutescens*). *Tropical Sci*. **40**: 58–62.

Park JeCheon, Park SungMin, Yoo KeunChang and Jeong CheonSoon (2001). Changes in postharvest physiology and quality of hot pepper fruits by harvest maturity and storage temperature. *J. Kor. Soc. Hort. Sci*. **42**: 289–94.

Park SungMin, Lee YounSu and Jeong CheonSoon (2001). Effect of preharvest foliar application of calcium chloride on shelf-life of red sweet pepper 'Ace'. *Kor. J. Hort. Sci. Tech*. **19**: 12–16.

Peixoto JR, Maluf WR and Campos VP (1999). Evaluation of lines, $F_1$ hybrids and cultivars of sweet pepper for resistance to *Meloidogyne* spp. Pesquisa. *Agropecuária Brasileira*. **34**: 2259–65.

Perucka I and Materska M (2004). The effect of $Ca_2^+$ on the content of vitamin C, provitamin A and xanthophylls in hot pepper fruits. Annales *Universitatis Mariae Curie-Sklodowska*, Sectio E, *Agricultura*. **59**: 1933–9.

Ramamoorthy V, Raguchander T and Samiyappan R (2002). Enhancing resistance of tomato and hot pepper to *Pythium* diseases by seed treatment with fluorescent pseudomonads. *Euro. J. Plant Pathol*. **108**: 429–41.

Ramirez-Luna E, Castillo-Aguilar C de la C, Aceves-Navarro E and Carrillo-Avila E (2005). Effect of products containing growth regulators on flowering and fruit set in 'Habanero' hot pepper. Revista Chapingo. *Serie Horticultura*. **11**: 93–98.

Roditakis NE, Lykouressis DP and Golfinopoulou NG (2001). Color preference, sticky trap catches and distribution of western flower thrips in greenhouse cucumber, sweet pepper and eggplant crops. *Southwestern Entomologist*. **26**: 227–37.

Ros C, Guerrero MM, Martínez MA, Barcelo N, Martínez MC, Rodriguez I, Lacasa A, Guirao P and Bello A (2005). Resistant sweet pepper rootstocks integrated into the management of soil borne pathogens in greenhouse. *In*: Vanachter A (*ed.*), *Acta Hortic.* No. **698**: 305–10.

Ros C, Guerrero MM, Martinez MA, Lacasa A and Bello A (2006). Integrated management of *Meloidogyne* resistance in sweet pepper in greenhouses. *In*: Castane C and Sanchez JA (*eds.*) Bulletin OILB/SROP. **4**: 319–24.

Sala FC, Costa CP da, Echer M de M, Martins MC and Blat SF (2004). Phosphite effect on hot and sweet pepper reaction to *Phytophthora capsici*. *Scientia Agricola* **61**: **5**: 492–5.

Salama GM and Zake MH (2000). Fertilization with manures and their influence on sweet pepper of plastic-houses. *Ann. Agric. Sci., Moshtohor*. **38**: 1075–85.

Salas MC, Urrestarazu M and Castillo E (2003). Effect of cultural practices on a sweet pepper crop in a mild winter climate. *In*: Malfa G la, Lipari V, Noto G and Leonardi C (*eds.*), *Acta Hortic*. No. **614(1)**: 301–6.

Saleh SMM, Medany MA, El-Behairy UA, Wadid MM and Raggal T (2002). Effect of colour of greenhouse cover on sweet pepper (*Capsicum annum* L.) growth and production. *Egypt. J. Hort*. **29**: 347–75.

Santos HS and Goto R (2004). Sweet pepper grafting to control phytophthora blight under protected cultivation. *Horticultura Brasileira*. **22**: 45–49.

Sasikala S, Natarajan S and Kumaresan GR (2004). Standardization of package of practices to promote growth and yield in sweet pepper (*Capsicum annuum* L. var. *grossum* Sendt.) hybrid under protected conditions. *South Indian Hort*. **52**: 164–9.

Sasikala S, Natarajan S, Kavitha M and Tamilselvi C (2007). Influence of growing media, irrigation regime, integrated nutrient management and mulching on the performance of sweet pepper (*Capsicum annuum* L. cv. grossum) hybrid under polyhouse condition. *International J. Agric. Sci*. **3**: 56–59.

Servin-Villegas R, García-Hernandez JL, Murillo-Amador B, Tejas A and Martinez-Carrillo JL (2006). Stability of insecticide resistance of silverleaf whitefly (Homoptera: Aleyrodidae) in the absence of selection pressure. *Folia Entomologica Mexicana*. **45**: 27–34.

Sharma P, Kadu LN and Sain SK (2005). Biological management of dieback and fruit rot of chilli caused by *Colletotrichum capsici* (Syd.) Butler and Bisby. *Indian J. Plant Prot*. **33**: 226–30.

Shin Ryoung, Han JungHeon, Lee GilJe and Peak KyungHee (2002). The potential use of a viral coat protein gene as a transgene screening marker and multiple virus resistance of pepper plants coexpressing coat proteins of cucumber mosaic virus and tomato mosaic virus. *Transgenic Res*. **11**: 215–9.

Shin Ryoung, Park JeongMee, An JongMin, Paek KyungHee (2002). Ectopic expression of *Tsi1* in transgenic hot pepper plants enhances host resistance to viral, bacterial, and oomycete pathogens. *Mol. Plant-Microb. Interac*. **15**: 983–9.

Shin YoungAn, Lee Um, YoungCheol and Park SuHyung (2006). Effects of seed spacing and depth and planting date on yield of once-over harvested hot pepper (*Capsicum annuum* L.) planted by direct seeding. *Kor. J. Horti. Sci. Tech*. **24**: 8–12.

Shirai T and Hagimori M (2004). A multiplication method of sweet pepper (*Capsicum annuum* L.) by vegetative propagation. *J. Jap. Soc. Hort. Sci.* **73**: 259–65.

Sid A, Ezziyyani M, Egea-Gilabert C and Candela ME (2004). Selecting bacterial strains for use in the biocontrol of diseases caused by *Phytophthora capsici* and *Alternaria alternata* in sweet pepper plants. *Biologia Plantarum*. **47**: 569–74.

Singh B and Kumar Mahesh (2006). Performance of sweet pepper varieties under semi-climate controlled greenhouse conditions of Northern India. *In*: Rezuwan K, Ibni Hajar Rukunuddin and Nor Raizan A H (*eds.*), *Acta Hortic*. No. **710**: 355–58.

Singh B, Kumar M and Singh V (2006). Nylon mesh screens reduce incidence of leaf curl virus and improve yield in sweet pepper. *J. Vegetable Sci*. **12**: 65–70.

Singh D, Kaur Sandeep, Dhillon TS, Singh Parminder, Hundal JS and Singh GJ (2004). Protected cultivation of sweet pepper hybrids under net-house in Indian conditions. *In*: Cantliffe DJ, Stoffella PJ and Shaw NL (*eds.*), *Acta Hortic*. No. **659(1)**: 515–21.

Singh R (2005). Integrated weed management in sweet pepper (*Capsicum annuum* L.) in semi-arid irrigated conditions of Punjab. *Haryana J. Agron*. **21**: 117–19.

Souza VL de and CaféFilho AC (2003). Effect of chemical control on the progress of sweet pepper powdery mildew under greenhouse conditions. *Summa Phytopathologica*. **29**: 317–22.

Sugita T, Yamaguchi K, Sugimura Y, Nagata R, Yuji K, Kinoshita T and Todoroki A (2004). Development of SCAR markers linked to $L^3$ gene in *Capsicum*. *Breeding Sci*. **54**: 111–15.

Sugita T, Kinoshita T, Kawano T, Yuji K, Yamaguchi K, Nagata R, Shimizu A, Chen LanZhuang, Kawasaki S and Todoroki A (2005). Rapid construction of a linkage map using high-efficiency genome scanning/AFLP and RAPD, based on an intraspecific, doubled-haploid population of *Capsicum annuum*. *Breeding Sci*. **55**: 287–95.

Supanjani and Lee KD (2006). Hot pepper response to interactive effects of salinity and boron. *Plant Soil and Environ*. **52**: 227–33.

Supanjani Han HyoShim, Jung JaeSung and Lee KyungDong (2006). Rock phosphate-potassium and rock-solubilising bacteria as alternative, sustainable fertilizers. *Agronomy for Sustainable Development*. **26**: 233–40.

Tebayashi S, Horibata Y, Mikagi E, Kashiwagi T, Mekuria D B, Dekebo A, Ishihara A and Kim ChulSa (2007). Induction of resistance against the leafminer, *Liriomyza trifolii*, by jasmonic acid in sweet pepper. *Bioscience, Biotech. Biochem*. **71**: 1521–26.

Teulon DAJ, Hollister B, Butler RC and Cameron EA (1999). Colour and odour responses of flying western flower thrips: wind tunnel and greenhouse experiments. *Entomologia Experimentalis et Applicata*. **93**: 9–19.

Tsai YL, Chen MJ, Hsu ST, Tzeng DDS and Tzeng KC (2004). Control potential of foliar *Pseudomonas putida YLFP14* against bacterial spot of sweet pepper. *Plant Pathol. Bulletin*. **3**: 191–200.

USDA Nutritional Database. 2007. Accessible at www.nal.usda.gov. 15 Feb.

Wang DeYuan, Wang YongFei, Yin QuiMiao and Li Ying (2002). Determination of hot pepper "Yuejiao No. 1" $F_1$ hybrid seed purity by RAPD markers. *Capsicum & Eggplant Newslett*. **21**: 29–32.

Wang ShuBin, Liu JinBing and Pan BaoGui (2006). Genetic effects of the male sterile cytoplasmic gene in hot (sweet) pepper. *Acta Agriculturae Shanghai* **22**: 14–17.

Weintraub P, Kleitman S, Shapira N, Argov Y and Palevsky E (2006). Efficacy of *Phytoseiulus persimilis* versus *Neoseiulus californicus* for controlling spider mites on greenhouse sweet pepper. *In*: Castane C and Sanchez JA (*eds*.), Bulletin OILB/SROP. **29**: 121–25.

Wiethoff J, Meyhöfer R and Poehling HM (2002). Use of combinations of natural enemies for biological control of *Myzus persicae* (Sulzer) (Hom.: Aphididae) on sweet pepper in greenhouses. *Gesunde Pflanzen*. **54**: 126–37.

Xu ZhiHong (2006). Plant biotechnology and its application in horticulture in China. *In*: Fari MG, Holb I and Bisztray GD (*eds*.), *Acta Hort*. No. **725**(Vol. 1): 49–53.

Yadav NR and Dashora LK (2003). Shelf-life of sweet pepper (*Capsicum annuum* L.) cv. 'California Wonder' as influenced by benzyladenine and vapor gard. *Adv. Hort. Forestry* **9**: 215–21.

Yi SoYoung, Kim JeeHyub, Joung YoungHee, Lee SanghYeob, Kim WooTaek, Yu SeungHun and Choi Doil (2004). The pepper transcription factor *CaPF1* confers pathogen and freezing tolerance in *Arabidopsis*. *Plant Physiol*. **136**: 2862–74.

Yoo TaeHyoung, Park ChangJin, Ham ByungKook, Kim KiJeong and Paek KyungHee (2004). Ornithine decarboxylase gene (*CaODC1*) is specifically induced during TMV-mediated but salicylate-independent resistant response in hot pepper. *Plant Cell Physiol.* **45(10)**: 1537–42.

Zong-ming Sheu, Jaw-rong Chen, and Tien-cheng Wang (2007). Application of ITS-RFLP analysis for identifying *Colletotrichum* species associated with pepper anthracnose in Taiwan, p. 35–36. Oh Dae-Geun and Ki-Taek Kim (*eds*.). Abstr. First International Symposium on Chilli Anthracnose, National Horticultural Research Institute, Rural Development of Administration, Republic of Korea. 17–19 September.

# 4

# Advances in Capsicum Biotechnology

V.A. PARTHASARATHY, D. PRASATH AND K. NIRMAL BABU

## 1. INTRODUCTION

Chilli and paprika (*Capsicum annuum* L.) are one of the major vegetables and spice crops of the world. It is the source of natural pungent compounds (capsaicin), colouring compounds (capsorubin) and vitamin C. The genus *Capsicum* originated in South America with two centres of domestication, one in Central America and the other in the Andean region of South America. In addition to its importance as a food and spice, the capsaicinoid compounds, responsible for pungent sensation of the fruits are widely used for diverse medicinal applications. While pain control is the most familiar application, reports on various pharmacological applications of these molecules from weight loss to cancer are available. *Capsicum* is best known for its capsaicinoids, a family of more than 25 related alkaloid analogs produced in epidermal cells of the placenta or dissepiment of the fruit that account for the pungent or 'hot' sensation when consumed, a trait evolved to deter mammalian herbivory. Capsicums vary in their pungency level and other flavour characteristics. Pungent peppers are called chillies, while non-pungent varieties are called 'sweet pepper' or 'paprika'. Carotenoid and anthocyanin pigments are responsible for fruit colour and for nutritional value of *Capsicum* fruit. The most common colours of pepper fruits are green, red, yellow, orange, chocolate and purple. The predominant red pigments are capsanthin and capsorubin, the yellow and orange pigments are lutein, β-carotene (provitamin A), zeaxanthin, violaxanthin and antheraxanthin. Among vegetables, peppers are ranked first in antioxidant content with very high levels of vitamin C (Palevitch and Craker, 1995).

The major breeding objectives are selection for yield, fruit colour and intensity, size, shape, degree of pungency, pericarp thickness, flowering time, fruit set at extreme temperature, concentration of fruit set, growth habit, adaptation to mechanical or hand harvesting, fruit quality traits and disease resistance.

Indian Institute of Spices Research, Calicut-673 012, Kerala (India).

## 2. BIOTECHNOLOGY IN *CAPSICUM*

The past few years have witnessed a quantum jump in utilization of biotechnological tools to achieve breeding objectives through commercial propagation, development of novel varieties/new breeding lines *via* somaclonal variation, anther culture, protoplast fusion and recombinant DNA technology for developing resistant lines for biotic and abiotic stresses. Industrial production of flavour compounds and secondary metabolites through bioreactor technology is another important field of approach.

### 2.1. Micropropagation and Plant Regeneration

Efficient micropropagation and plant regeneration protocols are prerequisites for any genetic manipulation initiative. Many reports are available on micropropagation and plant regeneration in *Capsicum* from shoot buds, shoot tips, nodes and axillary buds (Agarwal 1988; Cao and Jiu 1993; Ezura *et al.,* 1993; Christopher and Rajam 1994; Brinzel *et al.,* 1996). Reports on effect of culture conditions, culture media and genotype on *in vitro* induction of adventitious buds from various explants (Table 4.1) are also available (Buyukalaca and Mavituna 1996; Christopher and Rajam 1996; Kato *et al.,* 1996).

**Table 4.1.** *In vitro* response of various genotypes for micropropagation in *Capsicum*

| Genotype | Explant | Media* | Response | References |
|---|---|---|---|---|
| Jwala Sakti | Shoot tips | 4 mg/l BA + 2 mg/l Kin<br>2 mg/l NAA<br>3 mg/l BA+ 1 mg/l $GA_3$<br>1 mg/l IBA | Multiple shoots<br>Callus<br>Shoot buds<br>Rooting | Soniya and Nair (2004) |
| Bydagi Dabbi, Arka Lohit | Hypocotyl cotyledon | 2 mg/l BA + 1 mg/l IAA | Shoot regeneration | Mathew (2002) |
| Pusa Jwala, G4 | Cotyledon | 9–18 μM Zeatin + 2.89 μM $GA_3$<br>2.22 μM BAP + 2.89μM $GA_3$<br>4.90 μM IBA | Elongated shoot lets Elongation of shoot buds<br>Rooting | Shivegowda *et al.* (2002) |
| Soroksari | Shoot tip | MS+ 4.4 μM BA or 9.1 μM Zeatin | Shoot regeneration | Jasna Berljak (1999) |
| Land races | Shoot bud | 3 μM BAP+0.9 μM IAA or<br>5 μM BAP+ 1.5 μM IAA | Shoot proliferation | Mohamed (1998) |
| G2 | Apical, axillary buds | 3 mg/l BAP + 1 mg/l IAA<br>2 mg/l NAA + 0.5 mg/l Kin | Shoot proliferation<br>Rooting | Gupta *et al.* (1998) |
| G4, Bhiwapuri, Sweet and Cayenne pepper | Hypocotyl, leaf, cotyledon | 5.7 μM IAA + 22.2 μM BAP5.7 μM IAA + 13.3 μM BAP | Shoot proliferation<br>Rooting | Christopher and Rajam (1996) |

**Table 4.1.** *Contd.*

| Genotype | Explant | Media* | Response | References |
|---|---|---|---|---|
| *C. annuum* var. *longum* | Micro shoot | B5+300 ppm chlormequat | Rooting | Ma *et al.* (1991) |
| Pico Piquillo | Shoot bud | 1 mg/l IAA + 2 mg/l BAP<br>0.1 mg/l NAA +<br>0.05 mg/l IBA | Shoot proliferation<br>Rooting | Arroyo and Revilla (1991) |
| *C. annuum* | Seedling explant | 3 $mgl^{-1}$BA, 1 $mgl^{-1}$IBA | Plant regeneration through callus | Anu *et al.* (2004) |

*Source*: Parthasarathy and Nirmal (2001).

* All the media are Murashige and Skoog's (MS) medium unless otherwise specified. B5; Gamborg's medium.

*Capsicum annuum* is considered a recalcitrant species and regeneration *via* adventitious organogenesis, which usually occurs at low frequencies, is genotype dependant (Wolf *et al.*, 2001). Anu *et al.* (2004) reported efficient plant regeneration from callus and variations among somaclones in *C. annuum*.

## 2.2. Somatic Embryogenesis

Somatic embryogenesis in *Capsicum* spp. has been induced directly and indirectly *via* callus. Various explants such as cotyledon, leaves and embryo have been used for indirect embryogenesis. But for direct embryogenesis only embryos have been used (Buyukalaca and Mavituna, 1995a; Buyukalaca and Mavituna, 1996; Mavituna and Buyukalaca, 1996). Binzel *et al.* (1996) reported direct somatic embryogenesis without any intervening callus from immature zygotic embryos. Kintzios *et al*. (1998) indicated that both the intensity of light as well as the duration of incubation under illumination or darkness significantly affected somatic embryo induction and development. Harini and Sita (1993) also used immature zygotic embryos for direct embryogenesis and developed a protocol for *in vitro* regeneration of plants. Regenerated plants showed cytological and morphological uniformity.

Synthetic seeds have been developed by encapsulating matured somatic embryos in a calcium alginate gel (Buyukalaca and Mavituna, 1995b). Germination of the embryo was best with 3 per cent sodium alginate gel.

## 2.3. Anther Culture

Anther culture is the easiest way to produce haploids and dihaploids with many desirable traits, pest and disease resistance. George and

Narayanasamy (1973) reported development of haploid embryos and plantlets from immature anthers of *C. annuum* var. *grossum* through androgenesis. The success depends on genotypes (Novak, 1974; Mityko and Fari, 1997), pretreatment (Morrison *et al.* 1986; Vaulx *et al.* 1982; Mak and Maheswary, 1994) and supplementation of media. Young mother plants and frequent subculturing was reported to increase the androgenic yield in different *C. annuum* genotypes (Kristiansen and Anderson, 1993). The best embryoid development was reported with MS medium supplemented by carrot extract and activated charcoal (Vagera, 1984; Pandeva and Zagorsha, 1986). 2, 4-D was effective in inducing haploid callus when used in combination with NAA and BAP. Haploid callus and embryoids were induced more frequently when the anthers were at the late-uninucleate stage (Harn *et al.,* 1975). Morphogenetic growth and development of globular embryoids depended on the presence of FeEDTA in the medium and about 10 per cent of the plants were fertile (Vagera and Havranek. 1985). Efficiency of anther culture was dependent on donor genotype and induction of embryoids from microspores was best on MS medium containing 0.1 mg Kin and 0.1 mg 2, 4-D/l or medium containing 0.1 mg/l 2, 4-D and 0.1 mg NAA/l (Yoon *et al.,* 1991). Combinations of Kin and IAA produced significantly higher percentage of callus under continuous dark incubation than under alternate light/dark cycles (Mythili and Thomas, 1995).

Thidiazuron (TDZ) has also been reported to be very effective in morphogenesis and direct embryoid induction from callus (Pandeva *et al.,* 1990). Luz *et al.* (1998) reported that the most effective concentration of TDZ was 4.5 mM. Besides, cytokinins and phenyl urea derivatives, macro-elements and carbon source have also been effective in induction of androgenesis. Increase in number of embryos and plants could be attained by flushing cultures with air enriched with $CO_2$ at 900 µl/l (Dolcet-Sanjuan *et al.,* 1997). Depending on the genotype, haploid regeneration from anther culture in *Capsicum* is generally between 3 and 20 per cent, although success up to 50 per cent has been reported (Munyon *et al.,* 1989; Poulos, 1994). Induction of pollen embryogenesis in *C. annuum* was reported by Gonzalez *et al.* (1996) and Regner (1996). Occurrence of unreduced gametes and ploidy restoration in haploid peppers was reported by Yan *et al.* (2000).

## 2.4. Production of Secondary Metabolites

Biotechnology can be utilized to exploit the potential of spices for bioproduction of useful plant metabolites. The use of tissue culture for the biosynthesis of secondary metabolites, particularly in plants of pharmaceutical significance holds an interesting alternative to control production of plant constituents. Plant cells cultured *in vitro* produce wide range of primary and secondary metabolites of economic value. Production of flavour compounds and secondary metabolites *in vitro* using immobilized cells is an ideal system for chilli. Production of capsaicin has been reported using such system (Ravishankar *et al.,* 1993, 1995; Johnson *et al.,* 1996; Venkataraman and Ravishankar, 1997). Johnson *et al.* (1996) reported biotransformation of ferulic acid vanillylamine to capsacin and vanillin in immobilized cell cultures of *C. frutescens.*

Plant cell culture has been employed in *Capsicum* for the production of secondary metabolite like capsaicin (Chavez-Moctezuma and Lozoya-Gloria, 1966; Yoeman *et al.,* 1980; Ravishankar *et al.,* 2003). Yeoman *et al.* (1980) manipulated culture media to increase the production of capsaicin from *C. frutescens* and reported an increase in the capsaicin by adding precursor vanillylamine and isocapric acid, eliminating sucrose from the culture media and decreasing the level of N to 5 per cent in the MS solution. Nitrates and phosphates stress enhances the capsaicin production in immobilized cells by 13 and 5 fold, respectively, in comparison with free-cell culture systems (Ravishankar *et al.,* 1998). Mathematical modeling of capsaicin production in immobilized cells of *Capsicum* was studied by Suvarnalatha *et al.* (1993) to optimize the physical parameters, such as the bead strength of calcium alginate used for immobilization and the medium constituents for enhanced yield.

Jhonson *et al.* (1991) found that elicitor curdlan was most effective in eliciting capsaicin synthesis. Immobolized cells responded more effectively than placental tissues for curdlan treatment. A 6 to 7 fold increase was recorded in the capsaicin accumulation upon precursor biotransformation. The feeding of intermediate precursors to *Capsicum* cell cultures not only increased the capsaicin accumulation but also reduced the time required to produce high amount of capsaicin (Jhonson *et al.,* 1990, 1991). Immobilized placenta administered with intermediates of capsaicin pathway resulted in large amount of capsaicin accumulation (Jhonson and Ravishankar, 1996). During biotransformation studies to increase capsaicin yields, it was found that low capsaicin producing capsicum cell cultures formed vanillin when fed with phenylpropanoid compounds like protocatechuic acid, caffeic acid, ferulic acid, vanillylamine, coniferyl aldehyde and veratraldehyde (Rao, 1998).

Though the feasibility of *in vitro* production of spice principles has been demonstrated, methodology for scaling up and reproducibility need to be developed before it can reach commercial levels. Once standardized, this technology has tremendous potential in industrial production of capsaicin.

### 2.5. Protoplast Culture

The 'protoplast' is a naked cell and the absence of cell wall makes the protoplast suitable for a variety of manipulations that are not normally possible with intact cells and hence protoplast is an important tool for parasexual modification of genetic content of cells. To date, *Capsicum* has been a difficult genus for isolation and culture of protoplasts. Reports indicate that even the calli obtained from either the cultured protoplasts (Donato *et al.*, 1989) or directly from tissue explants (Ochoia and Ireta-Mereno, 1990) are recalcitrant. Successful isolation and culture of protoplasts have been reported in *Capsicum*. Organogenesis and plant regeneration from isolated protoplasts are available in chillies (Fari and Czako, 1981; Saxena *et al.*, 1981; Agarwal, 1988). Plant regeneration has been reported from protoplasts of *Capsicum* cv. California Wonder (Prakash *et al.*, 1997). Recently Kim *et al.* (2004a) isolated mesophyll protoplasts from pepper and introduced the maize transposable element, *Ac/Ds* by poly ethylene glycol (PEG) treatment to examine its mobility in pepper nuclei. The results showed that both *Ac* and *Ds* elements were highly mobile in pepper protoplasts.

## 3. MOLECULAR BREEDING

### 3.1. Genetic Mapping

Progress has been made in increasing yield and quality traits in chillies (Greenleaf, 1986; Poulos, 1994). Genes for resistance to TMV, TSWV, potyviruses, CMV, nematodes, bacterial leaf spot disease (*Bs2*), *Xanthomonas campestris, Phytophthora capsici* and powdery mildew have been identified in several *Capsicum* species and are being utilized in breeding programmes (Paran *et al.*, 2007). Morphological markers were initially used to study linkages. The most comprehensive linkage study using morphological markers was conducted by Pochard (1977) using trisomic analyses and eight genes were located to specific chromosomes. Three genes, *C*, now known as *Pun1* (presence of capsaicin in the fruit), *xantha 3* and *xantha* 8 were found to be linked to each other in the trisomic line JA. Two genes, *L* (resistance to TMV) and *MoA* (modifier of anthocyanin accumulation) were found to be linked to each other in the trisomic line BR. The genes *up* (erect fruit), *A* (anthocyanin accumulation) and *y* (now *Ccs*) (yellow fruit color) were found to be unlinked but were assigned to the trisomic lines, NO, RO and IN, respectively. Linkage between partial resistance to CMV and susceptibility to TMV was also reported (Pochard *et al.*, 1983).

### 3.2. Molecular Characterization

In recent past, there is an increased emphasis on use of molecular markers for characterization of the genotypes, genetic fingerprinting, identification, cloning of important genes, marker assisted selection, and in understanding of inter relationships at molecular level. Preliminary information on isoenzymes and breeding experiments divided the genus *Capsicum* into three clusters of taxa, each with one domesticated taxon (McLeod *et al.,* 1983). Later, Wang and Ma (1987) by studying the pattern of peroxidase isozymes assigned 8 species into 4 groups. Andrzejewski (1992) suggested the application of isozymes (superoxide dismutase, glutamate oxaloacetate transaminase, phosphoglucomutase, isocitrate dehydrogenase and shikimate dehydrogenase) as genetic markers in the breeding of *C. annuum, C. baccatum* and *C. chacoense*.

Subsequently, DNA finger printing of *Capsicum* accessions using randomly amplified polymorphic DNA (RAPD) was demonstrated by Hears *et al.* (1996). Lefebvre *et al.* (1993) studied nuclear restriction fragment length polymphism (RFLP) between pepper cultivars of *C. annuum*. Ninety accessions of *C. annuum* germplasm from diverse origin were investigated and compared using inert simple sequence repeat (ISSR) and RAPD markers (Wang and Fan, 1998). Both the markers detected high polymorphism and ISSR primer SCRI1418 was most polymorphic for finger printing. It is reported that amplified fragment length polymorphism (AFLP) primers are four times more efficient than RAPD primers in their ability to detect polymorphism (Paran *et al*., 1998). AFLP has also been employed in finger printing Guatemala's *Capsicum* genetic resources (Guzman *et al*., 2005).

Retrotransposon-based sequence-specific amplification polymorphism (SSAP), AFLP and simple sequence repeats (SSR) marker systems were used by Tam *et al.* (2005) to assess comparative genetic diversity within tomato and pepper collections. SSAP showed about four to nine fold more diversity than AFLP and had the highest number of polymorphic bands per assay and the highest marker index. The results of the entire three markers for pepper showed general agreement with pepper types. Additionally, retrotransposon sequences isolated from one species can be used in related solanaceous genera.

Determination of $F_1$ hybrid purity is a vital component of hybrid seed production for both commercial and breeding purposes. PCR-based RAPD analysis has been shown to be effective in testing hybridity of *Capsicum* as well as genetic purity of seeds (Ballester and deVicente 1998; Ilbi 2003; Wang *et al.,* 2002; Mongkolporrn *et al.*, 2004). Sequence characterized amplified region (SCAR) markers were also useful for purity testing of $F_1$ hybrid seeds in *Capsicum* (Jang *et al.,* 2004).

### 3.3. Gene Tagging, Mapping and Marker Assisted Selection (MAS)

Most *Capsicum* species are diploid with 12 pairs of chromosomes. A set of trisomic lines were produced and used for the chromosomal assignment of several mutations (Pochard, 1977). Various types of chromosomal rearrangements are prevalent in the genus both within and between species (Pickersgill, 1997; Onus and Pickersgill, 2004). The genome size of pepper was estimated by flow-cytometry at 7.65 pg/nucleus for *C. annuum* and 9.72 pg/nucleus for *C. pubescens* (Belletti *et al.,* 1998). Genome size in nucleotides is estimated to be about 3,000 Mbp (Arumuganathan and Earle, 1991). Much work has been done in molecular aspects of *Capsicum* especially, in molecular characterization, preparation of molecular maps and isolation of genes.

### 3.3.1. *Gene Tagging*

Monogenic resistance genes for viruses [potyviruses, tomato spotted wilt virus (TSWV), tobacco mosaic virus (TMV), and cucumber mosaic virus (CMV)], nematodes and bacteria were used as targets for mapping and developing linked markers for use in marker-assisted selection (MAS).

#### 3.3.1.1. *Bacterial Spot*

Bacterial spot caused by *Xanthomonas campestris* is an important disease of pepper. At least four dominant genes, *Bs1*, *Bs2*, *Bs3* and *Bs4* have been identified. Additionally, a non-hypersensitive resistance controlled by two recessive genes, *bs5* and *bs6*, was identified by Jones *et al*. (2002). Tai *et al*. (1999) identified molecular markers tightly linked to the *Bs2* gene. Although, the *Bs3* gene and the markers were not mapped in pepper, the AFLP fragments were cloned and mapped in tomato chromosome 2 within an interval of 1.6 centi Morgan (cM) between *TG33* and *TG31*.

#### 3.3.1.2. *Potyvirus*

Several genes for potyvirus resistance are known in *Capsicum* (Kyle and Palloix, 1997). Caranta *et al.* (1999) developed Cleaved amplified polymorphic sequence (CAPS) marker for the *Pvr4* locus for pyramiding potyvirus resistance genes in pepper. Arnedo-Andres *et al.* (2002) developed RAPD and SCAR markers linked to the *Pvr4* locus for resistance to *PVY* in *Capsicum*. Yeam *et al*. (2005) generated CAPS markers for three recessive viral resistance alleles used widely in pepper breeding, *pvr1*, $pvr1^1$, and $pvr1^2$. These markers are based on single nucleotide polymorphisms (SNPs) within the coding region of the *pvr1* locus encoding an *eIF4E* homolog on chromosome 3. The *pvr2* locus was mapped on chromosome 4 (Murphy *et al.,* 1998). This locus is now known to encode the eukaryotic translation initiation factor, *eIF4E* (Ruffel *et al*., 2002; Kang *et al*., 2005). Another recessive potyvirus resistance locus *pvr6* was mapped nearer to RFLP marker *TG57* in chromosome 3 (Caranta *et al.,* 1996). Resistance to pepper veinal mottle virus (PVMV) was observed when *pvr6/pvr6* occurred in a *pvr12/pvr12* background. A cluster of at least two dominant potyvirus resistance genes is located in chromosome 10 and it contains the dominant *Pvr4* and *Pvr7* genes (Caranta *et al.*, 1999; Grube *et al.*, 2000b; Arnedo-Andres *et al.*, 2002).

3.3.1.3. *Tomato Spotted Wilt Virus (TSWV)*

RAPD markers corresponding to hypersensitive reaction to tomato spotted wilt virus in *Capsicum*, governed by single dominant gene $T_{SW}$ was determined by Moury *et al.* (2000). RAPD markers were also used for mapping of $T_{SW}$ locus for resistance to tospo virus in *Capsicum* (Jahn *et al.*, 2002). In addition to the potyvirus resistance genes, the genomic region in chromosome 10 contains a dominant gene, *Tsw* that confers resistance to TSWV that belongs to the tospoviruses group (Jahn *et al.*, 2000; Moury *et al.*, 2000). This region, therefore, comprises the first cluster of dominant resistance genes in pepper.

3.3.1.4. *Tobacco Mosaic Virus (TMV)*

Resistance to TMV is conferred by the *L* gene for which a series of alleles have been identified (Boukema, 1980). The gene *L* was first assigned to chromosome BR by trisomic analysis (Pochard, 1977). Subsequently, *L*1 was mapped in *C. annuum* to chromosome 11 in the vicinity of the tomato RFLP marker *TG36* (Lefebvre *et al.*, 1995; Ben-Chaim *et al.*, 2001a). The *L*4 allele originated from *C. chacoense* was tagged by a RAPD marker, which was also converted to a SCAR marker at a distance of 1.5 cM from *L* (Matsunaga *et al.*, 2003).

3.3.1.5. *Root-Knot Nematode*

Several genes conferring resistance to root-knot nematodes have been reported in pepper. *Me*3 and *Me*4 were shown to be linked at approximately 10 cM and mapped to one of two chromosomes 7 or 12 (Djian-Caporalino *et al.*, 2001). An important nematode resistance gene in pepper that is commonly used in commercial cultivars is *N* (Thies and Fery, 2000).

3.3.1.6. *Fruit Pungency*

The ability to produce capsaicinoids is determined by a single dominant gene *C*, also known as *Pun1*. The *C* locus was found as loosely linked to RFLP markers from chromosome 2 in several mapping populations (Tanksley *et al.*, 1988; Lefebvre *et al.*, 1995; Ben-Chaim *et al.*, 2001b). Further mapping experiments identified the tomato RFLP marker *TG205* as co-segregating with *C* and in addition a PCR-based CAPS marker tightly linked to *C* was developed (Blum *et al.*, 2002).

Blum *et al.* (2002, 2003) mapped *C* locus (for pungency) using mapping populations developed by *C. frutescens* (pungent) × *C. annuum* var. *grossum* (non-pungent). Quantitative trait loci (QTL) interval analysis for individual and total capsaicinoid content identified a major QTL, termed *cap*, which explained 34–38 per cent of the phenotypic variation for this trait in two growing environments (Blum *et al.*, 2003). Pungency co-segregated with fruit colour and it was also reported that the deletion of capsanthin-capsorubin synthase gene (*CCS*) by induced mutation would lead to yellow fruit colour and deletion of specific band in an RFLP profile (Lefebvre *et al.*, 1998).

### 3.3.1.7. *Fruit Colour*

Carotenoids determine mature fruit colour, while anthocyanins and chlorophyll determine the colour of immature fruits. The inheritance of mature fruit colour was originally reported to be controlled by three independent genes *Y*, *C*1 and *C*2 (Hurtado-Hernandez and Smith, 1985). Evaluation of candidates for association with fruit colour loci suggested that the gene coding for capsanthin-capsorubin synthase (*CCS*) is a candidate for *Y* because complete co-segregation of *Y* and *CCS* was observed in several independent populations (Lefebvre *et al.*, 1998; Popovsky and Paran, 2000). A similar approach identified complete co-segregation between *C*2 and the gene coding for phytoene synthase (*PSY*; Thorup *et al.*, 2000; Huh *et al.*, 2001).

Brown colour in mature pepper fruit results from the simultaneous accumulation of red carotenoids and chlorophyll pigments as a consequence of impaired chlorophyll catabolism during ripening. This mutation is controlled by a single recessive gene, *chlorophyll retainer* (*cl*) (Smith, 1950). A double recessive mutant at *y* and *cl* results in a green mature colour. The *cl* locus was mapped recently to chromosome 1 (Efrati *et al.,* 2005). Purple immature fruit colour is controlled by the single dominant gene *A* which was mapped to chromosome 10 (Ben Chaim *et al.,* 2003a) and subsequently shown to co-segregate with the *Petunia* gene *Anthocyanin2* (*An2*), an *R2R3 MYB* transcription factor that regulates anthocyanin biosynthesis (Borovsky *et al.,* 2004).

### 3.3.1.8. *Fruit Texture*

Ripe fruits of wild peppers undergo rapid texture deterioration (termed soft flesh). The soft flesh and deciduous nature is controlled by a single dominant gene *S* mapped to chromosome 10. The soft flesh trait was found to co-segregate with the polygalacturonase (*PG*) gene from tomato (Rao and Paran, 2003). The complete linkage of *PG* and *S*, its increased expression in soft flesh fruits compared to bell peppers indicates that *PG* is a candidate gene for *S*.

#### 3.3.1.9. *Plant Architecture*

One of the most popular phenotypes utilized for ornamental pepper breeding is the *fasciculate* (*fa*) mutation. The *fa* mutation is controlled by a single recessive gene and it is characterized by the formation of clusters of flowers and fruits caused by reduced internodal length after initiation of flowering and by reduced flowering time (Paran *et al.*, 2007). The *fa* locus was mapped to chromosome 6 and co-segregated with the Self Pruning (*SP*) gene that controls inflorescence architecture in tomato (Paran, 2006).

## 3.4. QTL Mapping

### 3.4.1. *Cucumber Mosaic Virus (CMV)*

CMV is one of the most important viruses infecting pepper. Double haploid progenies from the cross of Perennial and the bell pepper variety Yolo Wonder were evaluated for a component of CMV resistance (Caranta *et al.*, 1997b). Two QTL regions were detected on linkage groups 6 and 12 with additive effects, explaining 24 per cent and 19 per cent of the phenotypic variation. One digenic interaction between the main effect QTL in chromosome 12 and *TG66* in chromosome 3 that contributed 33 per cent of the phenotypic variation. A similar cross between Perennial and bell type variety, Maor was analyzed for whole plant resistance to CMV by Ben-Chaim *et al.* (2001a) and 4 main effect QTLs were detected in chromosomes 4, 6, 11 and 13. The QTL with the largest effect was *cmv11.1* linked to the RFLP marker TG36 explaining 16 per cent to 33 per cent of the phenotypic variation for the trait. *cmv11.1* is also tightly linked in repulsion to *L* that confers resistance to TMV as originally observed by Pochard *et al.* (1983). Partial resistance to CMV was studied in double haploid progenies from the cross of Vania and H3 (Caranta *et al.*, 2002). Four main effect QTLs were detected, of which the QTL, *cmv 12.1* from Vania, had the largest effect on the resistance.

### 3.4.2. *Phytophthora capsici*

A comparative QTL study was performed in three mapping populations segregating for resistance originating from distinct sources, Vania, Perennial and Criollo de Morelos (CM) 334 (Thabuis *et al.*, 2003). A QTL with the largest effect (>50% of the explained variation) was detected in chromosome 5 in all three populations by two inoculation methods (root and stem tests) and four resistance components. This QTL consistently exhibited the highest level of resistance to *P. capsici* in germplasm (Quirin *et al.*, 2005). In addition to this locus on chromosome 5, a QTL was detected on chromosome 10 in two of the three crosses. Sugita *et al.* (2006) used a doubled-haploid (DH) population obtained by anther culture of an $F_1$ hybrid between a line susceptible to *Phytophthora capsici* 'K9 11' (*C. annuum*) and a line resistant to *P. capsici* 'AC 2258' (*C. annuum*) to identify 3 QTLs for *P. capsici* resistance. Of the 16 linkage groups (LGs), covering a total distance of 1100.5 cM, 3 QTLs were detected on LG1, LG6 and LG7. The QTL, *Phyt*-1, detected on LG7 explained 82.7 per cent of the phenotypic variance. The nearest marker was an AFLP marker, M10E3-6. The second QTL, *Phyt*-2, was found on LG1 explaining 6.4 per cent of the phenotypic variance. The nearest RAPD marker was RP13-1. The other QTL designated as *Phyt*-3, which was found on LG6, explained 5.6 per cent of the phenotypic variance. The nearest AFLP marker was M9E3-11. It was confirmed that the lines with a high resistance could be efficiently selected by using two markers, M10E3-6 and RP13-1, simultaneously. The presence of both *Phyt*-1 and *Phyt*-2 under homozygous conditions may enable to breed resistant cultivars of sweet pepper. The molecular markers identified could be useful for marker-assisted selection (MAS) in order to breed sweet pepper cultivars for resistance to *P. capsici* using 'AC 2258' as a source of resistance.

### 3.4.3. *Potyviruses*

The doubled haploid progenies from the cross of the resistant accession, Perennial and the susceptible cultivar Yolo Wonder were analyzed for resistance to two isolates of PVY and to potyvirus E (Caranta *et al.*, 1997a). A total of 11 chromosomal regions had significant effect on the resistance. Only one QTL in the region containing *pvr1* (formerly *pvr2*) on chromosome 4 was detected for all three viruses. Two additional QTLs were also detected for the two PVY strains.

### 3.4.4. *Powdery Mildew*

QTL analysis for powdery mildew resistance in the double haploid progenies from the cross of H3 and the susceptible cultivar Vania revealed the presence of seven genomic regions involved with the resistance (Lefebvre *et al.*, 2003). The most significant QTL was *Lt-6.1* detected on chromosome 6.

### 3.4.5. *Anthracnose*

Voorrips *et al.* (2004) reported one major QTL (B1) for resistance to two *Colletotrichum* races that causes the anthracnose on linkage group B in an $F_2$ population derived from an interspecific cross of *C. annuum* and *C. chinense*. Three minor QTLs were also detected.

### 3.4.6. *Fruit Size*

Two QTL studies were performed in crosses of the large blocky cultivar, Maor with the small-fruited accessions, Perennial (*C. annuum*) and BG 2816 (*C. frutescens*) (Ben-Chaim *et al.*, 2001b; Rao *et al.*, 2003). Five and eight QTLs were detected in the cross of Maor × Perennial and Maor × BG 2816, respectively but none of the QTLs had a major effect larger than 20 per cent of the explained phenotypic variation.

### 3.4.7. *Fruit Shape*

Three and five QTLs for fruit shape were detected in the crosses of Maor × Perennial and Maor × BG 2816, respectively. A major QTL, *fs3.1*, which accounted for about 67 percent of the phenotypic variation in Maor × Perennial and 24 percent of the phenotypic variation in Maor × BG 2816 was detected. This indicates the *fs3.1* is a major QTL that differentiates elongated from blocky fruits in *Capsicum*. A second major fruit shape QTL, *fs10.1*, was detected on chromosome 10 in a cross of *C. annuum* (5226) line × *C. chinense* (159234) line (Ben-Chaim *et al.,* 2001b, 2003b). This QTL explained up to 44 per cent of phenotypic variation for fruit shape and linked to the *A* locus, confirming the linkage between fruit shape and fruit colour genes (Peterson, 1959). The comparative QTL mapping of fruit weight and shape in pepper and tomato was done by Paran *et al.* (2004) and they reported that both convergent and divergent selections were operated on fruit weight in pepper. QTL mapping for fruit shape and size in chromosome 2 and 4 in *Capsicum* was done by Zygier *et al.* (2005).

The *C* locus that is required for the production of pungent fruit (Blum *et al.*, 2002) was mapped and a major QTL affecting the level of pungency identified (Blum *et al.*, 2003). QTLs for major fruit quantitative traits such as fruit weight, fruit shape, pericarp thickness and maturity were also identified (Ben Chaim *et al.*, 2001b; Ben Chaim *et al.*, 2003a; Ben Chaim *et al.,* 2003; Rao *et al.*, 2003b).

### 3.4.8. *Pungency*

Quantitative variation in capsaicinoids (capsaicin and dihydrocapsaicin) content was examined in a cross of the sweet blocky parent Maor and the pungent wild accession BG 2816 (*C. frutescens*). Three RAPD markers were identifies and mapped on chromosome 7. A major QTL termed *cap* contributed to the increased level of pungency and it explained up to 38 per cent of the phenotypic variation for this trait. SCAR markers for capsaicinoid synthase gene responsible for pungency was designed by Lee *et al.* (2005) and Minamiyama *et al.* (2005). The map location of pepper anthocyanin genes were studied by various workers (Ben Chaim *et al.,*

2001; Rao and Paran, 2003). The two anthocyanin loci were linked to a major QTL, *fs10.1*, for fruit shape index, that also segregated in the $F_2$ population (Chaim *et al.,* 2003). The linkage relationship in pepper resembles similar in potato, in which anthocyanin and tuber shape genes were found to be linked to each other.

### 3.4.9. *Fertility Restoration*

Male sterility in *Capsicum* is an important system for the production of commercial hybrid Seeds. QTL analysis for fertility restoration of nuclear-cytoplasmic male sterility was performed in a population of double haploid progenies of Yolo Wonder and Perennial, fertility restorer lines crossed to a cytoplasmic genic male sterile line. One major QTL for fertility restoration was detected in chromosome 6. Four additional QTLs with minor effect were also detected at which alleles from both parents contributed to fertility restoration (Wang *et al.*, 2004). Two RAPD markers linked to a major fertility restorer gene were detected by Zhang *et al.* (2000). A similar work by Kumar *et al.* (2002) details the RAPD protocol for tagging of fertility restorer and male sterility genes in chilli.

## 3.5. Construction of Linkage Maps

The first study in which isozymes were used for linkage mapping in *Capsicum* was reported by Tanksley (1984) who mapped 14 isozymes in an interspecific cross of *C. annuum* and *C. chinense*, of which nine were arranged in four linkage groups. The first RFLP map of pepper consisting of molecular markers was reported by Tanksley *et al.* (1988) using a pair of segregating populations in tomato and pepper. The syntenic relationship between the pepper and tomato genomes was assessed based on common markers. This indicated conserved gene repertoire but considerable rearrangement in gene order.

Pepper maps were constructed using $F_2$ or $BC_1$ populations from interspecific crosses of *C. annuum* × *C. chinense* (Tanksley *et al.,* 1988; Prince *et al.,* 1993; Livingstone *et al.,* 1999; Kang *et al.,* 2001; Lee *et al.*, 2004). Two maps, one based on $BC_2$ population from a cross of *C. annuum* (Maor) × C. *frutescens* (BG2816) using RFLP markers (Rao *et al.*, 2003) and the second based on $F_2$ population from *C. annuum* (NuMex RNaky) × *C. frutescens* (14–6) cross using public and proprietary SSR marker sets anchored by RFLP and cloned genes were also constructed (Ben Chaim, 2006). In addition to the interspecific maps, intraspecific maps from crosses within *C. annuum* have also been constructed using AFLP and RFLP markers (Ben Chaim *et al.,* 2001a). A map derived from double haploids (DH) lines was reported by Lefebvre *et al.* (1995, 2002).

The most comprehensive comparative map of pepper was developed by Livingstone *et al.* (1999). This map included about 1000 loci including many of pepper origin (RFLP, RAPD and AFLP markers). In addition, several hundred loci were detected by RFLP with tomato probes, allowing considerably improved resolution in assessment of syntenic relationships between tomato and pepper. The most saturated and integrated map of pepper was constructed by merging segregation data from six individual inter- and intra-specific maps (Paran *et al.*, 2004). This integrated map has a total of 2262 loci covering 1,832 cM distributed in 13 linkage groups. Map integration improved the average marker density throughout the genome to 1 marker per 0.8 cM. Because of uneven marker distribution, 15 gaps of at least 10 cM between adjacent markers still remain in the map. Despite the use of numerous markers, a few small unlinked linkage groups remained, indicating that the pepper map is still incomplete and the parents involved in mapping crosses differ by undefined chromosomal rearrangements

Two interspecific SSR based maps based on $F_2$ populations derived from a *C. frutescens* × *C. annuum* (FA 03) and *C. annuum* × *C. chinense* (AC 99) were created. All available marker sets were mapped on the same population creating relatively dense maps. The FA 03 map contains a total of approximately 728 markers covering 1,358 cM grouped to 12 linkage groups and associated with the 12 chromosomes of pepper. The AC 99 map currently contains a total of about 450 markers covering 1,304 cM, 150 SSR loci are common to both maps and can be used to describe colinearity. Selected tomato RFLP markers with known locations on the pepper genome were also used to anchor these maps to previous linkage maps of both pepper and tomato/potato.

### 3.6. Comparative Mapping in the Solanaceae

Comparative mapping in plant genomes originated in the Solanaceae, through the early work of Tanksley and colleagues (Bernatzky and Tanksley, 1986; Bonierbale *et al.*, 1988; Tanksley *et al.*, 1988; Tanksley *et al.*, 1992; Prince *et al.*, 1993; Livingstone *et al.*, 1999; Doganlar *et al.*, 2002a). Orthology of loci or identity by descent from a common ancestral sequence was determined based on map position. For example, if a tomato marker was within a block of markers, all showing conserved synteny between tomato and pepper, then that marker was orthologous to the corresponding marker in pepper. Comparative mapping is the side-by-side comparison of gene order and other genomic features in related individuals. Comparative mapping can aid in the identification of gene function and choice of additional markers for linkage mapping, marker-assisted selection and positional cloning. Additional genetic and physical markers from a dense map of a related species can be selected for application in the species of interest, provided an area of conserved synteny is under investigation. It also provides insights into genome evolution. Knowledge of the location and function of a morphological gene may provide insights into the function of a putative gene in the homologous region in pepper (*Capsicum* spp.).

In the solanaceae family, tomato genomics is currently the driving force behind much of the progress in the study of genomics for the entire family. The tools and markers developed for tomato often work well across the family. High density molecular and genetic linkage maps are available in tomato (Tanksley *et al.*, 1992), potato (Bonierbale *et al.*, 1988; Gebhardt *et al.*, 1991; van Os *et al.*, 2006), *Capsicum* (Livingstone *et al.*, 1999; Kang *et al.*, 2001; Lee *et al.,* 2004; Paran *et al.*, 2004; Minamiyama *et al.*, 2006; Yi *et al.*, 2006), eggplant (Doganlar *et al.*, 2002a), *Petunia* (Strommer *et al.*, 2000; Strommer *et al.*, 2002) and tobacco (Julio *et al.*, 2006; Bindler *et al.*, 2007). Mapping in pepper has revealed extensive chromosomal rearrangements between pepper and tomato genomes (Tanksley *et al.*, 1988; Prince *et al.,* 1993). Additional maps in potato have been constructed, including a comparative map between potato and *Arabidopsis* (Gebhardt *et al.*, 2003) showing 90 putative syntenic blocks. Paran *et al.* (2004) developed a multi-lab integrated map for pepper with 2,262 markers, which is currently the standard for pepper mapping.

The ultimate form of comparative mapping is increasingly becoming feasible, with significant sequencing efforts being devoted to many members of the solanaceae. Sequencing of the tomato nuclear genome is in progress by the members of the International Tomato Genome Sequencing Project. In addition, both the tomato mitochondrial and chloroplast genomes are being sequenced by LAT-SOL and EUSOL, two consortia of countries from Latin America and Europe. Sequencing efforts are also taking place in pepper (Lee *et al.*, 2004a; Yi *et al.,* 2006) and other members of solanaceae family.

### 3.7. Marker Assisted Breeding

The only marker assisted selection programmes documented in pepper is for resistance to *Phytopthora capsici*, although allele-specific molecular markers and markers linked to a number of useful traits have been reported. A SCAR marker was developed as a breeding tool to assist in selection of the major resistance QTL, *Phyto.5.2*, by screening diverse resistant and susceptible germplasm with RAPD primers (Quirin *et al.*, 2005). An RAPD marker associated with resistance was cloned and converted to SCAR marker that mapped back to an inferred position for *Phyto.5.2* on chromosome 5. The wide range differentiation of resistant and susceptible genotypes by this marker may provide an excellent tool for selecting peppers highly resistant to *P. capsici*.

## 3.8. Map Based Cloning, Characterization and Isolation of Genes

Intensive plant genomic sequencing data has provided new opportunities to study gene functions and their isolation. Several genes have been recently cloned and characterized (Chung *et al.*, 2003; Oh *et al.*, 2003; Kim *et al.*, 2004b; Yi *et al.*, 2004). Tai and Staskawicz (2000) have constructed a yeast artificial chromosome (YAC) library of pepper (*C. annuum*). This library was constructed from the cultivar Early Calwonder-123 R, which contained three resistance genes against *Xanthomonas campestris* and was estimated to have 3x coverage of the pepper genome. At least three additional large-insert libraries have been constructed in bacterial artificial chromosome (BAC) vectors. The first library had 10 × genome coverage of a *C. annuum* doubled haploid line homozygous for the *pvr1* (Ruffel *et al.*, 2004). The second library contained 12 genome equivalents from *C. annuum* Criollo de Morelos 334 (CM 334) and to date has been used in studies of several target regions including the *L*, *Y* and *C2* loci (Yoo *et al.*, 2003). The third library of 8.5 genome equivalents was constructed from the Mexican *C. frutescens* BG 2816 accession (Jahn, 2006) and was used for the isolation of the *FASCICULATE* gene (Paran, 2006). These libraries originate from different backgrounds and provide a rich resource for *Capsicum* genomics that can be utilized for the isolation of genes and regulatory elements that control agriculturaly important traits and for comparative sequence analysis of specific genomic regions with other solanaceous plants.

The *Bs2* gene that confers resistance to the bacterial spot pathogen *Xanthomonas campestris* was isolated by positional cloning approach (Tai *et al.*, 1999a, b). Molecular markers linked to *Bs2* were identified by screening near-isogenic lines that differ for the resistance gene. Tightly linked AFLP markers that flank *Bs2* were found which allowed physical mapping. These markers were used to screen a YAC library of pepper and identified positive clones that contained the appropriate region (Tai and Staskawicz, 2000). The *Bs2* locus was fine-mapped, isolated and confirmed. Transgenic expression of *Bs2* in tomato and tobacco conferred resistance to *X. campestris*, indicating that all downstream elements necessary for resistance may be broadly present in solanaceous plants. *Bs2* is the first resistance gene cloned in pepper by a map based cloning approach. Studies have shown that *avrBs2*, an avirulence gene of this pathogen triggers disease resistance in pepper plants containing the *Bs2* resistance gene and contribute to bacterial virulence on susceptible host plants (Gassman *et al.*, 2000). The *avrBs3* gene of the same organism has been found to be specific for avirulence in pepper (Bonas *et al.*, 1993).

Huh *et al.* (2001) utilized candidate gene approach to identify phytoene synthase (*PS*) as the locus for mature fruit colour in red pepper (*Capsicum* spp). A PCR-based approach was used to isolate resistance gene analogs

(RGAs) in pepper with primers corresponding to the nucleotide binding site of the *RPS2* (*Arabidopsis*), *N* (tobacco) and *L6* (flax) genes and to the kinase domains of the *Fen* and *Pto* (tomato) genes. The PCR products were cloned, sequenced and localized on the interspecific pepper maps. Cloned resistance genes (*Pto*, *Cf2*, *N*) and pathogenesis-related (PR) protein genes were used as probes for heterologous RFLP mapping (Pflieger *et al.*, 1998).

It has been observed that hot pepper (*C. annuum*) exhibits a hypersensitive response (HR) against infection by many tobacco viruses. A clone (*CaPR-4*) encoding a putative pathogenesis related protein 4 was isolated by Park *et al.* (2001) by differential screening of the cDNA library prepared for resistant pepper plant leaves inoculated with tobacco mosaic virus (TMV) pathotype PO. This study was extended to demonstrate *Ca PR-4* gene expression in pepper plants by various signal molecules, such as jasmonic acid and other abiotic elicitors. Such information has been useful in developing systemic acquired resistance (SAR) approaches for disease management. Similarly, the expression of SAR study would shed light on the defense system of *Capsicum* for future use in the genetic engineering of plant to confer resistance.

Sarowar *et al.* (2007) investigated the function of *CAF1* in plants, by over expression of the pepper *CAF1* (*CaCAF1*) in tomato and virus-induced gene silencing (VIGS) of the gene in pepper plants. Over expression of *CaCAF1* in tomato resulted in significant growth enhancement, with increasing leaf thickness, and enlarged cell size by more than two-fold when compared with the control plants. The electron microscopic analysis revealed that the *CaCAF1*-transgenic tomato plants had thicker cell walls and cuticle layers than the control plants. In addition to developmental changes, overexpression of *CaCAF1* in tomato plants resulted in enhanced resistance against the oomycete pathogen *Phytophthora infestans*. In contrast, VIGS of *CaCAF1* in pepper plants caused significant growth retardation and enhanced susceptibility to the pepper bacterial spot pathogen *Xanthomonas axonopodis* pv. *vesicatoria*. These results suggest the roles for plant *CAF1* in normal growth and development, as well as in defense against pathogens.

There are relatively few studies on the tissue-specific expression in *Capsicum* fruits. A recent report by Sung *et al.* (2001) has shown that a regulatory role for flower and fruit development in *C. annuum* may be developed by an interaction of protein products through MADS-box genes, *canMADS1* (isolated from floral bud) and *CanMADS6*. This was found by using the *OSMADS1* rice MADS-box gene as a probe.

Various genes coding for enzymes in capsaicinoid biosynthesis pathway have been isolated (Curry *et al.,* 1999; Kim *et al.,* 2001; Aluru *et al.,* 2003). Capsaicin biosynthesis involves condensation of vanillylamine and 8-methyl nonenoic acid, brought about by capsaicin synthase (*CS*). Characterization of capsaicin synthase and identification of its gene (*csy1*) for pungency factor (capsaicin) in pepper was reported by Prasad *et al.* (2006). CS gene (*csy1*) was cloned and full-length cDNA (981 bp) was sequenced. The deduced amino acid sequence of capsaicin synthase from full-length cDNA was 38 kDa. Functionality of *csy1* through heterologous expression in recombinant *Escherichia coli* was also demonstrated. These findings have implications in the regulation of capsaicin levels in *Capsicum* genotypes.

Other genes of pepper that have been isolated include *polygalcturonase* (Rao and Paran, 2003), capsanthin capsorubin synthase (*CCS*) (Lefebvre *et al.*, 1998), phytoene synthase (Thorup *et al.*, 2000; Huh *et al.*, 2001), *Pun1* (Stewart *et al.*, 2005) and *pvr1* (Ruffel *et al.*, 2002; Kang *et al.*, 2005).

## 4. GENETIC TRANSFORMATION IN *CAPSICUM*

Although excellent progress has been made in obtaining transgenic plants from many species of the solanaceae, *Capsicum* has lagged behind due to the unavailability of an efficient regeneration protocol (Liu *et al.,* 1990; Ebida and Hu, 1993). Despite several articles describing systems for the regeneration of both chilli and sweet peppers cultivars, elongation of shoot buds is a formidable job (Ochoa-Alezo and Ireta-Moreno, 1990). Wang *et al.* (1991) reported the recovery of plantlets from cultures of sweet pepper hypocotyls and cotyledons with GUS gene expression.

Zhu *et al.* (1996) reported transgenic sweet pepper plants from *Agrobacterium*-mediated transformation. They reported the regeneration of fertile transgenic sweet pepper (*C. annuum* var. *grossum*) plants at a relatively high rate from various explants that were co-cultivated with *A. tumefaceins* strain GV3111-SE harbouring a plasmid that contained the cucumber mosaic virus coat protein (*CMV-CP*) gene. Similarly, *Agrobacterium*-mediated transformation has been reported in *Capsicum* of a high-pungent variety with GUS and *NPT II* gene insertion (Manoharan *et al.*, 1998). There are some reports on generation of transgenic sweet pepper plants (Zhu *et al.*, 1996) for viral coat protein gene and herbicide gene, respectively, and of chilli pepper for reporter gene (Shivegowda *et al*, 2002) and for resistance to cucumber mosaic virus (Dong *et al*., 1992; 1995; Kim *et al.*, 1997). Regeneration of transgenic pepper plants resistant to TMV and CMV was reported by Ping *et al.* (1999).

Mesophyll protoplasts of pepper *C. annuum* were transformed by Jeon *et al.* (2007) using PEG-mediated transformation with a pCAMBIA1302 vector carrying a maize transposable element, *Ac* (activator), a selection marker *HPT* (hygromycin phospho-transferase), and a GFP-coding region driven by the *35S* promoter in the presence of PEG. They studied the transposition of the *Ac/Ds* element with a long term goal of developing a gene tagging system and a method for transposon mutagenesis in pepper.

## 5. THE SOLANUM GENOME NETWORK (SGN)

The solanum genome network (SGN) provides tremendous information on various aspects of the solanaceous genome. The information on all linkage groups and their relative sizes, an abstract with background information, and details about the map including marker counts and types of markers are available. A comparison linkage group will display the comparison between the two linkage groups. In addition to the map data, SGN stores detailed information about molecular markers. This includes sequences for RFLP markers and primers, sequence source (GenBank accession or SGNESTor unigene sequence identity) and PCR reaction conditions for CAPS, SSR and other PCR based markers. SGN also houses the master database for the COS (Fulton *et al.*, 2002b) and COS II markers. Currently, there are over 2,800 COS II markers available that have been developed to work on a maximum number of *Asterid* species (Wu *et al.*, 2006). For the COS II, the database stores the alignment of the sequences that were used to develop the markers, the experimental PCR conditions for each accession that the markers were applied to, and sequence reads from the PCR products. The data for the different accessions is submitted by users who have used the primers in the species and accessions that are documented. Using the SGN marker search, markers can be searched with a large number of search criteria, such as constraining the search to a species, chromosome, chromosome interval, marker type, LOD scores, association to a physical map, and more.

The *Capsicum annuum* EST data and additional microarray data are available at http://plant.pdrc.re.kr:7777/. Two inter-specific SSR-based maps were developed based on $F_2$ populations derived from a *C. frutescens* × *C. annuum* (FA03) and *C. annuum* × *C. chinense* (AC 99) have been created.

## 6. CONCLUSION

Micropropagation, plant regeneration and protoplast technology is available in *Capsicum* spp. Development of dihaploids through anther culture has been effectively utilized in *Capsicum* for molecular mapping and gene expression studies. Considerable progress has been made for molecular characterization and in pepper mapping, but the pepper map is not yet complete. Future mapping in pepper will rely on integration of sequence information from other solanaceous plant species. The international

Solanaceae Genome Project (SOL) and the sequence of the tomato genome as a public reference genome for other solanaceous taxa will enhance the future mapping experiments in pepper. More efficient positional cloning and candidate gene approaches will help in identification of genes that control pepper development and production. Functional genomics, such as determination of expression profiles by microarrays, metabolic profiling and screening mutant populations are currently underway in pepper. These complementary approaches will allow the identification of many new genes and their function in *Capsicum*.

Application of recombinant-DNA technology for production of resistant types to biotic and abiotic stress has great potential in the genus *Capsicum*. Though successes were reported in *in vitro* production of capsaicin, capsanthin and capsorubin, production, techniques are to be refined and scaled up before their commercialization. Metabolic engineering of *Capsicum*, for enhanced capsaicin or carotenoid pathway gene has been demonstrated in several systems, including rice (Ye *et al*., 2000). However, there is a need to develop a high pigment and low pungency product from *Capsicum*, which will be of value for pharamaceuticals, consumers and as a food colourant. Studies on a placenta-specific expression of capsaicinoids and a fruit-wall specific expression of carotenoids will be of great help. Such developments need to be completed for disease resistance, which would result in the overall improvement of the *Capsicum* for pre- and post-harvest applications for augmenting qualitative and quantitative output of the products from *Capsicum* species.

## REFERENCES

Agarwal S (1988). Shoot tip culture of pepper and its micropropagation. *Curr. Sci.* **57**: 1347–8.

Aluru MR, Mazourek M, Landry LG, Curry J, Jahn M and O'Connell MA (2003). Differential expression of fatty acid synthase genes, Acl, Fat and Kas, in *Capsicum* fruit. J. *Exp. Bot*. **54**: 1655–44.

Andrews J (1995). The Domesticated *Capsicum*, (New edn.) University of Texas Press, Austin, TX.

Anu A, Nirmal Babu K and Peter KV (2004). Variations among somaclones and its seedling progeny in *Capsicum annuum*. *Plant Cell Tissue Organ Cul.* **76**: 261–7.

Arnedo-Andres MS, Gil Ortega R, Luis Arteaga M and Hormaza JI (2002). Development of RAPD and SCAR markers linked to the *Pvr4* locus for resistance to PVY in pepper (*Capsicum annuum* L.). *Theor. Appl. Genet.* **105(6–7)**: 1067–74.

Arroyo R and Revilla MA (1991). *In vitro* plant regeneration from cotyledon and hypocotyl segments in two bell pepper cultivars. *Plant Cell Rep*. **10**: 414–6.

Arumuganathan K and Earle ED (1991). Nuclear DNA content of some important plant species. *Plant Mol. Biol. Rep*. **9**: 208–19.

Belletti P, Marzachi C and Lanteri S (1998). Flow cytometric measurement of nuclear DNA content in *Capsicum* (Solanaceae). *Plant Syst. Evol.* **209**: 85–91.

Ben Chaim A and Paran I (2000). Genetic analysis of quantitative traits in pepper (*Capsicum annuum*). *J. Am. Soc. Hort. Sci.* **125**: 66–70.

Ben Chaim A, Borovsky E, De Jong W and Paran I (2003b). Linkage of the *A* locus for the presence of anthocyanin and *fs10.1*, a major fruit shape QTL in pepper. *Theor. Appl. Genet.* **106**: 889–94.

Ben Chaim A, Borovsky E, Rao G U, Tanyolac B and Paran I (2003a). *fs3.1*: a major fruit shape QTL conserved in *Capsicum*. *Genome* **46**: 1–9.

Ben Chaim A, Grube R, Lapidot M, Jahn M and Paran I (2001a). Identification of quantitative trait loci associated with resistance to cucumber mosaic virus in *Capsicum annuum*. *Theor. Appl. Genet.* **102**: 1213–20.

Ben Chaim A, Paran I, Grube R, Jahn M, van Wijk R and Peleman J (2001b). QTL mapping of fruit related traits in pepper (*Capsicum annuum*). *Theor. Appl. Genet.* **102**: 1016–28.

Berzal-Herranz A, de la Cruz A, Tenllado F, Diaz-Ruiz JR, Lopez L, Sanz AI, Vaquero C, Serra MT and Garcia-Luque I (1995). The *Capsicum L3* gene-mediated resistance against the tobamoviruses is elicited by the coat protein. *Virology* **209**: 498–505.

Bindler G, van der Hoeven R, Gunduz I, Plieske J, Ganal M, Rossi L, Gadani F and Donini P (2007). A microsatellite marker based linkage map of tobacco. *Theor. Appl. Genet.* **114**: 341–9.

Binzel ML, Sankhla N, Joshi S and Sankhla D (1996). Induction of direct somatic embryogenesis and plant regeneration in pepper (*Capsicum annuum* L.). *Plant Cell Rep.* **15**: 536–40.

Blum E, Liu K, Mazourek M, Yoo EY, Jahn M and Paran I (2002). Molecular mapping of the *C* locus for presence of pungency in *Capsicum*. *Genome* **45**: 702–5.

Blum E, Mazourek M, O'Connell MA, Curry J, Thorup T, Liu K, Jahn M and Paran I (2003). Molecular mapping of capsaicinoid biosynthesis genes and quantitative trait loci analysis for capsaicinoid content in *Capsicum*. *Theor. Appl. Genet.* **108**: 79–86.

Boiteux LS and De-Avila AC (1994). Inheritance of a resistance specific to tomato spotted wilt Tospovirus in *Capsicum chinense* 'PI 159236'. *Euphytica* **75**: 139–42.

Bonas U, Conrads-Strauch J and Balbo (1993). Resistance in tomato to *Xanthomonas campestris* pv. *vesicatoria* is determined by alleles of the pepper-specific avirulence gene *avrBs3*. *Mol. Gen. Gent. Apr.* **238(1–2)**: 261–9.

Bonierbale MW, Plaisted RL and Tanksley SD (1988). RFLP maps-based on a common set of clones reveal modes of chromosomal evolution in potato and tomato. *Genetics* **120**: 1095–1103.

Borovsky Y, Oren-Shamir M, Ovadia R, De Jong W and Paran I (2004). The *A* locus that controls anthocyanin accumulation in pepper encodes a *MYB* transcription factor homologous to *Anthocyanin2* of *Petunia*. *Theor. Appl. Genet.* **109**: 23–29.

Bosland PW and Votava EJ (2000). Peppers: Vegetable and Spice Capsicums. CABI, New York.

Bosland PW (1992). Chiles: a diverse crop. *Hort. Technol.* **2**: 7–10.

Boukema IW (1980). Allelism of genes controlling resistance to TMV in *Capsicum* L. *Euphytica* **29**: 433–9.

Buyukalaca S and Mavituna F (1996). Somatic embryogenesis and plant regeneration of pepper in liquid media. *Plant Cell Tissue Organ Cul.* **46(3)**: 227–37.

Buyukalaca S and Mavituna F (1995a). Large scale production of pepper somatic embryos using bioreactors. *Acta Hort.* **412**: 58–63.

Buyukalaca S.and Mavituna F (1995b). Artificial seeds of pepper somatic embryos. *Acta Hort.* **412**: 106–10.

Cao DS and Jiu SR (1993). *In vitro* plant regeneration of sweet pepper. *Acta Hort. Sinica* **20**: 171–75.

Caranta C, Lefebvre V and Palloix A (1997b). Polygenic resistance of pepper to potyviruses consists of a combination of isolate specific and broad spectrum quantitative trait loci. *Mol. Plant. Micr. Interact.* **10**: 872–8.

Caranta C, Palloix A, Gebre-Selassie K, Lefebvre V, Moury B and Daubeze AM (1996). A complementation of two genes originating from susceptible *Capsicum annuum* lines confers a new and complete resistance to pepper vienal mottle virus. *Phytopathol.* **86**: 739–43.

Caranta C, Pflieger S, Lefebvre V, Daubeze AM, Thabuis A and Palloix A (2002). QTLs involved in the restriction of cucumber mosaic virus (*CMV*) long-distance movement in pepper. *Theor. Appl. Genet.* **104**: 586–91.

Caranta C, Thabuis A and Palloix A (1999). Development of CAPS marker for the *Pvr4* locus: a tool for pyramiding potyvirus resistance genes in pepper. *Genome* **42(6)**: 1111–6.

Caranta CA, Palloix A and Lefebvre V (1997a). QTLs for a component of partial resistance to cucumber mosaic virus in pepper: restriction of virus installation in host-cell. *Theor. Appl. Genet.* **94**: 431–8.

Caterina MJ, Leffler A, Malmberg AB, Martin WJ, Trafton J, Petersen-Zeitz KR, Koltzenburg M, Basbaum AI and Julius D (2000). Impaired nociception and pain sensation in mice lacking the capsaicin receptor. *Science* **288**: 306–13.

Chaim AB, Borovsky Y, De Jong W and Paran I (2003). Linkage of the *A* locus for the presence of anthocyanin and *fs10.1*, a major fruit-shape QTL in pepper. *Theor. Appl. Genet.* **106**: 889–94.

Chavez-Moetezuma MP and Lozoya-Gloria E (1996). Biosynthesis of the sesquiterpene phytoalexin capsidiol in elicited root cultures of chili pepper (*Capsicum annuum*). *Plant Cell Rep.* **15(5)**: 360–6.

Christopher T and Rajam MV (1996). Effect of genotype, explant and medium on *in vitro* plant regeneration of red pepper (*Capsicum annuum* L.) *Plant Cell Tissue Organ Cult.* **46**: 245–50.

ChristopherT and Rajam MV (1994). *In vitro* clonal propagation of *Capsicum* spp. *Plant Cell Tissue Organ Cult.* **38(1)**: 25–29.

Chung E, Kim S-Y, Yi SY and Choi D (2003). *Capsicum annuum* dehydrin, an osmotic-stress gene in hot pepper plants. *Mol. Cells* **15**: 327–32.

Comai L and Henikoff S (2006). TILLING: practical singlenucleotide mutation discovery. *Plant J.* **45**: 684–94.

Comai L, Young K, Till BJ, Reynolds SH, Greene EA, Codomo CA, Enns LC, Johnson JE, Burtner C, Odden AR and Henikoff S (2004). Efficient discovery of DNA polymorphisms in natural populations by Ecotilling. *Plant J.* **37**: 778–86.

Cook AA and Guevara YG (1984). Hypersensitivity in *Capsicum chacoense* to race 1 of the bacterial spot pathogen of pepper. *Plant Dis.* **68**: 329–30.

Csillery G, Szarka E, SardiE, Mityko J, Kapitany J, Nagy B and Szarka J (2004). The unity of plant defense: genetics, breeding and physiology. *In*: Proc. of the 12th Meeting on Genet. Breed. *Capsicum* and Eggplant, Noordwijkerhout, The Netherlands, pp. 147–53.

Curry J, Aluru M, Mendoza M, Nevarez J, Melendrez M and O'Connell MA (1999). Transcripts for possible capsaicinoid biosynthetic genes are differentially accumulated in pungent and non-pungent *Capsicum* spp. *Plant Sci.* **148**: 47–57.

de la Cruz A, Lopez L, Tenllado F, Diaz-Ruiz JR, Sanz AI, Vaquero C, Serra MT and Garcia-Luque I (1997). The coat protein is required for the elicitation of the *Capsicum L2* genemediated resistance against the tobamoviruses. *Mol. Plant Microbe Interact.* **10**: 107–13.

Deshpande RB (1933). Studies in Indian chillies. 3. Inheritance of some characters in *Capsicum* L. *Indian J. Agri. Sci.* 3.

Diaz I, Moreno R and Power JB (1988). Plant regeneration from protoplasts of *Capsicum annuum. Plant Cell Rep.* **7**: 210–2.

Djian-Caporalino C, Pijarowski L, Fazari A, Samson M, Gaveau L, O'Byrne C, Lefebvre V, Caranta C, Palloix A and Abad P (2001). High-resolution genetic mapping of the pepper (*Capsicum annuum* L.) resistance loci *Me*3 and *Me*4 conferring heat-stable resistance to root-knot nematodes (*Meloiogyne* spp.). *Theor. Appl. Genet.* **103**: 592–600.

Doganlar S, Frary A, Daunay M-C, Lester RN and Tanksley SD (2002). Comparative genetic linkage map of eggplant (*Solanum melongena*) and its implications for genome evolution in the Solanaceae. *Genetics* **161**: 1697–711.

Doganlar S, Frary A, Daunay M-C, Lester RN and Tanksley SD (2002). Conservation of gene function in the solanaceae as revealedby comparative mapping of domestication traits in eggplant. *Genetics* **161**: 1713–26.

Donato MD, Perucco E and Mozzetti C (1989). Protoplasts culture and callus proliferation from cotyledons of *Capsicum annuum* L. *Adv. Hortic Sci.* **3**: 17–20.

Dong CZ, Jiang CX., Feng LX, Li SD., Gao ZH., Guo JZ and Zhu DW (1995). Transgenic tomato and pepper plants containing CMV sat-RNA cDNA. *Acta Hort.* **402**: 78–86.

Dong CZ, Jiang CX, Feng LX and Guo JZ (1992). Transgenic pepper plants (*Capsicum annuum* L.) containing CMV sat-RNA cDNA. *Acta Hort. Sinica* **19**: 180–4.

Ebida IAA and Hu C (1993). *In vitro* morphogenetic responses and plant regeneration. from pepper. (*Capsicum annuum* L.). *Plant cell Rep.* **13**: 107–10.

Efrati A, Eyal Y and Paran I (2005). Molecular mapping of the chlorophyll retainer (*cl*) mutation in pepper (*Capsicum* spp.) and screening for candidate genes using tomato ESTs homologous to structural genes of the chlorophyll catabolism pathway. *Genome* **48**: 347–51.

Ezura H, Nishimiya S and Kasumi M (1993). Efficient regeneration of plants independent of exogenous growth regulators in bell pepper (*C. annuum* L.). *Plant Cell Rep.* **12**: 676–80.

Fari M and Czako M (1981). Relationships between position and morphogenic responses of pepper hypocotyl explants cultured *in vitro*. *Sci. Hortic.* **15**: 207.

Fulton TM, Bucheli P, Voirol E, Lopez J, Petiard V and Tanksley SD (2002a). Quantitative trait loci (QTL) affecting sugars, organic acids and other biochemical properties possibly contributing to flavour, identified in four advanced backcross populations of tomato. *Euphytica* **127**: 163–77.

Fulton TM, Grandillo S, Beck-Bunn T, Fridman E, Frampton A, Lopez J, Petiard V, Uhlig J, Zamir D and Tanksley SD (2000). Advanced backcross QTL analysis of a *Lycopersicon esculentum* × *Lycopersicon parviflorum* cross. *Theor. Appl. Genet.* **100**: 1025–42.

Fulton TM, van derHoeven R, Eanetta NT and Tanksley SD (2002b). Identification, analysis, and utilization of conserved ortholog set markers for comparative genomics in higher plants. *Plant Cell* **14**: 1457–67.

Gassmann W, Dahlbeck D, Chesnokova O, Minsavage GV, Jones JB and Staskawicz, BJ (2000). Molecular evolution of virulence in natural field strains of *Xanthomonas campestris* pv. *vesicatoria*. *J. Bacteriol.* **182(24)**: 7053–9.

Gebhardt C, Ritter E, Barone A, Debener T, Walkemeier B,Schachtschabel U, Kaufman H, Thompson RD, Bonierbale MW, Ganal MW, Tanksley SD, Salamini F (1991). RFLP maps of potato and their alignment with the homoeologous tomato genome. *Theor. Appl. Genet.* **83**: 49–57.

Gebhardt C, Walkemeier B, Henselewski H, Barakat A, DelsenyM and Stuber K (2003) Comparative mapping between potato (*Solanum tuberosum*) and *Arabidopsis thaliana* reveals structurally conserved domains and ancient duplications in the potato genome. *Plant J.* **34**: 529–41.

George L and Narayana Swamy S (1973). Haploid *Capsicum* through experimental androgenesis. *Protoplasma.* **78**: 467–70.

Gilchrist EJ and Haughn GW (2005). TILLING without a plough: a new method with applications for reverse genetics. *Curr. Opin. Plant Biol.* **8**: 211–5.

Gilchrist EJ, Haughn GW, Ying CC, Otto SP, Zhuang J, Cheung D, Hamberger B, Aboutorabi F, Kalynyak T, Johnson L, Bohlmann J, Ellis BE, Douglas CJ and Cronk QC (2006). Use of Ecotilling as an efficient SNP discovery tool to survey genetic variation in wild populations of *Populus trichocarpa*. *Mol. Ecol.* **15**: 1367–78.

Gonzalez-Melendi P, Testillano PS, Ahamadian P, Fadon B and Risueno MC (1996). New *in situ* approaches to study the induction of pollen embryogenesis in *Capsicum annuum*. *European J. Cell Biol.* **69(4)**: 373–86.

Greenleaf WH (1986). Pepper breeding. *In*: Bassett MJ (*ed.*) Breeding vegetable crops. AVI, Westpost, CT, pp. 67–134.

Grube RC, Blauth JR, Arnedo-Andres MS, Caranta C and Jahn M (2000b). Identification and comparative mapping of a dominant potyvirus gene cluster in *Capsicum*. *Theor. Appl .Genet.* **101**: 852–9.

Grube RC, Radwanski ER and Jahn M (2000a). Comparative genetics of disease resistance within the solanaceae. *Genetics* **155**: 873–87.

Gupta CG, Lakshmi N and Srivalli T (1998) Micropropagation studies on a male sterile line of *Capsicum annuum* L. *Capsicum and Eggplant Newslett.* **17**: 42–45.

Guzman FA, Ayala H, Azurdia C, Duque MC. and Carman de Vicente M (2005). AFLP assessment of genetic diversity of *Capsicum* genetic resources in Guatemala: home gardens as an option for conservation. *Crop Sci.* **45**: 363–70.

Harini I and Sita GL (1993). Direct somatic embryogenesis and plant regeneration, from immature embryos of chilli (*Capsicum annuum* L.). *Plant Sci.* **89**: 107–12.

Harn C, Kim MZ, Choi KT and Lee YI (1975). Production of haploid callus and embryoid from the cultured anther of *Capsicum annuum*. *Kor. J. Plant Tissue Cult.* **3**: 1–7.

Harvell KP and Bosland PW (1997). The environment produces a significant effect on pungency of chiles (*Capsicum annuum* L.). *HortScience* **32**: 1992.

Hibberd AM, Bassett MJ and Stall RE (1987). Allelism tests of three dominant genes for hypersensitive resistance to bacterial spot of pepper. *Phytopathol.* **77**: 1304–7.

Huh JH, Kang BC, Nahm SH, Kim S, Lee MH and Kim BD (2001). A candidate gene approach identified phytoene synthase as the locus for mature fruit color in red pepper (*Capsicum* spp.). *Theor. Appl. Genet.* **102(4)**: 524–30.

Hurtado-Hernandez H and Smith PG (1985). Inheritance of mature fruit colour in *Capsicum annuum* L. *J. Hered.* **76**: 211–3.

Ilbi H (2003). RAPD markers assisted varietal identification and genetic purity test in pepper, *Capsicum annuum*. *Sci. Hortic.* **97**: 211–8.

Jahn M, Paran I, Hoffmann K, Radwansky E, Livingstone K, Grube R, Aftergoot E, Lapidot M and Moyer J (2000). Genetic mapping of the *Tsw* locus for resistance to the tomato spotted wilt virus in *Capsicum* spp. and its relationship to the *Sw-5* gene for resistance to the same pathogen in tomato. *Mol. Plant Microbe Interact.* **13**: 673–82.

Jang I, Moon JH, Yoon JB, Yoo JH, Yang TJ, Kim YJ and Park HG (2004). Application of RAPD and SCAR markers for purity testing of $F_1$ hybrid seed in chilli. *Mol. Cells* **18**: 295–9.

Jasna Berljak (1999). *In vitro* plant regeneration from pepper (*Capsicum annuum* L.). *Phyton* (Austria), **39(3)**: 289–92.

Jeon JM, Ahn NY, Son BH, Kim CY, Han C, Kim GD, Gal SW and Lee SH (2007). Efficient transient expression and transformation of PEG-mediated gene uptake into mesophyll protoplasts of pepper (*Capsicum annuum* L.) *Plant Cell Tissue Organ Cult.* **88**: 225–32.

Johnson TS and Ravishankar GA (1996). Biotransformation of ferulic acid and vanillylamine to capsaicin and vanillin in immobilised cell cultures of *Capsicum frutescens. Plant Cell Tissue Organ Cult.* **44**: 117–26.

Johnson TS, Ravishankar GA and Venkataraman LV (1990). *In vitro* capsaicin production by immobilised cells and placental tissues of *Capsicum annuum* L. grown in liquid medium. *Plant Sci.* **70**: 223–9.

Johnson TS, Ravishankar GA and Venkataraman LV (1991). Elicitation of capsaicin production in freely suspended cells and immobilised cell cultures of *Capsicum frutescens. Food Biotechnol.* **5**: 197–205.

Johnson TS, Ravishanker GA and Venkataraman LV (1996). Biotransformation of ferulic acid vanillyamine to capsacin and vanillin in immobilised cell cultures of *Capsicum frutescens. Plant Cell Tissue Organ Cult.* **44(2)**: 117–23.

Jones JB, Minsavage GV, Roberts PD, Johnson RR, Kousik CS, Subramanian S and Stall RE (2002). A non-hypersensitive resistance in pepper to the bacterial spot pathogen is associated with two recessive genes. *Phytopathol.* **92**: 273–7.

Jordi Ballester and Carmen de Vincente M (1998). Determination of $F_1$ hybrid seed purity in pepper using PCR-based markers. *Euphytica* **103**: 223–6.

Kang BC, Nahm SH, Huh JH, Yoo HS, Yu JW, Lee MH and Kim BD (2001). An interspecific (*Capsicum annuum* L. × *C. chinense*) $F_2$ linkage map in pepper using RFLP and RAPD markers. *Theor. Appl. Genet.* **102(4)**: 531–9.

Kang BC, Yeam IH, Frantz DJ, Murphy JF and Jahn MM (2005). The *pvr1* locus in pepper encodes a translation initiation factor *eIF4E* that interacts with *Tobacco etch virus* VPg. *Plant J.* **42**: 392–405.

Kato K, Matsumoto M, Shimoda Y, Yamasaki S and Shimamura F (1996). Effect of culture conditions on *in vitro* induction of adventitious buds from leaf explants in pepper. Scientific Report of the Faculty of Agriculture, Okayama University No. **85**: 39–44.

Kim BS and Hartmann RW (1985). Inheritance of a gene (*Bs3*) conferring hypersensitive resistance to *Xanthomonas campestris* pv *vesicatoria* in pepper (*Capsicum annuum*). *Plant Dis.* **69**: 233–5.

Kim CM, Je BI, Koo JC, Piao HL, Park SJ, Jeon JM, Kim MK, Park SH, Park JY, Lee EJ, Chung WS, Lee KH, Kang KY, Lee SH and Han C-d (2004a). Studies on the mobilization of a maize transposable family, *Ac / Ds*, in pepper using *in vivo* transient assay system. *J. Plant Biol.***47**: 1–7.

Kim M, Kim S, Kim S and Kim BD (2001). Isolation of cDNA clones differentially accumulated in the placenta of pungent pepper by suppression subtractive hybridization. *Mol. Cells* **11**: 213–9.

Kim S, An CS, Hong and YN, Lee KW (2004b). Cold inducible transcription factor, *CaCBF*, is associated with a homeodomain leucine zipper protein in hot pepper (*Capsicum annuum* L.). *Mol. Cells* **18**: 300–8.

Kim SJ, Lee SJ, Kim BD and Paek KH (1997). Satellite-mediated resistance to cucumber mosaic virus in transgenic plants of hot pepper (*Capsicum annuum* L.). *Plant Cell Rep.* **16**: 825–30.

Kintzios SE, Hiureas G, Shortsiaanitis E, Sereti E, Blouhos P, Manos C, Makri O, Taravira N, Drossopoulos JB, Holevas CD and Drew RA (1998). Effect of light on the induction, development and maturation of somatic embryos from various horticultural and ornamental species. *Acta Hort.* **461**: 427–32.

Kumar S, Singh V, Kumar S, Singh M, Rai M and Kalloo G (2002). RAPD protocol for tagging of fertility restorer and male sterility genes in chilli (*Capsicum annuum* L.). *Veg. Sci.* **29**: 101–5.

Kyle MM and Palloix A (1997). Proposed revision of nomenclature for potyvirus resistance genes in *Capsicum*. *Euphytica* **97**: 183–8.

Lee JM, Nahm SH, Kim YM and Kim BD (2004). Characterization and molecular genetic mapping of microsatellite loci in pepper. *Theor. Appl. Genet.* **108**: 619–27.

Lee S, Kim SY, Chung E, Joung YH, Pai HS, Hur CG and Choi D (2004). EST and microarray analyses of pathogen responsive genes in hot pepper (*Capsicum annuum* L.) non-host resistance against soybean pustule pathogen (*Xanthomonas axonopodis pv. glycines*). *Funct. Integr. Genome* **4**: 196–205.

Lefebvre V, Daubeze AM, Rouppe van der Voort J, Peleman J, Bardin M and Palloix A (2003). QTLs for resistance to powdery mildew in pepper under natural and artificial infections. *Theor. Appl. Genet.* **107**: 661–6.

Lefebvre V, Kuntz M, Camara B and Palloix A (1998). Capsanthin capsorubin synthase gene: a candidate gene for the *y* locus controlling the red fruit colour in pepper. *Plant. Mol. Biol.* **36**: 785–9.

Lefebvre V, Palloix A, Caranta C and Pochard E (1995). Construction of an intraspecific integrated linkage map of pepper using molecular markers and double haploid progenies. *Genome* **38**: 112–21.

Lefebvre V, Pflieger S, Thabuis A, Caranta C, Blattes A, Chauvet JC, Daubeze AM and Palloix A (2002). Towards the saturation of the pepper linkage map by alignment of three intra-specific maps including known function genes. *Genome* **45**: 839–54.

Lippert LF, Smith PG and Bergh BO (1966). Cytogentics of Vegetable Crops. Garden Pepper, *Capsicum* spp. *Bot. Rev.* **32**: 24–55.

Liu W, Parrot WA, Hildebrand D, Collins GB and Williams EG (1990). *Agrobacterium* induced gall formation in bell pepper (*C. annuum* L.) and formation of shoot like structures expressing induced genes. *Plant Cell Rep.* **9**: 360–4.

Livingstone KD, Lackney VK, Blauth J, van Wijk R and Jahn MK (1999). Genome mapping in *Capsicum* and the evolution of genome structure in the solanaceae. *Genetics* **152**: 1183–1202.

Matsunaga H, Saito T, Hirai M, Nunome T and Yoshida T (2003). DNA markers linked to pepper mild mottle virus (PMMoV) resistant locus (*L*4) in *Capsicum*. J. *Japanese Soc. Hort. Sci.* **72**: 218–20.

Mcleod MJ, Guttman SI and Eshbaugh WH (1983). *In*: Isozymes in Plant Genetics and Breeding, Tanksley SD and Orton TJ; (*eds*). Part B, Elsevier, Amsterdam, pp. 189–201.

Michelmore RW, Paran I and Kesseli RV (1991). Identification of markers linked to disease resistance genes by bulked segregant analysis: a rapid method to detect markers in specific genomic regions by using segregating populations. *Proc. Natl. Acad. Sci. USA.* **88**: 9828–32.

Minamiyama Y, Tsuro M and HiraiM (2006). An SSR-based linkage map of *Capsicum annuum*. *Mol. Breed.* **18**: 157–69.

Moscone EA, Lambrou M, Hunziker AT and Ehrendorfer F (1993). Giemsa C-banded karyotyes in *Capsicum* (Solanaceae). Plant Syst. Evol. **186**: 213–29.

Moury B, Pflieger S, Blattes A, Lefebvre V and Palloix A (2000). A CAPS marker to assist selection of tomato spotted wilt virus (TSWV) resistance in pepper. *Genome* **43**: 137–42.

Murphy JF, Blauth JR, Livingstone KD, Lackney VK and Jahn M (1998). Genetic mapping of the *pvr1* locus in *Capsicum* spp. and evidence that distinct potyvirus resistance loci control responses that differ at the whole plant and cellular levels. *Mol. Plant Micr. Interact.* **11**: 943–51.

Ochoia AN and Ireta-Mereno L (1990). Culture differences in shoot forming capacity of hypocotyl tissues of chilli. *Sci. Hortic.* **42**: 21–28.

Oh B-J, Ko MK, Kim KS, Kim YS, Lee HH, Jeon WB and Im KH (2003). Isolation of defense-related genes differentially expressed in the resistance interaction between pepper fruits and the anthracnose fungus *Colletotrichum gloeosporioides*. *Mol. Cells* **15**: 349–55.

Onus AN and Pickersgill B (2004). Unilateral incompatibility in *Capsicum* (Solanaceae): occurrence and taxonomic distribution. *Ann. Bot.* (London) **94**: 289–95.

Palevitch D and Craker LE (1995). Nutritional and medical importance of red pepper (*Capsicum* spp.). *J. Herbs Spices Med. Plants* **3**: 55–83.

Paran I (2003). Marker assisted utilization of exotic germplasm. *In*: Nguyen HT and Blum A (*eds.*) Physiology and biotechnology integration for plant breeding, Marcel Dekker, New York.

Paran I, Aftergoot E and Shifriss C (1998). Variation in *Capsicum annuum* revealed by RAPD and AFLP markers. *Euphytica* **99**: 167–73.

Paran I, Ben-Chaim A, Byoung-Cheor, Kang, and Molly Jahn (2007). Capsicums, pp. 209–23 in C. Kole (*ed.*). Genome mapping and molecular breeding in plants, Volume 5, Vegetables, Springer-Verlag Berlin Heidelberg.

Paran I, van der Voort JR, Lefebvre V, Jahn M, Landry L,vanSchriek M, Tanyolac B, Caranta C, Ben Chaim A, Livingstone K, Palloix A and Peleman J (2004). An integrated genetic linkage map of pepper. *Mol. Breed.* **13**: 251–61.

Parrella G, Ruffel S, Moretti A, Morel C, Palloix A and Caranta C (2002). Recessive resistance genes against potyviruses are localized in colinear genomic regions of the tomato (*Lycopersicon* spp.) and pepper (*Capsicum* spp.) genomes. *Theor. Appl. Genet.* **105**: 855–61.

Parthasarathy VA and Nirmal V (2001). *In*: Biotechnology of Horticultural Crops, (Parthasarathy VA, Bose TK. and Das P; *eds.*) Naya Prokash, Calcutta, Vol. 2. pp. 146–73.

Peterson PA (1959). Linkage of fruit shape and colour genes in *Capsicum*. *Genetics* **44**: 407–19.

Pfieleger S, Lefebvre V, Caranta C, Blattes A, Goffinet B, Palloix A (1999). Disease resistance gene analogs as candidates for QTLs involved in pepper-pathogen interactions. *Genome* **42**: 1100–10.

Pfieleger S, Palloix A, Caranta C, Blattes A and Lefebvre V (2001). Defense response genes colocalize with quantitative disease resistance loci in pepper. *Theor. Appl. Genet.* **103**: 920–9.

Pflieger S, Veronique Lefebvre, Carol C, Radwankski E, Livingston K, Jahn MK and Palloix. (1998). *In*: Intl. Plant and Animal Genome VI conference, Jan 18–22, San Diego, CA, USA.

Pickersgill B (1997). Genetic resources and breeding of *Capsicum* spp. *Euphytica* **96**: 129–33.

Pierre M, Noel L, Lahaye T, Ballvora A, Veuskens J, Ganal M and Bonas U (2000). High resolution genetic mapping of the pepper resistance locus *Bs3* governing recognition of the *Xanthomonas campestris* pv *vesicatora AvrBs3* protein. *Theor. Appl. Genet.* **101**: 255–63.

Ping YUB, Lei S, Xing JW, Ping LX, Zhong XZ and Jing JM (1999). Regeneration of transgenic pepper plants resistant to TMV and CMV and its resistance identification. *Acta Agriculturae Boreali Sinica* **14(93)**: 103–8.

Pochard E (1977).Localization of genes in *Capsicum annuum* L. by trisomic analysis. *Ann. Amelior. Plantes* **27**: 255–66.

Pochard E, Dumas de Valuix R and Florent A (1983). Linkage between partial resistance to CMVand susceptibility to TMV in the line Perennial: analysis on androgenetic homozygous lines. *Capsicum* and *Eggplant Newlett.* **2**: 34–35.

Popovsky S and Paran I (2000). Molecular analysis of the *Y* locus in pepper: its relation to capsanthin capsorubin synthase and to fruit colour. *Theor. Appl. Genet.* **101**: 86–89.

Poulos JM (1994). Pepper breeding (*Capsicum* spp.): achievements, challenges and possibilities. Plant Breed. Abstr. **64(2)**: 143–55.

Prakash AH, Rao KS and Kumar MU (1997). Plant regeneration from protoplasts of *Capsicum* cv. California Wonder. *J. Biosci.* **22(3)**: 339–44.

Prasad BCN, Kumar V, Gururaj HB, Parimalan R, Giridhar P and Ravishankar GA (2006). Characterization of capsaicin synthase and identification of its gene (*csy1*) for pungency factor capsaicin in pepper (*Capsicum* sp.), *Proc. Natl. Acad. Sci. USA* **103(36)**: 133315–20.

Prince JP, Pochard E, and Tanksley SD (1993). Construction of a molecular linkage map of pepper and a comparison of synteny with tomato. *Genome* **36**: 404–17.

Quirin EA, Ogundiwin EA, Prince JP, Mazourek M, Briggs MO, Chlanda TS, Kim KT, Falise M, Kang BC and Jahn MM (2005). Development of sequence characterized amplified region (SCAR) primers for the detection of *Phyto.5.2*, a major QTL for resistance to *Phytophthora capsici* Leon. in pepper. *Theor. Appl. Genet.* **110**: 605–12.

Rao GU and Paran I (2003a). Polygalacturonase: a candidate gene for the soft flesh and deciduous fruit mutation in *Capsicum*. *Plant Mol. Biol.* **51**: 135–41.

Rao GU, Ben Chaim A, Borovsky E and Paran I (2003b). Mapping of yield related QTLs in pepper in an inter-specific cross of *Capsicum annuum* and *C. frutescens*. *Theor. Appl. Genet.* **106**: 1457–66.

Ravishankar GA, Sudhakar JT and Venkataraman LV (1993). Biotechnological approach of *in vitro* production of capsaicin. *In*: Proc. Natl. Sem. Post-Harvest Tech. Spices. Trivandrum: 75–82.

Ravishankar GA, Suresh B, Giridhar P, Ramachandra Rao and Sudhakar Jhonson T (2003). *In*: Capsicum-The genus *Capsicum* (De, AK, *ed.*). Medicinal and aromatic plants–industrial profiles; v.33. Taylor and Francis, London.

Regner F (1996). Anther and microspore culture in *Capsicum*, *In*: Jain SM, Sopory SK and Veilleux RE (*eds.*) *In vitro* haploid production in higher plants. Vol. 3 (pp. 77–89). Kluwer Academic Publishers, The Netherlands.

Ruffel S, Caranta C, Palloix A, Lefebvre V, Caboche M and Bendahmane A (2004). Structural analysis of the eukaryotic initiation factor 4E gene controlling potyvirus resistance in pepper: exploitation of a BAC library. *Gene* **338**: 209–16.

Ruffel S, Dussault MH, Palloix A, Moury B, Bendahmane A, Robaglia C and Caranta C (2002). A natural recessive gene against potato virus Y in pepper corresponds to the eukaryotic initiation factor 4E (*eIF4E*). *Plant J.* **32**: 1067–75.

Sadder M and Weber G (2002). Comparison between genetic and physical maps in *Zea mays* L. of molecular markers linked to resistance against *Diatraea* spp. *Theor. Appl. Genet.* **104**: 908–15.

Sarowar S, Oh HW, Cho HS, Baek KH, Seong ES, Joung YH, Choi GJ, Lee S and Choi D. (2007). *Capsicum annuum* CCR4-associated factor *CaCAF1* is necessary for plant development and defence response. *Plant J.* **51(5)**: 792–802.

Saxena P, Gill R, Rashid A and Maheshwari CS (1981). Isolation and culture of protoplasts of *Capsicum annuum* L. and their regeneration into plants flowering *in vitro*. *Protoplasma* **108**: 357–60.

Shivegowda ST, Mythili JB, Saiprasad GVS, Gowda TKS, Anand L and Gowda R (2002). *In vitro* regeneration and transformation in chilli pepper (*Capcisum annuum* L.). *J. Hort. Sci.Biotech.* **77(5)**: 629–34.

Singh J (1992). Breeding multiple resistant lines in Chilli Pepper at PAU, Ludhiana. *In*: Proc. of convention Genet. Breed. Capsicum and Eggplant, Rome, 7–10 September 1992, pp. 127–31.

Slade AJ, Fuerstenberg SI, Loeffler D, Steine MN and Facciotti D (2005). A reverse genetic, non-transgenic approach to wheat crop improvement by TILLING. *Nat. Biotechnol.* **23**: 75–81.

Smith PG (1950). Inheritance of brown and green mature colour in peppers. *J. Hered.* **41**: 138–40.

Stewart CS, Kang BC, Liu K, Mazourek M, Yoo EY, Moore SL, Kim BD, Paran I and Jahn M (2005). The *Pun1* gene in pepper encodes a putative acyltransferase. *Plant J.* **42**: 675–88.

Strommer J, Gerats AGM, Sanago M, Molnar SJ and Gerats T (2000). A gene-based RFLP map of petunia. *Theor. Appl. Genet.* **100**: 899–905.

Strommer J, Peters J, Zethof J, de Keukeleiere P and Gerats T (2002). AFLP maps of *Petunia hybrida*: building maps when markers cluster. *Theor. Appl. Genet.* **105**: 1000–9.

Sugita T, Yamaguchi K, Kinoshita T, Yuji K, Sugimura Y, Nagata R, Kawasaki S and Todoroki A (2006). *Breed. Sci.* **56**: 137–45.

Tai T and Staskawicz BJ (2000). Construction of yeast artificial chromosome library of pepper (*Capsicum annuum* L.) and identification of clones from the *Bs2* resistance locus. *Theor. Appl. Genet.* **100(1)**: 112–7.

Tai TH, Dahlbeck D, Clark ET, Gajiwala P, Pasion R, Whalen M, Stall RE and Staskawicz BJ (1999b). Expression of the *Bs2* pepper gene confers resistance to bacterial spot disease in tomato. *Proc. Natl. Acad. Sci. USA* **96**: 14153–8.

Tai TH, Dahlbeck D, Stall RE, Peleman J and Staskawicz BJ (1999a). High resolution genetic physical mapping of the region containing the *Bs2* resistance gene of pepper. *Theor. Appl. Genet.* **99**: 1201–6.

Tam SM, Mhiri C, Vogelaar A, Kerkveld M, Pearce SR and Grandbastien MA (2005). Comparative analyses of genetic diversities within tomato and pepper collections detected by retrotransposon-based SSAP, AFLP and SSR. *Theor. Appl. Genet.* **110(5)**: 819–31.

Tanksley SD (1984). Linkage relationships and chromosomal locations of enzyme-coding genes in pepper (*Capsicum annuum* L.). *Chromosoma* **8**: 352–60.

Tanksley SD, Bernatzky R, Lapitan NL and Prince JP (1988). Conservation of gene repertoire but not gene order in pepper and tomato. *Proc. Natl. Acad. Sci. USA* **85**: 6419–23.

Tanksley SD, Ganal MW, Prince JP, De Vicente MC, Bonierbale MW, Broun P, Fulton TM, Giovannoni JJ, Grandillo S, Martin GB, Messeguer R, Miller JC, Miller L, Paterson AH, Pineda O, Roder MS, Wing RA, Wu W and Young ND (1992). High-density molecular linkage maps of the tomato and potato genomes. *Genetics* **132**: 1141–60.

Tewksbury JJ and Nabhan GP (2001). Directed deterrence by capsaicin in chilies. *Nature* **412**: 403–4.

ThabiusA, Palloix A, Pflieger S, Daubeze AM, Caranta C and Lefebvre V (2003). Comparative mapping of *Phytophthora* resistance loci in pepper germplasm: evidence for conserved resistance loci across Solanaceae and for a large genetic diversity. *Theor. Appl. Genet.* **106**: 1473–85.

Thabuis A, Lefebvre V, Bernard G, Daubeze AM, Phaly E, Pochard E, Palloix A (2004b). Phenotypic and molecular evaluation of a recurrent selection program for a polygenic resistance to *Phytophthora capsici* in pepper. *Theor. Appl. Genet.* **109**: 342–51.

Thabuis A, Palloix A, Servin B, Daubeze AM, Signoret P, Hospital F and Lefebvre V (2004a). Marker-assisted introgression of *Phytophthora capsici* resistance QTL alleles into a bell pepper line: validation of additive and epistatic effects. *Mol. Breed.* **14**: 9–20.

Thies JA and Fery RL (2000). Characterization of resistance conferred by the *N* gene to *Meloidogyne arenaria* races 1 and 2, *M. hapla*, and *M. javanica* in two sets of isogenic lines of *Capsicum annuum* L. *J. Am. Soc. Hort. Sci.* **125**: 71–75.

Thorup T, Tanyolac B, Livingstone K, Popovsky S, Paran I and Jahn M (2000). Candidate gene analysis of organ pigmentation loci in the solanaceae. *Proc. Natl. Acad. Sci. USA* **97**: 11192–7.

van Os H, Andrzejewski S, Bakker E, Barrena I, Bryan GJ,Caromel B, Ghareeb B, Isidore E, de Jong W, van Koert P, Lefebvre V, Milbourne D, Ritter E, van der Voort JN, Rousselle-Bourgeois F, van Vliet J, Waugh R, Visser RG, Bakker J and van Eck HJ (2006). Construction of a 10,000-marker ultradense genetic recombination map of potato: providing a framework for accelerated gene isolation and a genome-wide physical map. *Genetics* **173**: 1075–87.

Vasil IK and Vasil V (1980). Isolation and culture of protoplasts. *In*: Perspectives in plant cell and tissue culture, Vasil IK, (*ed.*). International Rev. Cytology (Supplement) IIB: 10–19.

Venkataraman LV and Ravishanker GA (1997). Biotechnological approaches for production of saffron and capsaicin—a perspective. *In*: Edison S, Ramana KV, Sasikumar B, Nirmal Babu K and Santhosh J Eapen (*eds.*). Biotechnology of spices, medicinal and aromatic plants, Indian Society for Spices, Calicut, India, p. 156–65.

Voorrips RE, Finkers R, Sanjaya L and Groenwold R (2004). QTL mapping of anthracnose (*Colletotrichum* spp.) resistance in a cross between *Capsicum annuum* and *C. chinense*. *Theor Appl. Genet.* **109**: 1275–82.

Wang LH, Zhang BX, Lefebvre V, Huang SW, Daubeze AM and Palloix A (2004). QTL analysis of fertility restoration in cytoplasmic male sterile pepper. *Theor. Appl. Genet.* **109**: 1058–63.

Wu F, Mueller LA, Crouzillat D, Petiard V and Tanksley SD (2006). Combining bioinformatics and phylogenetics to identify large sets of single copy, orthologous genes (*COS II*) for comparative, evolutionnary and systematics studies: a test case in the Euasterid plant clade. *Genetics* **174**: 1407–20.

Wu JL, Wu C, Lei C, Baraoidan M, Bordeos A, Madamba MR, Ramos-Pamplona M, Mauleon R, Portugal A, Ulat VJ, Bruskiewich R,Wang G, Leach J, Khush G and Leung H (2005). Chemical- and irradiation-induced mutants of indica rice IR64 for forward and reverse genetics. *Plant Mol. Biol.* **59**: 85–97.

Yamada Y and Fujita Y (1983). *In*: Hand book of Plant Cell Culture, Vol. 1 (Evans DA, Sharp WP, Ammirato PV and Yamada T, *eds.*). Mcmillan, Yew York, pp. 717–28.

Yeoman MM, Meidzybrodzka MB, Lindsey K and Lauchal WR. (1980). *In*: Plant Cell culture: Results and Perspectives (Sala F, Parisi B, Cella R and Ciffiri O, *eds.*) Elsevier, North Hall Publ. Amsterdam, pp. 327–343.

Yi G, Lee IM, Lee S, Choi D and Kim BD (2006). Exploitation of pepper EST-SSRs and an SSR-based linkage map. *Theor. Appl. Genet.* **114**: 113–30.

Yi SY, Kim J-H, Joung YH, Lee S, Kim W-T, Yu SH and Choi D (2004). The pepper transcription factor *CaPF1* confers pathogen and freezing tolerance in *Arabidopsis*. *Plant Physiol.* **136**: 2862–74.

Yoo Y, Kim S, Kim YH, Lee CJ and Kim BD (2003). Construction of a deep coverage BAC library from *Capsicum annuum*, CM 334. *Theor. Appl. Genet.* **107**: 540–3.

Zygier S, Chaim AB, Efrati A, Kaluzky G, Borovsky Y and Paran I (2005). QTLs mapping for fruit size and shape in chromosomes 2 and 4 in pepper and a comparison of the pepper QTL map with that of tomato *Theor. Appl. Genet.* **111**: 437–45.

# 5

# Breeding for Virus Resistance

K. Madhavi Reddy and M. Krishna Reddy

## 1. INTRODUCTION

Chilli (*Capsicum annuum* L.) is an important commercial crop of India grown for its green fruits as vegetable and ripe dried form as a spice. India stands first in chilli cultivation covering 45 per cent of chilli growing areas of the world, but the productivity of dry chilli is lower (0.9 t/ha) as compared to the world average (2.0 t/ha). There is a tremendous demand for Indian chillies in the international market that provides a wide scope to increase the export. Viral diseases are major constraints, which contribute to low yield and inferior fruit quality (Avilla *et al.*, 1997). Up to 100 per cent loss due to viruses has rendered chilli growing uneconomical, deserting the entire fields prior to harvest. More than 45 viruses are known to infect peppers (*Capsicum* spp.) all over the world, 21 of which occur in India. Among them, Chilli veinal mottle virus (ChiVMV) of potyvirus group, Cucumber mosaic virus (CMV) of cucumovirus group, Chilli leaf curl virus (ChLCV) of begomovirus and Groundnut bud necrosis virus (GBNV) of tospovirus group are some of the most devastating viruses, which cause major set-backs in chilli production in India.

## 2. STATUS OF IMPORTANT VIRUSES

### 2.1. Chilli Veinal Mottle Virus (ChiVMV)

The development of chilli pepper cultivars with genetic resistance to potyviruses is one of the most practical, economical and environmentally secure strategies for reducing losses caused by this group of viruses.

#### 2.1.1. *International Status*

Three major resistance genes effective against different Potato virus Y (PVY) pathotypes have been reported in different pepper accessions. Several

Indian Institute of Horticultural Research, Hessaraghatta, Lake Post, Bangalore.

QTLs for PVY resistance have been mapped and one major QTL was detected in the vicinity of the *pvr2* locus in an Indian chilli, Perennial, suggesting a possible allelic relationship between major genes and QTLs (Caranta *et al.*, 1997). In 1973, Zitter and Cook reported an extreme resistance to PVY pathotype-0 and tobacco etch virus (TEV) in the *C. annuum* line 'Avelar', in addition to a monogenic recessive and partial resistance to pepper mottle virus (PepMoV). Partial resistance to PepMoV was attributed to the *pvr3* locus and it was shown to have no discernible effect on TEV.

Association of several potyvirus resistance factors could generate new resistances against other potyviruses. A doubled haploid (DH) line completely resistant to pepper veinal mottle virus (PVMV) was recovered from the $F_1$ hybrid between two susceptible *C. annuum* lines, Perennial, and Florida VR2, as both of these lines were susceptible to PVMV suggesting a complementary action of genes originating from Perennial and Florida VR2. A CAPS (cleaved amplified polymorphic sequence) marker tightly linked to *Pvr4* locus has been developed. The *Pvr4* resistance gene in pepper confers a complete resistance to the 3 pathotypes of PVY and to a PepMoV. In order to use this gene in a marker assisted selection (MAS) programme and to permit the pyramiding of several potyvirus resistance genes in the same cultivar, tightly linked AFLP markers were identified by bulk segregant analysis (BSA) method. Eight AFLP markers were mapped in an interval from 2.1±0.8 to 13.8±2.9 cM around this locus. The closest co-dominant AFLP marker was converted into CAPS marker using data from the alignment of the two-allele sequence. CAPS marker was further characterized to be used in MAS programmes (Caranta *et al.*,1999). Dominant gene *Pvr7* [from *C. chinense* Jacq. PI 159236 confers resistance to the PepMoV Florida strain (V1182)] is tightly linked to the dominant potyvirus resistance gene *Pvr4* with observed recombination frequencies of 0.012 to 0.016. The results indicated that *Pvr4, Pvr7* and *Tsw* (a gene conferring resistance to tomato spotted wilt virus), comprise the first identified cluster of dominant disease resistance genes in *Capsicum* (Grube *et al.,* 2000).

The *pvr1* locus in *Capsicum* encodes a translation initiation factor *elF4E* that interacts with Tobacco etch virus VPg. Two new genes, $pvr2^1$ and $pvr2^2$, previously known to be *elF4E* with narrower resistance spectra are alleles at the *pvr1* locus. Based on these data and current nomenclature guidelines, $pvr1^1$ and $pvr1^2$ alleles were redesigned. The *pvr2* locus in pepper, conferring recessive resistance against strains of potato virus Y (PVY) also corresponds to a eukaryotic initiation factor 4E (*elF4E*) gene. RFLP analysis on the PVY-susceptible and resistant pepper cultivars, using an *elF4E* cDNA from tobacco as probe, revealed perfect map co-segregation between a polymorphism in the *elF4E* gene and the *pvr2* alleles, $pvr2^1$ (resistant to PVY-0) and $pvr2^2$ (resistant to *PVY*-0 and 1). CAPS markers for three recessive viral resistance alleles, *pvr1*, *pvr1*$^1$ and *pvr1*$^2$ were generated (Yeam *et al.*, 2005). These markers are based on single nucleotide polymorphisms (SNPs) within the coding region of the *pvr1* locus encoding an *elF4E* homolog on chromosome 3. These three markers define a system of indirect selection for potyvirus resistance in *Capsicum* based on genomic sequences.

### 2.1.2. *National Status*

Earlier monogenic recessive sources of resistance against PVY in *C. angulosum* EC 97758, *C. microcarpum,* Perennial and S 41-1 were described (Singh and Chenulu, 1985). In contrast to the global status, ChiVMV is the most prevalent potyvirus in India.

Isolates collected from chilli growing areas of Karnataka and Tamil Nadu were identified as ChiVMV based on their host range, serological relationships, electron microscopy and phylogenetic analysis of coat protein sequences. Molecular variability present among the isolates was studied by RT-PCR (real time PCR) technique coupled with sequence analysis of coat protein region and 3′ non-coding region (NCR) using primers specific to Nib region and coat protein of ChiVMV. Comparison of nucleotide sequences of isolates with already reported sequences of ChiVMV, showed 91–93 per cent nucleotide identity. The biological differences among the six isolates were noticed in infecting test hosts. Mechanical inoculation of ChiVMV isolates onto 25 genotypes of chilli pepper under screen house conditions helped to identify three immune and ten highly resistant genotypes as confirmed by symptoms and enzyme linked immunosorbant assay (ELISA) testing. The various levels of symptom and severity have been reported in six isolates of ChiVMV (Krishna Reddy *et al.,* 2004).

PCR amplification and restriction enzyme selection for CAPS marker were used to identify *pvr* gene linked to chilli veinal mottle potyvirus resistance. Amplification with Pvr-S primer followed by *Bsr1* restriction digestion of $F_2$, $F_3$, $BC_1r$ and $BC_1s$ individual populations have showed clear polymorphism between homozygous resistance (RR), heterozygous resistance (Rr) and susceptible lines indicating the resistance in Perennial chilli variety is tightly linked to *pvr1*$^1$ locus that is located on chromosome 4. Further, comparison of nucleotide sequence of coding region has indicated that 99.1 to 99.5 per cent protein and 99.1 to 99.8 per cent nucleotide identity with genotype carrying *pvr1*$^1$ genes of the *elF4E* (Madhavi Reddy *et al.,* 2007).

## 2.2. Cucumber Mosaic Virus (CMV)

Cucumber mosaic virus (CMV), member of the cucumovirus group, is one of the most devastating viruses in pepper all over the world. CMV is transmitted mainly by aphids in a non-persistent manner, the most common aphids are *Myzus persicae* and *Aphis gossypi*.

### 2.2.1. *International Status*

There have been many reports of genetic resistance to CMV, especially in *C. annuum*. However, classification, inheritance patterns and transfer of resistance from many of these sources have been complicated by isolates specificity or screening difficulties. The best characterized sources of resistance *C. annuum* 'Perennial', a pungent Indian chilli, has been described as having monogeneic recessive, partially dominant or polygenic recessive inheritance by different researchers (Lapidot *et al.,* 1997; Pochard and Daubeze, 1989; Pochard *et al.,* 1986; Singh and Thakur, 1979). Further it remains unclear whether the response to resistance to CMV in Perennial is tolerance or resistance. These inconsistencies may be explained, at least in part, by the use of different viral isolates, environmental conditions and inoculation techniques by different researchers. Regardless of its source, the inconsistency observed in the performance (resistance to CMV) of Perennial is a significant shortcoming of this genotype. The characterization of additional or improved sources of resistance to CMV in pepper would facilitate the development of commercial pepper varieties with enhanced CMV resistance (Grube *et al.*, 2000). Rusco and Csillery (1980) used Perennial as the resistant donor and Fulton's CMV strain for selection and breeding and suggested two modes of inheritance: monogenic recessive, and partially dominant. In France, Perennial was considered to be tolerant to CMV, since slight mosaic symptoms appeared after inoculation at different stages of plant development. Perennial, however, had the ability to recover from CMV mosaic symptoms. The tolerance was determined to be recessive and polygenic.

A tool for marker assisted selection (MAS) to facilitate the transfer of CMV resistance was developed in *Capsicum*. QTL analysis for resistance to CMV was performed in an intraspecific *C. annuum* population. A total of 180 $F_3$ families were derived from a cross between the susceptible bell-type cultivar 'Maor' and the resistant small-fruited Indian line 'Perennial' and inoculated with CMV in experiments carried out in the USA and Israel using two different virus isolates. Four QTLs were found to be associated with resistance to CMV. Two digenic interactions involving markers with and without an individual effect on CMV resistance were also detected. The QTL controlling the largest percentage (16–33%) of the observed phenotypic variation (*cmv11.1*) was detected in all three experiments and was also involved in one of the digenic interactions. Markers in two genomic regions that were identified as linked to QTL for CMV resistance were also linked to QTL for fruit weight, confirming additional breeding observations of an association between resistance to CMV (originating from Perennial) and small fruit weight (Ben Chaim *et al.*, 2001).

QTLs involved in the restriction of CMV to long-distance movement in pepper has been mapped. Partial restriction of CMV to long-distance movement originating from the *C. annuum* inbred line 'Vania' was assessed in a doubled-haploid progeny using two screening methods. The first method allowed to assess the resistance of adult plants decapitated above the 4th leaf and inoculated the 3rd leaf using a common CMV strain, while the second method allowed to assess CMV resistance to long-distance movement on seedlings inoculated using an atypical CMV strain. In both the tests, the behavior of the $F_1$ hybrid between 'Vania' and the susceptible line 'H 3' indicated that partial resistance is inherited as a dominant trait (Caranta *et al.,* 2002).

### 2.2.2. *National Status*

The CMV resistance of 'Perennial' and of three resistant accessions was reported to be monogenic recessive (Singh and Thakur, 1977). Segregation studies were carried out following two inoculations, and the symptom-less plants remained virus-free until the end of the season. Bansal *et al.* (1992) demonstrated that 'Perennial' and three other accessions were practically immune to CMV in Punjab. Singh (1973) screened several chilli lines against chilli mosaic under field conditions and found that lines, Puri Red, Puri Orange, G-2, Kondivenum and Suryamukhi were resistant. Tiwari and Anand (1977) reported that Pusa Jwala, a selection from NP 46 A × Puri Red was found resistant to mosaic disease of chilli. Konai and Nariani (1980) screened 30 chilli varieties against four viruses (TMV, CMV, PVX and TLCV) to determine the multiple resistance and found that none of the varieties tested were found immune to all the four viruses. An unidentified variety, Delhi Local was found to be immune to CMV and PVX, but tolerant to TMV and TLCV. Varieties Pant C1 and Pant C2 were, however, found tolerant to all the four viruses. Gahukar and Nariani (1980) screened a large number of chilli varieties against three strains of CMV. Tiwari and Vishwanathan (1988) found that variety, PSP-11 (*C. frutescens*) as resistant to CMV, PVX and TMV. Singh and Shukla (1990) reported that out of 148 varieties of *Capsicum* screened for resistance to mosaic and leaf curl diseases, 122 lines were found susceptible and 26 were disease-free. Bal *et al.* (1995) studied genetics of resistance to mosaic and leaf curl viruses in chilli. Punjab Lal, a multiple disease resistant cultivar, was crossed with two susceptible lines, *viz.,* Ludhiana Local Selection and Hungarian Sweet Yellow and their six basic generations were scored for different viruses under field conditions. Genetic analysis indicated that susceptibility to mosaic as well as leaf curl was dominant and resistance controlled by monogenic recessive genes.

Singh and Singh (1998) screened 11 chilli varieties against CMV and found Punjab Lal (S- 118-2), Bengal Green-1 and Perennial as resistant. Six genotypes, *viz*., SS-4, Lorai, Tiwari, S-20-1, Bengal Green Excel, and Shahkoti, were found moderately resistant (25–50% prevalence) with mild symptoms, while Ludhiana Local and Pusa Jwala were highly susceptible (100% prevalence) with severe symptoms. Nine chilli pepper lines and their corresponding $F_1$ hybrids were screened against CMV through sap inoculation under screen house conditions and two parents, *viz*., VR 42 and VR 55 and a hybrid VR 42 × VR 55 were found resistant to CMV, whereas the susceptible check, Arka Gaurav expressed 83 per cent infection (Prasad *et al*., 2001). Host range studies, electron microscopy and partial molecular characterization could confirm CMV incidence and distribution on chilli from northern Telangana region of Andhra Pradesh (Jagadeeshwar *et al.,* 2006).

### 2.3. Leaf Curl Virus (LCV)

Leaf curl disease of chilli caused by white fly (*Bemisia tabaci*) transmitted (WFT) geminivirus has emerged as a serious problem in most of the chilli growing areas of the world. Incidences up to 100 per cent in farmers' fields are a common phenomenon. The typical leaf curl affected plants showed symptoms of upward curling, puckering, crinkling and reduced leaf size, shortened internodes, and stunted growth. Such plants produced either less or no fruits. Due to regular epidemics of this disease in the recent years, farmers have withdrawn chilli cultivation or replaced with other crops in some parts of India.

#### 2.3.1. *International Status*

The occurrence of chilli leaf curl disease has been reported from USA (Stenger *et al*., 1990), Nigeria (Alegbejo, 1999) and several other countries like Pakistan, Bangladesh and Indonesia (Fauquet and Stanley, 2003; Green and Kim, 1991). Pepper mild tigre virus (PMTV) is another whitefly-transmitted virus, which has been reported (Brown and Nelson, 1989) to cause a severe disorder in commercial pepper plantings in Mexico. In whitefly-inoculated chilli peppers, bright yellow interveinal chlorosis, mild stunting and fruit malformations have been observed. Yield losses of 50–80 per cent due to PMTV infection have been reported in commercial plantings. Besides pepper, the host range of the virus includes *Datura stramonium, Nicotiana tabacum, Xanthi* and *Solanum lycopersicum.* The chino del tomate virus (CdTV), reported in Mexico (Brown and Nelson, 1988), mainly infect tomato, where it may affect 100 per cent of the plants in the field. The disease is characterized by curling and rolling of leaves, thickening of veins, yellow mosaic, stunting and a reduction in fruit set on tomato. On pepper, the virus causes only mild symptoms, such as mottle, mosaic and vein clearing, and yield reduction is minimal. Differences in host range and symptomatology suggest that the virus is distinct from the tobacco leaf curl in India, Japan and Sudan (Newton and Peiris, 1953; Osaki and Inouye, 1978; Rataul and Brar, 1989), and tomato yellow dwarf virus in Ceylon and Japan (Newton and Peiris, 1953; Osaki and Inouye, 1978). CdTV appears to be similar to Tomato yellow leaf curl virus (TYLCV) described in Africa (Cherif and Russo, 1983; Yassin and Nour, 1965) and the Middle East (Cohen and Nitzany, 1966; Makkouk, 1978).

### 2.3.2. *National Status*

The occurrence of chilli leaf curl disease has also been reported from India (Mishra *et al.,* 1963; Dhanraj and Seth, 1968; Singh and Singh, 1976; Raj *et al.*, 2005). There are various reports of leaf curl diseases affecting peppers in India (Vasudeva and Samraj, 1948; Gattani and Mathur, 1951; Vasudeva, 1954; Mishra *et al.*, 1963; Muniyappa and Veeresh, 1984). Losses up to 80 per cent have been reported in many parts of northern India (Singh *et al.*, 1979). The usual symptoms are stunting of the plants and heavy crinkling and rolling of the leaves with chlorosis or yellowing. Older leaves may become leathery and brittle. Earlier reports indicate that the disease is caused by tobacco leaf curl virus (TLCV) on their transmission by the whitefly to tobacco and tomato and typical symptoms produced on both these hosts (Pal and Tandon, 1937). Whitefly-transmission studies have indeed shown that an isolate of tobacco could be experimentally transmitted to *C. annuum* and 34 other plant species, including *S. lycopersicum, Beta vulgaris, Carica papaya, Cyamopsis tetragonoloba, Sesamum indicum, Phaseolus vulgaris* and *Petunia hybrida* (Valand and Muniyappa, 1992). Screening for resistance started in the late sixties (Dhanraj *et al.*, 1968). Most of the virus screening has been done under field conditions, assessing disease incidence and disease severity (Sharma and Singh, 1985; Memane *et al.*, 1987; Tewari and Viswanath, 1988). For disease rating, a coefficient of infection has been used, for which the percentage disease incidence was multiplied with a response value assigned to each observed disease severity grade. A good correspondence was usually obtained between field and greenhouse assessments (Mayee *et al.*, 1975). Since then, a number of lines with varying degrees of resistance have been identified. Some of these lines also exhibit resistance or tolerance to other viruses (Tiwari and Viswanath, 1986). Many multiple virus-resistant varieties have been developed at PAU, Ludhiana. Important multiple resistant lines are Perennial, BG1, Lorai and Punjab Lal (Thakur *et al.*, 1987; Singh and Kaur, 1990). The variety Pant C1, resistant to mosaic and leaf curl disease, has been released by GBPUA&T, Pantnagar. However, some of these multiple disease-resistant lines have undesirable agronomic characteristics such as small fruit size, late maturity, and low yield.

Leaf curl disease of chilli has emerged as a serious problem in the chilli growing areas of Delhi, Haryana, Punjab, Rajasthan, West Bengal and Uttar Pradesh states. Surveys conducted in 2005 and 2006 showed very high disease incidence of leaf curl disease (40–95%) in farmers' fields. Molecular genetic diversity of begomoviruses infecting chilli in North India is reported (Krishnareddy *et al.*, 2007). Cloning and sequencing of DNA-A from different samples has indicated that chilli is affected by three different begomoviruses. An isolate from Kalyani showed highest nucleotide identity with pepper leaf curl–Bangladesh virus (96.3%), three isolates from Rajasthan showed nucleotide identity of 88.9 to 91.1 per cent with chilli leaf curl virus—Multan and eight isolates, two each from Delhi, Haryana, Punjab and Uttar Pradesh showed nucleotide identity of 92.4 to 98.5 per cent with tomato leaf curl–New Delhi virus.

### 2.4. Tomato Spotted Wilt Virus (TSWV)

Tospovirus is also a very dangerous threat to chilli crop cultivation. Severe damage to *Capsicum* spp. occurs worldwide. Plants are highly susceptible to all stages of development, showing symptoms like necrosis of leaves and fruits. The behavior of the vector (thrips), feeding preferentially in flowers, complicates the picture by providing an alternative route to leaves for entry of the virus, and in mature plants sometimes fruits alone become infected.

#### 2.4.1. *International Status*

Tomato Spotted Wilt Virus (TSWV) is common on solanaceous crops in tropical and subtropical regions throughout the world; it may also occur in temperate regions (Ie, 1970; Allen and Broadbent, 1986). The *Tsw* resistance gene to tospovirus sero-group I was originated from *C. chinense* and have been introgressed into pepper cultivars (Costa *et al.*, 1995). The hypersensitive resistance to TSWV in pepper determined by a single dominant gene in several *C. chinense* genotypes. Boiteux (1995) confirmed the resistance to tospovirus and was conferred in two *C. chinense* accessions, *viz.*, PI 152225 and PI 159236 by a single dominant gene, which displayed resistance to a broad range of TSWV isolates. Black *et al.* (1996) showed that resistance was governed by a single dominant gene and is identical in both the *C. chinense* accessions. The genus tospovirus includes seven confirmed and five tentative species (Elliot *et al.*, 2000). The genetic bases of resistance to all these viruses are poorly known. *Tsw* is neither effective against GRSV and TCSV (Boiteux and De Avila, 1994), nor against two isolates of Impatiens necrotic spot virus (INSV) (Roggero *et al.*, 1999); both these isolates gave about 50 per cent systemic infection in the PI resistant accessions kept at 25/19°C, day/night. Research is needed to reach the goal of genetic control of tospovirus diseases in pepper.

In order to facilitate the selection of this resistance gene, four randomly amplified polymorphic DNA (RAPD) markers were found linked to the *Tsw* locus using bulked segregant analysis by Moury *et al.* (2000). A close RAPD marker was converted into a CAPS. This CAPS marker is tightly linked to *Tsw* and useful for marker-assisted selection (MAS) in a wide range of genetic intercrosses for the development of resistant pepper varieties resistant to TSWV. Livingstone *et al.* (1999) have published an improved genome map of pepper, and Jahn *et al.* (2000) compared the map of the *Tsw* locus of *C. chinense* with that of the *Sw-5* locus of tomato, while Grube *et al.* (2000) analyzed in detail the loci of resistant genes for three members of the solanaceae. This research stressed the potential use of comparative genome mapping in order to rapidly identify other genes showing similar function and specificity against other pathogens.

Molly Jahn *et al.* (2000) examined the relationship between *Tsw* and *Sw-5* through genetic studies of TSWV. The capacity of TSWV–A to overcome the *Tsw* gene in pepper and the *Sw-5* gene in tomato maps to different TSWV genome segments. Despite phenotypic and genotypic similarities of resistance in tomato and pepper, distinct viral gene products control the outcome of infection in plants carrying *Sw-5* and *Tsw*, and that these loci do not appear to share a common evolutionary ancestor.

#### 2.4.2. *National Status*

In India, so far TSWV was not detected and is mostly prevalent in Europe and America. In India the major tospovirus affecting chilli is groundnut bud necrosis virus (GBNV), which is primarily transmitted through thrips (*Thrips palmi*). The increasing problem of tospovirus has stimulated a search for sources of genetic resistance. GBNV is known to affect several crops in addition to chilli (Rao *et al.*, 1980; Krishnareddy and Varma, 1989). Recently, severe epidemics of tospoviruses causing serious yield losses have been observed from Andhra Pradesh (Guntur and Khamman), Karnataka (Bellary and Bagalkot) and Tamil Nadu (Coimbatore). In India, no work has been done on tospoviruses affecting chilli in the areas of screening techniques and identification of stable resistant sources. Due to its disease severity spreading in India in the recent past, there is a strong need to work for disease resistance against tospovirus.

## 3. BRIEF STATUS OF VIRUS RESISTANCE PROGRAMME IN CHILLI

Several virus resistance loci and QTLs have been mapped in *Capsicum* sp., *L* for resistance to tobacco mosaic virus and pepper mild mottle virus (Lefebvre *et al.,* 1995), *pvr1, pvr2, pvr6* for resistance to potyviruses (Caranta *et al.,* 1996), *Tsw* for resistance to tomato spotted wilt virus, potyviruses (Caranta *et al.,* 1997) and cucumber mosaic virus (Caranta *et al.,* 1997). 'Hot genomic regions' involved in several disease resistances and co-localization of major genes/QTLs were observed. These results lead to the characterization of resistance loci, using a candidate gene approach.

In contrast, in India little work has been done on virus resistance in chilli. Information on strain variation of major viruses affecting chilli is very limited in India. No systematic work on screening of chilli lines for specific strains of different viruses and also on the identification of genetic markers linked to virus resistance in chilli has been done so far. In INRA, France, a multiple virus resistant Indian chilli line, 'Perennial' was used extensively to study the reaction to different viruses (partial resistance to CMV, absolute resistance to PVY pathotype 0, potyvirus E and ChiVMV and a partial resistance to PVY pathotype 1, 2), inheritance of resistance, identification and characterization of the resistance genes (Caranta and Palloix, 1996; Caranta *et al*., 1996).

At IIHR, major chilli growing areas were surveyed and different isolates were collected from Andhra Pradesh, Karnataka and Tamil Nadu and biologically characterized. Artificial screening methodology, *i.e.* mechanical sap inoculation technique has been standardized and severe and widely prevalent isolates of ChiVMV and CMV have been identified. A number of chilli lines were screened against different isolates and horizontal resistant lines having resistance to CMV and ChiVMV strains have been identified. Advanced breeding lines with combined resistance to CMV and ChiVMV have been developed. The leaf curl symptoms were identified by PCR and whitefly transmission. Complete genome sequencing of three leaf curl viruses has been done. ChLCV isolates from Rajasthan were 87.3 per cent similar to ChLCV–Pakistan. ChLCV isolates from Varanasi were 86.1 per cent similar to ChLCV–Bangladesh. ChLCV isolates from Tamil Nadu were 89.0 per cent similar to ToLCSV–Sri Lanka. ChLCV isolates from Haryana has 96.5 per cent identity with tomato leaf curl New Delhi Virus (ToLCNDV). Screening of more than 140 chilli lines at two hot spots, *i.e.* one in Tamil Nadu and other in Haryana has resulted in identification of ChLCV resistant lines. At IIVR Varanasi, 307 genotypes were systematically screened against ChLCV (syn. pepper leaf curl virus) under screen house conditions and resistant lines, *viz.,* BS 35, GKC 29 and EC 497636 have been identified employing field screening, grafting and PCR methods.

## 4. GAPS IDENTIFIED

Breeding for yield and quality parameters have been the major driving forces in genetic improvement programmes in our country. In the recent past, importance of breeding for major diseases has been recognized.

Breeding for virus resistance is a major challenge to chilli pepper breeders. In order to speed up the resistance breeding process the major areas to be emphasized are:

(i) To understand the viruses infecting the crop, *i.e.* mainly through proper identification by characterizing the virus both biologically and molecularly,

(ii) Development of reliable diagnostic techniques of the major viruses affecting chilli,

(iii) Screening germplasm and advanced breeding materials under controlled conditions, and

(iv) Breeding for virus resistance by integrating phenotypic and molecular marker assisted approaches.

## 4.1. Approaches to Fill the Gaps

The following approaches may be followed in order to bridge the gaps:

- Isolation and characterization (biological and molecular) of major viruses affecting chilli; development of reliable disease screening methodologies and confirmation of resistance by ELISA, PCR and nucleic acid probes.
- India, being secondary centre of diversity for chilli, is endowed with excellent biodiversity. Systematic effort in identification of resistant sources will enable to identify new sources of resistance from both cultivated and related wild species.
- Cataloguing the isolates of CMV, ChiVMV, ChLCV and GBNV and resistance sources against existing isolates affecting chilli in different chilli growing areas in India. Identification of resistant genes against different isolates would help in pooling resistant genes for stable and durable resistance.
- Introgression of stable virus resistant genes from wild species through marker assisted selection will help in the development of useful pre-breeding lines and thus, in broadening the genetic base of the crop.
- Development of stable, multiple virus resistant chilli varieties/hybrids.

## REFERENCES

Bal SS, Singh J and Dhanju KC (1995). Genetics of resistance to mosaic and leaf curl viruses in chilli (*Capsicum annuum* L.). *Indian J. Virol.* **11(1)**: 77–79.

Ben Chaim A, Grube RC, Lapidot M, Jahn M and Paran I (2001a). Identification of quantitative trait loci associated with resistance to cucumber mosaic virus in *Capsicum annuum. Theor. Appl. Genet.* **102**: 1213–20.

Boiteux LS (1995). Allelic relationships between genes for resistance to tomato spotted wilt tospovirus in *Capsicum chinense. Theor. Appl. Genet.* **90**: 146–9.

Brown JK and Poulos BT (1989). Two whitefly-transmitted geminiviruses isolated from pepper affected with tigre disease. *Phytopathol.* **79**: 908.

Caranta C, Polloix A, Lefebvre V and Daubeze AM (1997). QTLs for a component of partial resistance to cucumber mosaic virus in pepper: restriction of virus installation in host-cells. *Theor. Appl. Genet.* **94**: 43–438.

Caranta C, Thabuis A and Palloix A (1999). Development of a CAPS marker for the *Pvr4* locus: A tool for pyramiding potyvirus resistance genes in pepper. *Genome* **42**: 1111–6.

Caranta C, Pflieger S, Lefebvre V, Daubeze AM, Thabuis A and Palloix A (2002). QTLs involved in the restriction of cucumber mosaic virus (CMV) long-distance movement in pepper. *Theor. Appl. Genet.* **104**: 586–91.

Cohen S and Nitzany FE (1966). Transmission and host range of the tomato yellow leaf curl virus. *Phytopathol.* **56**: 1127–31.

Dhanraj KS, Seth ML and Basal HC (1968). Reactions of certain chilli mutants and varieties to leaf curl virus. *Indian Phytopathol.* **21**: 342–3.

Grube RC, Blauth JR, Arnedo A, Caranta C and Jahn MK (2000). Identification and comparative mapping of a dominant potyvirus resistance gene cluster in *Capsicum. Theor. Appl. Genet.* **101**: 852–9.

Konai M and Nariani T K (1980). Screening of chilli varieties for virus resistance to mosaic and leaf curl viruses. *Indian Phytopathol.* **33**: 155.

Krishna Reddy M, Madhavi Reddy K, Lakshminarayan Reddy CN, Smitha R and Jalali S (2004). Molecular characterization and genetic variability of chilli veinal mottle virus and its reaction on chilli pepper genotypes. *In:* Proceedings of the XIIth EUCARPIA meeting on Genetics and Breeding of Capsicum and Eggplant, May 17–19, 2004, Noorwijkerhout, The Netherlands. R.E.Voorrips (*ed.*): 171–77.

Krishnareddy M, Venkataravanappa V, Madhavi Reddy K and S Jalali (2007). Molecular genetic diversity of begomoviruses infecting chilli in North India. Proceedings of International Conference on Emerging and Re-emerging Viral Diseases of the Tropics and Sub-tropics, Dec 11–14, 2007, New Delhi, pp. 140.

Madhavi Reddy K, Krishna Reddy M, Reddy L and Sadashiva AT (2004). Resistance of *Capsicum annuum* L. lines to chilli veinal mottle potyvirus isolate, inheritance and allelism studies for resistance genes. *In*: Proceedings of the XIIth EUCARPIA meeting on Genetics and Breeding of Capsicum and Eggplant, May 17–19, 2004, Noorwijkerhout, The Netherlands. R.E.Voorrips (*ed.*): pp. 190.

Madhavi Reddy K, Krishna Reddy M, Ravishankar KV, Patil GP and Savita GK (2007). Use of allele-specific CAPS marker in identification of recessive chilli veinal mottle virus (ChiVMV) resistance in chilli (*Capsicum annuum* L.). Proceedings of International Conference on Emerging and Re-emerging Viral Diseases of the Tropics and Sub-tropics, Dec 11–14, 2007, New Delhi, pp. 218.

Mishra MD, Raychaudhuri SP and Jha A (1963). Virus causing leaf curl of chilli (*Capsicum annuum* L.). *Indian J. Microbiol.* **2**: 73–76.

Moury B, Pflieger S, Blattes A, Lefebvre V, Palloix A (2000). A CAPS marker to assist selection of tomato spotted wilt virus (TSWV) resistance in pepper. *Genome* **43**: 137–42.

Muniyappa V and Veeresh GK (1984). Plant virus diseases transmitted by whiteflies in Karnataka. *Proc. Indian Acad. Sci.*, India **93**: 397–406.

Narasimha Prasad BC, Madhavi Reddy K and Sadashiva AT (2001). Development of $F_1$ hybrids with resistance to cucumber mosaic virus (CMV) in chilli (*Capsicum annuum* L.). *Capsicum & Eggplant Newslett.* **20**: 78–81.

Prasad BCN, Madhavi Reddy K and Sadashiva AT (2000). Development of $F_1$ hybrids with resistance to cucumber mosaic virus (CMV) in chilli (*Capsicum annuum* L.). *In*: Symposium on 'Emerging Trends in Plant Disease Management' Bangalore from 7–8th December, 2000.

Sharma OP and Singh J (1985). Reaction of different genotypes of pepper to cucumber mosaic and tobacco mosaic viruses. *Capsicum & Eggplant Newslett.* **4**: 47.

Shasikumar KT, Madhavi Reddy K, Krishna Reddy M and Sadashiva AT (2000). Development of multiple virus resistant hybrids in chilli (*Capsicum annuum* L.). *In*: Symposium on 'Emerging Trends in Plant Disease Management' Bangalore from 7–8th December, 2000.

Singh SJ (1973). Reaction of chilli varieties (*Capsicum annuum* L.) to mosaic and leaf curl diseases under field conditions. *Indian J. Hort.* **30**: 444–7.

Singh S and Chenulu VV (1985). Studies on resistance to virus diseases in *Capsicum* species: III. Inheritance of resistance to potato virus Y. *Indian Phytopathol.* **38(3)**: 479–83.

Singh DV and Lal SB (1964). Occurrence of viruses of tomato and their probable strains at Kanpur. *Proc. Nat. Acad. Sci.,* India. **34**: 179–87.

Singh SJ, Sastry KS and Sastry KSM (1979). Efficacy of different insecticides and oil in the control of leaf curl virus disease of chillies. *J. Plant Dis. Prot.* **86**: 253–86.

Singh BR and Shukla P (1990). Properties of new strain of cucumber mosaic virus from chilli. *Indian J Virol.* **6(1–2)**: 58–63.

Singh MJ and Singh J (1998). Sources of resistance to cucumber mosaic virus in chillies. *Plant Dis. Res.* **13(2)**: 184.

Stenger DC, Duffus JE and Villalon B (1990). Biological and genomic properties of geminivirus isolated from pepper. *Phytopathol.* **80**: 704–9.

Tiwari VP and Ramanujam S (1974). Grow Jwala, a disease resistant high yield chilli. *Indian Farming* **24(1)**: 20.

Tiwari VP and Vishwanathan SM (1998). Note on breeding for resistance in chillies – today and tomorrow. *New Botanist* **15(2–3)**: 185–6.

Valand GB and Muniyappa V (1992). Epidemiology of tobacco leaf curl virus in India. *Ann. Appl. Biol.* **120**: 257–67.

Vasudeva RS (1954). Report of the Division of Mycology and Plant Pathology. P. 79–89 *In*: Sci. Rept. Agric. Res. Inst., New Delhi 1952–1953. p. 79–89.

Vasudeva RS and Samraj J (1948). A leaf curl disease of tomato. *Phytopathol.* **38**: 364–9.

Venkata Ramana C, Madhavi Reddy K, Sadashiva AT and Krishna Reddy M (2005). Combining ability estimates in virus resistant and susceptible lines of chilli (*Capsicum annuum* L.). *J. Appl. Hort.*, **7(2)**: 108–12.

Yassin AM and Nour MA (1965). Tomato leaf curl disease, its effect on yield and varietal susceptibility. *Sudan Agric. J.* **1**: 3–7.

# 6

# Exploitation of Male Sterility Systems

M.S. Dhaliwal

## 1. INTRODUCTION

The commercial production of $F_1$ hybrid seeds of chilli has been accomplished using hand emasculation, genic (or genetic) male sterility (GMS) and cytoplasmic male sterility (CMS). Manual hybrid seed production, especially hand emasculation is time consuming and labour intensive. Male sterility can be used to reduce amount of labour needed and thus to reduce cost of hybrid seeds. Male sterility also increases the purity of the $F_1$ seeds since no self-pollination takes place. Male sterility was first reported by Martin and Crawford (1951) in *Capsicum frutescens.* They visualized potential of exploiting male sterility for hybrids development in chilli. Later, Peterson (1958) reported and documented male sterility in *C. annuum*. Initial reports indicated that the system was unstable and resulted in fertile pollen under cool conditions. Since then, considerable knowledge has been accumulated on the nature of the trait, the means of its identification and induction, inheritance of both genic and cytoplasmic-genic male sterility (C-GMS or CMS), its maintenance, and the potential for breeding hybrid cultivars. Today, stable C-GMS lines have been developed. Some public sector institutions and several internationally known seed companies use the genic mechanism *msms* on a large scale for producing hybrids both in hot and sweet peppers, whereas the cytoplasmic-genic source is currently used for breeding pungent hybrids (Shifriss, 1997).

## 2. EXPLOITATION OF GMS SYSTEM FOR HYBRIDS DEVELOPMENT

In GMS system, the nuclear genes determine the sterility mechanism. It is an important pollination control mechanism, which has been exploited commercially for hybrid seed production in chilli. Punjab Agricultural

Punjab Agricultural University, Ludhiana.

University (PAU), Ludhiana is the pioneer institution in India for developing and commercially exploiting GMS system in chilli.

## 2.1. Inheritance of Male Sterility

Male sterility in peppers has been extensively studied by Shifriss and others. In 1969, Shifriss and Frankel found a genetic male sterile plant in a population of the cultivar 'All Big' bell pepper. They discovered that a single recessive gene determined the character. The plants having recessive gene in homozygous state (*msms*) are male sterile, whereas those in heterozygous state (*Msms)* and homozygous dominant (*MsMs*) state are male fertile. In 1972, Shifriss and Rylsky described the discovery of a second gene encoding genetic male sterility. The second source came from California Wonder bell peppers. Again, the character was inherited as a single recessive gene. The authors suggested that the genetic male sterile plant described by Shifriss and Frankel (1969) should be designated as $ms_1$ and the second non-allelic gene $ms_2$. Patel *et al.* (1998) identified a naturally occurring male sterile plant at farmer's field and developed the line called ACMS-2. A single recessive gene again governed inheritance of sterility. Yu (1990) also reported monogenic recessive inheritance of genetic male sterility. In *Capsicum*, over 20 male sterile gene have been reported (Wang and Bosland, 2006).

## 2.2. Development of GMS Line 'MS 12'

Punjab Agricultural University, Ludhiana, India has developed the GMS line 'MS 12' by transferring male sterility gene present in French introduction '*ms-509*'. Locally developed variety 'Punjab Lal' was used as the recurrent parent. Following alternate methods of backcrossing and selfing, the *ms* gene was transferred into the genetic background of Punjab Lal, a multiple disease resistant variety (Singh and Kaur, 1986). Since then the public sector and the private seed companies for hybrids development in chilli have extensively used this source.

Since the GMS system is governed by recessive genes, the conventional breeding approach is to follow selfing after each backcross to recover recessive genes. This requires almost double the time than what is required for transferring a dominant gene. Time period can be reduced if modified backcross method is followed where selfing is resorted after 2–3 backcrosses, however, there are chances of loosing male sterility gene in the process. Following biotechnological approaches *viz*. marker assisted selection (MAS), the time required for MS (male sterie) gene transfer can be reduced to approximately half. The molecular markers, especially the dominant markers, discriminate between the heterozygote and the homozygote individuals (Collards *et al.*, 2005), thus eliminating the need to self backcross progenies.

### 2.3. Maintenance of GMS and Producing $F_1$ Hybrid Seed

The male sterility characteristic in GMS is maintained by backcrossing it with the heterozygote fertile. The progeny segregates into fertile (*Msms*) and sterile (*msms*) plants in the ratio of 1:1. Male sterile plants can be identified at the time of flowering. The easiest way is to see presence/absence of pollen by touching anthers of freshly opened flowers on thumbnail or on a black paper. Presence of whitish/creamy powdery mass (pollen) indicates that the plant is fertile and *vice-versa*. The other identification features are given in Table 6.1. The sterile and fertile plants are tagged using different colour. The heterozygous fertile plants act as maintainer and the seed is harvested from the male sterile plants only. For hybrid seed production, the female (male sterile) and the male lines are planted in the ratio of 2:1. The population density of the female parent is 200 per cent of the standard as 50 per cent of the plants that turned out to be male fertile are rouged-out at flowering. The identification process should be completed as soon as possible. After the identification process is over, the previously set fruits are removed. Pollination is done by the insects, such as bees, thrips, ants, etc. Seeds harvested from the female parent are the hybrid seeds.

**Table 6.1.** Identification of male sterile and fertile plants in female parent of MS 12

| Character | Male fertile | Male sterile |
|---|---|---|
| Pollen | Present | Absent |
| Anther colour | Light grey | Purple or yellow |
| Anther size | Plump | Shriveled |
| Dehiscence | Present | Absent |
| Natural fruit set | Heavy | Poor |

Exploiting GMS system available in MS 12, PAU, Ludhiana has developed two hybrids *viz.*, CH-1 (Hundal and Khurana, 1993) and CH-3 (Hundal and Khurana, 2001) in the year 1992 and 2000, respectively for commercial cultivation in Punjab. The hybrid CH-1 had become an instant success with Punjab growers and had been cultivated even in adjoining states of Haryana, Rajasthan and Uttrakhand.

One of the important limitations of GMS system is its inability to produce progeny of cent per cent male sterile plants. Improper identification of male sterile plants from the mixed population and laxity in following other precautions such as removal of preset fruits after the identification process is completed may further distort the ratio in favour of male fertile (MF) plants. However, availability of morphological markers at seedling stage would be of much use to establish pure stand of MS plants. In an attempt to increase the ratio of male sterile plants in a population, Shifriss and Pilovsky (1993) crossed two isogenic lines that differed for male sterility genes. The intension of this digenic cross was to produce a system in which a male sterile plant contained both previously described male sterile genes *viz.*, $ms_1$ and $ms_2$. This plant was then crossed to a fertile plant that was heterozygous for both the genes *i.e.* $ms_1ms_1ms_2ms_2 \times Ms_1ms_1Ms_2ms_2$. The resultant progeny from such a cross segregated in a ratio of three male sterile plants to one fertile plant. The implication for such a cross was that only a quarter of plants would have to be removed from the seed production block. Unfortunately the procedure required both parents to be maintained asexually and protected from viral contamination (Shifriss, 1997).

## 3. EXPLOITATION OF CMS SYSTEM FOR HYBRID DEVELOPMENT

Cytoplasmic male sterility (CMS) is another means by which hybrids may be produced in chilli. In CMS system, the sterility mechanism is determined by the interaction of cytoplasm and the nuclear genes. The main advantage of CMS system over the GMS system is that the progeny (sterile line × maintainer line) is 100 per cent MS and the labour used for identifying male sterile plants from the mixed population in the GMS system and removal of male fertile plants, is saved. This reduces the hybrid seed cost and it also ensures purity of the hybrid seed.

### 3.1. Inheritance of Cytoplasmic Male Sterility

In this system, the cytoplasmic factor 'S' interacts with recessive nuclear gene *rf* and produces a male sterile line of the genotype S *rfrf*. This male sterile or female line is known as 'A'-line in $F_1$ hybrid seed production programme. Its isogenic line with normal cytoplasm (N *rfrf*) is male fertile and is known as 'B'-line. This is used to maintain male sterility of the A-line. To produce hybrid seed, A-line is inter-planted with the pollinator, also known as the restorer or 'C'-line of the genotype N/S *RfRf*.

Most of the literature indicated that the sterility is governed by single recessive gene in combination with S-cytoplasm (Peterson, 1958; Shifriss and Frankel 1971; Shifriss and Guri, 1979). Some reports indicated that CMS could be unstable under different environmental conditions, depending on the genotype (Peterson, 1958; Novak *et al.*, 1971; Shifriss and Guri, 1979). The inter-cultivar variation in levels of male sterility *i.e.* quantity of pollen grains in cv. Bikura vs Zohar suggested that there were additional genes with accumulative effects that influenced the expression of sterility (Shifriss and Guri, 1979). Nowak *et al.* (1971), working with Peterson's source (PI 164835) of male sterility and broader genetic material found digenic inheritance. Variable degree of sterility, *i.e.* from complete male sterility to partial fertility was also observed. Lee *et al.* (2008) identified the partial restoration (*pr*) locus related to the fertility restoration of CMS. The trait was expressed only when the pepper plant had the sterile (S) cytoplasm and homozygous recessive (*prpr*) alleles. The homozygous dominant (*PrPr*) alleles determine complete fertility.

Since varieties known to be restorers or non-restorers in Peterson's S-cytoplasm behaved similarly in most cases when tested with Shifriss and Frankel's S-cytoplasm, and thus Peterson's S-cytoplasm may be identical to that of Shifriss and Frankel's S-cytoplasm. It should be noted that both Peterson's and Shifriss and Frankel's S-cytoplasms were found in material (PI 164835) collected from India. Shifriss and Frankel (1971) argued that it would not be surprising if future work shows that the various S-cytoplasms known in cultivated peppers are identical. Their hypothesis was based on the earlier report by Duvick (1959) who studied 100 S-type sources of male sterility in corn and classified into only two S-type cytoplasms. Assuming a single gene inheritance, of the six possible cytoplasmic-nuclear genotype combinations, only one is sterile (Table 6.2).

**Table 6.2.** Cytoplasmic–nuclear genotypes and their fertility status

| Cytoplasmic genotype | Nuclear genotype | Fertility | Line classification |
|---|---|---|---|
| S (Sterile) | *RfRf* | Fertile | R-line (restores fertility) |
| S | *Rfrf* | Fertile | R-line (restores fertility) |
| S | *rfrf* | Sterile | A-line (female parent) |
| N (Normal) | *RfRf* | Fertile | R-line (restores fertility) |
| N | *Rfrf* | Fertile | R-line (restores fertility) |
| N | *rfrf* | Fertile | B-line (maintainer line) |

## 3.2. Development of CMS Lines

Gabelman (1956) suggested induction of cytoplasmic-genic male sterility (C-GMS or CMS) in cross-pollinated crops. According to Duvick (1959), male sterility due to cytoplasmic-genic interaction is found or can be induced in nearly every species. According to Peterson (1958), pepper varieties can be divided into two groups regarding the source of S-cytoplasms, one group contains male sterility genes (*ms*, syn. *rf*) and the other group has dominant restorer genes (*MS*, syn. *Rf*). Since bell pepper varieties were found to possess *ms* genes, Shifriss and Frankel (1971) assumed that a stable cytoplasmic male sterility may possibly be synthesized from new sources of S-cytoplasm and *ms* genes already present in bell type varieties.

With this perception *i.e.* to search for new sources of cytoplasmic male sterility in peppers, Shifriss and Frankel (1971) made crosses using 23 small fruited hot pepper lines as female and 3 sweet pepper varieties as male parents. Among the $F_2$ populations raised, five segregated into male fertile and male sterile individuals in the ratio of 3:1. From two populations carried further (1005 and 1006), the sterility was found to be controlled by single recessive gene. In order to test a possible cytoplasmic–genic interaction in that material, several reciprocal crosses were made and reciprocal differences were observed. When 'Yolo Y' and 'Vinedale' were used as seed parents, only male fertile plants were obtained, however, when plants from the 1005 and 1006 series were used as seed parents, single gene (1:1) segregation occurred. Therefore, it was suggested that male sterility in populations 1005 and 1006 and their derivatives resulted from a combination of S–type cytoplasm carried by the original female plus an *ms* gene present in the pollen parents. It was discovered that all the hot pepper varieties used contained only fertility restorer gene (*MS*), while the sweet pepper varieties contained either sterility gene *ms* or both types of genes. These findings should encourage breeders to look for new sources of cytoplasmic male sterility in chilli peppers.

Shifriss and Guri (1979) developed several C–GMS lines in pepper and reported variation in level of male sterility among and within male sterile cultivars. They demonstrated that:

(a) Flowers of male sterile (partially fertile) cultivars produced both aborted and normal pollen grains throughout the year but in different quantities and proportions.

(b) An increase in male sterility (per cent of aborted pollen grains) with the approach of hot season was noticed in all cultivars.

(c) Bikura and Yolo Y demonstrated a higher level of male sterility than Zohar and the difference between these two groups was great during March–April.

(d) Degree of plant sterility was unaffected by plant age.

They further stated that Bikura was found to be a promising cultivar by providing stable male sterile phenotype. It recorded less than 1 per cent selfing and therefore, can serve as A-line using natural cross-pollination in hybrid seed production. In unstable cultivars typified by intra-plant variation, further progress towards utilization of cytoplasmic-genic male sterility may be achieved through the following breeding procedures:

(a) Selection for higher level of male sterility through maintainer plants.

(b) Incorporation of seedling marker genes into natural male sterile cultivars for early elimination of progenies resulting from selfing and

(c) Sampling environmental conditions that produce the highest level of male sterility.

In 1996, the Asian Vegetable Research and Development Centre (AVRDC), Taiwan received funding from the Seminis Company to transfer C-GMS system into the selected chilli and sweet pepper inbred lines (Anon, 2006). The C-GMS material was donated by Choong Ang Seed Company, Korea. Each year 10–15 tropically adapted chilli pepper inbred lines were test-crossed to a Korean C-GMS line to determine if they were maintainer (B-lines) or restorer (R-lines). Based on test-cross results, potential B-lines were backcrossed to the male sterile $F_1$ to begin the process of A-line conversion. Six stable maintainer B-lines (Tumpang, Tit Paris, 9907–9611, Kunja, Arunala and Cayenne Large Red Thick) were crossed to the source of male sterility and back-crossed five to six times to their respective maintainer lines (Table 6.3). Three of the maintainer (B-lines) and their respective CMS A-lines were crossed to restorer lines, 9852-100 and 9852-110, and the $F_1$ hybrids were evaluated for yield and horticultural traits (Table 6.4).

**Table 6.3.** Conversion of maintainer B-lines to cytoplasmic male sterile A-lines

| Variety | B-line | A-line | Pedigree |
|---|---|---|---|
| Tumpang | PBC 534 | CCA 4261 | Seungchon (CMS)/Tumpang |
| Tit Paris | PBC 380 | CCA 4757 | Seungchon (CMS)/PBC 380 |
| 9907–9611 | 9907-9611 | CCA 4758 | Seungchon (CMS)/9907–9611 |
| Arunalu | PBC 483 | CCA 4759 | Seungchon (CMS)/Arunalu |
| Kunja | PBC 362 | CCA 4916 | Seungchon (CMS)/Kunja |
| Cayenne Large Red Thick | PBC 293 | CCA 4917 | Suwon (CMS)/Cayenne Large Red Thick |

**Table 6.4.** Horticultural characters of maintainer B-lines and CMS A-lines crossed to restorer (C-line) 9852-100 and 9852-110

| Code # | Parent | Cross | Fruit length (cm) | Fruit weight (g) | Yield (g/plant) |
|---|---|---|---|---|---|
| CCA 4927 | PBC 380 × 9852-100 | B × C | 9.5 | 6.0 | 550 |
| CCA 4928 | 9907-9611 × 9852-100 | B × C | 9.7 | 4.8 | 780 |
| CCA 4929 | PBC 483 × 9852-100 | B × C | 9.1 | 4.5 | 560 |
| CCA 4930 | PBC 380 × 9852-110 | B × C | 9.4 | 5.3 | 630 |
| CCA 4931 | 9907-9611 × 9852-110 | B × C | 9.5 | 4.5 | 598 |

**Table 6.4.** *Contd.*

| Code # | Parent | Cross | Fruit length (cm) | Fruit weight (g) | Yield (g/plant) |
|---|---|---|---|---|---|
| CCA 4932 | PBC 483 × 9852-110 | B × C | 8.6 | 4.2 | 534 |
| CCA 4933 | CCA 4757 × 9852-100 | A × C | 8.0 | 4.6 | 584 |
| CCA 4934 | CCA 4758 × 9852-100 | A × C | 8.8 | 4.5 | 501 |
| CCA 4935 | CCA 4759 × 9852-100 | A × C | 9.6 | 4.6 | 751 |
| CCA 4936 | CCA 4757 × 9852-110 | A × C | 8.5 | 4.7 | 532 |
| CCA 4937 | CCA 4758 × 9852-110 | A × C | 8.2 | 3.8 | 540 |
| CCA 4938 | CCA 4759 × 9852-110 | A × C | 8.8 | 4.0 | 584 |

Observations on CMS A-lines and maintainer B-lines for horticultural characters indicated significant differences in the impact of S *versus* N cytoplasm in two of the three pairs, and across all three pairs in a combined analysis, particularly for fruit length and fruit weight (Table 6.5). Similarly, significant differences were detected between the two pollen parent lines in their impact on fruit length and fruit weight.

**Table 6.5.** Contrast in horticultural characters between cytoplasmic male sterile A-line *versus* maintainer B-line when used as seed parent in hybrid combinations

| Contrast | Fruit length | Fruit weight | Total yield |
|---|---|---|---|
| E 1 vs E 7 | ** | ** | NS |
| E 2 vs E 8 | ** | NS | ** |
| E 3 vs E 9 | NS | NS | ** |
| E 4 vs E 10 | ** | ** | NS |
| E 5 vs E 11 | ** | ** | NS |
| E 6 vs E 12 | NS | NS | NS |
| E 1, 2, 3 vs E 7, 8, 9 | ** | ** | NS |
| E 4, 5, 6 vs E 10, 11, 12 | ** | ** | NS |
| E 1, 2, 3, 4, 5, 6 vs. E 7, 8, 9, 10, 11, 12 | ** | ** | NS |

While significant differences due to pollen parent are expected, a noticeable influence of the cytoplasm factor (outside of the male sterility) was a surprise. Potential causes of the anomalous differences may include:

- Insufficient backcrossing to make the CMS A-line and maintainer B-line pairs more nearly isogenic.
- Inadequate replication, randomization, or other experimental design limitations.

- Genuine influence of the different cytoplasms on plant growth and development.

Wankhade *et al.* (2004) developed a C-GMS line NCL 3000A by backcross method. The line NCL 3000 was used as a recurrent parent and PBC 534 received from AVRDC, Taiwan, was used as a donor parent for CMS background. Eight backcrosses were made to transfer the nuclear genome of maintainer line NCL 3000 in sterile cytoplasmic background. They also made 80 crosses with NCL 3000A to identify fertility restorers and sterility maintainers for the C-GMS line. Out of 80 $F_1$s, fertility was restored in 11 (13.75%), partially restored in 33 (41.25%) and maintained sterility in 36 (40.00%) crosses. The authors suggested that the identified maintainer lines should be backcrossed as a recurrent parent to the sterile hybrids for the development of new C-GMS lines.

Due to limited availability of maintainer (*rf*) gene in hot pepper lines, process of nuclear diversification of CMS line is delayed, leading to restriction in choice of the parents to obtain heterotic hybrids (Shifriss, 1997). Molecular marker aided selection (MAS) using tightly linked marker(s) with restorer gene is expected to facilitate:

(i) Rapid screening of inbred line for *Rf*/*rf* gene without developing and evaluating test cross progenies,

(ii) Accelerated transfer of *rf* gene in hot pepper female parent (maintainer breeding), and

(iii) Accelerated transfer of *Rf* gene in sweet pepper male parent (restorer breeding) without progeny testing of selected plant(s) after each backcross generation.

### 3.2.1. *Genetics of Fertility Restoration and Distribution*

Kumar *et al.* (2007) conducted an experiment to study genetics of fertility restoration and to examine distribution of RAPD markers (OPW $19_{800}$ and OPP $13_{1400}$) linked with fertility restoration gene *Rf* in pepper inbreds reported by Zhang *et al.* (2000). Forty two hot and five sweet pepper inbred lines were crossed on a cytoplasmic male sterile (CMS) line CCA 4261 and $F_1$s were evaluated for fertility restoration under open field conditions. The results revealed that most of the hot pepper lines possess *Rf* gene. All the five sweet pepper lines had maintainer alleles. Among the 37 restorer lines, OPW $19_{800}$ and OPP $13_{1400}$ marker bands were present only in 17 and 10 lines, respectively. The presence of these two markers often did not coincide with presence of *Rf* gene. Therefore, search for more widely distributed *Rf* gene associated markers would be required to exercise marker assisted selection for restorer and maintainer plants. Nevertheless,

Kumar *et al.* (2007) suggested that these RAPD markers might have case specific utilization in CMS based hybrid breeding programme. The absence of both these markers in all the five sweet pepper maintainer lines may be useful during transfer of *Rf* gene in these sweet pepper lines (restorer breeding). Albeit practical use of RAPD markers in plant breeding has been questioned due to their poor reproducibility across the lab (Mohan *et al.*, 1997), they have efficiently and successfully been utilized for hybrid seed purity testing of pepper (Ilbi, 2003).

Shifriss (1997) and Zhang *et al.* (2000) suggested linkage of genes controlling sweet pepper fruit traits (non-pungent and larger fruit size) with the *rf* gene and linkage of genes controlling hot pepper traits (pungency and smaller fruits) with *Rf* gene. Lee *et al.* (2008) developed a cleaved amplified polymorphic sequence (CAPS) marker closely linked to partial restoration (*pr*) locus, related to fertility restoration of CMS. Out of the eight *Pr*-linked amplified fragment length polymorphism (AFLP) markers identified, E-AGC/M-GCA122 and E-TCT/M-CCG116 were the closest to the locus, estimated at about 1.8 cM in genetic distance. E-AGC/M-GCA122 was converted into CAPS marker, PR-CAPS, based on the internal and flanking regions of the AFLP fragment. The PR-CAPS marker could be useful in selecting fully fertile lines (*Pr*/*Pr*) and eliminating partially fertile (*pr*/*pr*) and potential (*Pr*/*pr*) lines in segregating populations during the development of new restorer inbred lines.

### 3.3. Exploitation of CMS for Hybrid Development

Commercial utilization of CMS for hybrid seed production starts with the development of isogenic lines incorporating the recessive *rfrf* maintainer gene; one version carries the S-cytoplasm (A-line) while the other carries the N-cytoplasm (B-line). Pollination of the sterile A-line by the male fertile B-line maintains the sterility of A-line. B-line is propagated through self-pollination. For $F_1$ hybrid development, A-line is crossed with C-line that has ability to restore fertility of the resultant hybrids. Both the lines (A and C) should have good combining ability to produce superior performing $F_1$ hybrid. Daskaloff and Mihailov (1988) suggested a unique method of producing $F_1$ hybrid seed based on CMS combined with lethal gene and a female sterile pollenizer. The CMS female line Zlaten medal '(S) *rfrf ll*' possessed a conditional lethal (CMSL) gene that can be easily inactivated with a specific treatment. 'Barzana *cfs*' is a conditional female sterile line used as a pollenizer. The '*cfs*' pollenizer plants were characterized by a permanent abundant flowering during the whole vegetative period. At the end of the vegetative period, the mutant plants form fruits. The mean seed yield per plant is about 1.5 g (15% of the control), which is satisfactory for easy maintenance of the line. A female sterile pollenizer in pepper hybrid seed production could be of great use, because to stimulate flowering of the male parent, it is necessary to remove the young fruits 2–3 times, which involves additional labour expenses.

The CMSL line 'Zlaten medal (S) *rfrf ll*' can be maintained by self-pollination during the winter months so that no maintenance is required. A specific treatment must be applied to inhibit the expression of the lethal gene in order to obtain normally growing female plants. In the hybrid seed production block the female parent and the pollenizer must be planted in alternate rows. Pollination is done by insects *viz.*, bees, thrips, ants etc. If the CMSL line expresses phenotypic instability of sterility, all non-hybrid $F_1$ plants obtained by self-pollination will be killed after germination due to the action of the lethal gene. All surviving plants will be $F_1$ hybrids.

## 4. TESTING PURITY OF $F_1$ HYBRID SEED

Purity of $F_1$ hybrid seed can be judged either by using morphological or molecular markers. For this purpose, recessive marker genes should be present in the male sterile line whereas, the pollinator should possess the corresponding dominant genes. Hence, progenies that show the recessive traits are considered to have resulted from selfing in A-line. Shifriss and Guri (1979) used erect fruit habit (*pp*) of male sterile lines 'Bikura' and 'Zohar' with pollinators 'Yolo Y' and 'Maor' carrying the dominant pendent trait (*PP*) to study purity of $F_1$ hybrid seeds. In another hybrid combination, 'Yellow Yoto Y' male sterile had the recessive yellow fruit colour at maturity (*yy*) while the pollinator parent had the dominant allele for red fruit colour (*YY*). In both the hybrids developed by PAU (CH-1 and CH-3), erect fruit bearing habit of the female parent MS-12 is used as the morphological marker. However, availability of morphological markers at seedling stage would be more useful to establish pure stand of hybrid crop. Kumar *et al.* (2007) suggested use of RAPD markers (OPW $19_{800}$ and OPP $13_{1400}$) for hybrid purity testing in the cross CCA 4261 × Pusa Jwala, since both the markers are male specific.

## REFERENCES

Anonymous (2006). Cytoplasmic-genic male sterile pepper. *In*: Annual Report 2005–06, AVRDC—The World Vegetable Center, Taiwan, ROC, pp. 18–21.

Collard BCY, Jahufer MZZ, Brouwer JB and Pang ECK (2005). An introduction to markers, quantitative trait loci (QTL) mapping and marker aided selection for crop improvement: the basic concepts. *Euphytica* **142**: 169–96.

Daskaloff S and Mihailov L (1988). A new method for hybrid seed production based on cytoplasmic male sterility combined with lethal gene and a female sterile pollenizer in *Capsicum annuum* L., *Theor. Appl. Genet.* **76**: 530–2.

Duvick DN (1959). The use of cytoplasmic male sterility in hybrid seed production. *Econ. Bot.* **13**: 167–95.

Gableman WH (1956). Male sterility in vegetable breeding. *In*; Genetics and Plant Breeding, Brookhaven Symp. Biol. **9**: 113–22.

Hundal JS and Khurana DS (1993). 'CH-1'—A new hybrid of chilli. *Prog. Farming*, **29**: 11–13.

Hundal JS and Khurana DS (2001). A new hybrid of chilli 'CH-3'—suitable for processing. *J. Res. Punjab agric. Univ.* **39(2)**: 326.

Kumar Sanjeet, Singh Vineeta, Singh Major, Rai Shubha, Kumar Sanjeev, Rai SK and Rai Mathura (2007). Genetics and distribution of fertility restoration associated RAPD markers in inbreds of pepper. *Sci. Hortic.* **111**: 197–202.

Lee JD, Yoon JB and Park HG (2008). A CAPS marker associated with the partial restoration of cytoplasmic male sterility in chilli pepper (*Capsicum annuum* L.). *Mol. Breed.* **21(1)**: 95–104.

Ilbi H (2003). RAPD markers assisted varietal identification and genetic purity test in pepper, *Capsicum annuum*. *Sci. Hortic.* **97**: 211–8.

Martin J and Crawford JH (1951). Several types of sterility in *Capsicum frutescens*. *Proc. Amer. Soc. Hort. Sci.* **57**: 335–8.

Mohan M, Nair S, Bhagwat A, Krishna TG, Yano M, Bhatia CR and Sasaki T (1997). Genome mapping, molecular markers and marker assisted selection in crop plants. *Mol. Breed.* **3**: 87–103.

Novak F, Betlach J and Dubovsky J (1971). Cytoplasmic male sterility in sweet pepper. Phenotype and inheritance of male sterile character. *J. Plant Breed.* **65**: 129–40.

Patel JA, Shukla MR, Doshi KM, Patel SA, Patel BR, Patel SB and Patel AD (1998). Identification and development of male sterile line in chilli. *Veg. Sci.* **25**: 145–8.

Peterson PA (1958). Cytoplasimically inherited male sterility in *Capsicum*. *Am. Nat.* **92**: 111–9.

Shifriss C and Frankel R (1969). A new male sterility gene in *C. annuum* L. *J. Am. Soc. Hort. Sci.* **94**: 385–7.

Shifriss C and Frankel R (1971). New source of cytoplasmic male sterility in cultivated peppers. J. *Hered.* **62**: 254–6.

Shifriss C and Rylsky I (1972). A male sterile (*ms-2*) gene in California Wonder pepper (*Capsicum annuum* L.). *Hort. Sci.* **7(1)**: 36.

Shifriss C and Guri A (1979). Variation in stability of cytoplasmic-genic male sterility in *Capsicum annuum*. *J. Am. Soc. Hort. Sci.* **104**: 94–96.

Shifriss C and Pilovsky Meir (1993). Digenic nature of male sterility in chilli (*Capsicum annuum* L,). *Euphytica* **67(1–2)**: 111–2.

Shifriss C (1997). Male sterility in pepper (*Capsicum annuum* L.). *Euphytica* **93(1)**: 83–88.

Singh J and Kaur S (1986). Present status of hot pepper breeding for multiple disease resistance in Punjab. Proc. of VI EUCARPIA meeting on Genetics and Breeding of Capsicum and Eggplant, Zaragoza, Spain, October 21–24, pp. 111–4.

Wankhade RR, Rajput JC, Halakude IS, and Sawarkar NW (2004). Identification of fertility restorers and sterility maintainers for CGMS line in chilli (*Capsicum annuum* L.). *Indian J. Genet.* **64(4)**: 337–8.

Yu IO (1990). The inheritance of male sterility and its utilization for breeding in pepper. Ph. D. dissertation, Kyung Hee University, Korea, p. 69.

Zhang BX, Huang S, Yang G and Guo J (2000). Two RAPD markers linked to a major fertility restorer gene in pepper. *Euphytica* **113**: 155–61.

# 7

# Hybrid Development for Yield and Quality

K. MADHAVI REDDY

## 1. INTRODUCTION

Chilli or hot pepper (*Capsicum annuum* L.; $2n = 24$) is one of the most valuable commercial crops grown in India and belongs to the family solanaceae. Chilli is rich in vitamins A, C, E and P. Capsaicin is responsible for pungency and has good medicinal value. Capsanthin is the most important pigment of chilli, used as a natural food colour. Hot peppers are native to Central and South America. Portuguese traders introduced them to India, Indonesia, and other parts of Asia around 450–500 years ago and due to the long history of cultivation, out-crossing nature and popularity of the crop, large genetic diversity including local landraces have evolved. In hot chilli, great range of variability for several morphological attributes occurs throughout India. India is the largest producer and consumer of chilli among other major producers in the world. India contributes about 36 percent to the total world production, and is at the top in terms of international trade, exporting 20 percent of its total production. Since last decade, chilli production in India is moving northwards on increasing demand from diversified sectors and changing consumption patterns. Dry chilli production rose by nearly 43 percent from 8.7 lakh tonnes in 1997–98 to 12.5 lakh tonnes in 2007–08. The cardinal factors driving this significant increase in production are the use of high yielding hybrids in place of varieties, increase in average yield from 1035 kg/ha to 1736 kg/ha, favourable weather conditions, and changing consumption pattern (Anon., 2008).

India has immense potential to export different types of chillies required by various markets around the world. It is the leader in exports, with around 25 per cent share in world trade, followed by China with 24 per cent share in total global exports and is a serious competitor to India. According to the Spices Board, the total export of chillies from India

Indian Institute of Horticultural Research, Hessaraghatta, Lake Post, Bangalore.

in 2007–08 touched a record high of 2.09 lakh tonnes, valued at Rs. 1097.59 crores. It exports in different forms like chilli powder, dried chilli, pickled chillies and chilli oleoresins. Indian chilli exports are mainly influenced by domestic demand and uneven production, which is interrupted by erratic monsoon, drought and yield factor. Because of its extensive cultivation in India, cucumber mosaic virus, chilli veinal mottle virus, anthracnose fruit rot, and *Phytophthora* blight are most wide spreading constraints infecting hot pepper production causing more than 60 per cent yield reduction depending on the stage of infection and cultivars grown. The development of hot pepper $F_1$ hybrid along with resistance to major diseases is one of the most practical, economical and environmentally secure strategies for reducing yield losses. Marker assisted selection (MAS) provides a potential for increasing selection efficiency by allowing for earlier selection and reducing plant population size in breeding process of chilli.

## 2. HYBRID DEVELOPMENT IN CHILLI

Despite of high seed cost, there is increasing demand for hybrids because of high yield, greater adaptability, uniformity and resistance to biotic and abiotic stresses. Moreover, hybrid/heterosis breeding is comparatively easy to vegetable breeders as it is easy to incorporate resistant genes for biotic and abiotic stresses and also horticultural traits in $F_1$ hybrid and also the right of the bred variety is protected in terms of parental lines. Chilli peppers express considerable amount of heterosis (20–50%) for yield, hence are amenable for exploitation of hybrid vigour as $F_1$ hybrids. Further, there is a great demand and huge market for chilli $F_1$ hybrid seeds in India.

### 2.1. Why $F_1$ Hybrids in Chillies?

- Earliness
- High productivity
- High fruit weight
- High dry recovery
- Resistance to biotic and abiotic stresses.

Chilli peppers grown from hybrid seed are highly uniform and usually higher yielding. In chilli, percent share of hybrid varieties is very low *i.e.* approximately 2.5 to 3 per cent. The reason may be probably due to small flower size and low seed yield per an act of pollination. Though chilli is considered as a self-pollinated crop the extent of cross-pollination reported is 60–90 per cent, may be due to exerted nature of stigma. Therefore, to maintain purity of the parental lines a minimum of 400–500 m isolation distance is recommended.

## 2.2. Methods for the Development of $F_1$ Hybrids in Chillies

The methodology of hybrid seed production described in several books and articles for many other crops are suited for chilli pepper as well. Broadly, following points are considered for the production of hybrids/hybrid seeds:

- Production of inbred lines
- Testing of combining ability
- Improvement of inbred/parental lines
- Hybrid seed production
  — Emasculation and pollination
  — Use of male sterility (GMS/CMS).

### 2.2.1. *Cost-Effective $F_1$ Seed Production Using Male Sterile Lines*

Several systems to produce hybrid seed are possible, including the use of genetic male sterile (GMS) plants and cytoplasmic-genic male sterile (C-GMS) plants. Unfortunately, the production of today's pepper hybrids commonly relies on making manual crosses between the two parents—a very labour intensive and expensive process.

#### 2.2.1.1. *Genetic Male Sterility*

Genetic male sterility is one means by which hybrid seed may be produced. The sterile plants are used as the female parent of a hybrid cross. The male sterile characteristic is often inherited as a single recessive gene (*ms*). The use of genetic male sterility is limited in hybrid seed production due to the inefficiency of producing and maintaining a population of male sterile plants. In order to produce more male sterile plants one must cross a fertile plant heterozygous for the male sterile trait to the male sterile plant, and then only half the progeny from this cross will be male sterile.

In an attempt to increase the ratio of male sterile plants in a population, Shifriss and Pilovsky (1993) crossed two isogenic lines that differed for male sterility genes. The intention of this digenic cross was to a produce a system in which a male sterile plant contained both previously described male sterile genes, *i.e.* $ms_1$ and $ms_2$. This plant was then crossed to a fertile plant that was heterozygous for both genes, *i.e.* $ms_1ms_1ms_2ms_2 \times Ms_1ms_1Ms_2ms_2$. The resultant progeny from such a cross was segregated in a ratio of three male sterile plants to one fertile plant. The implication for such a cross was that only a quarter of plants would have to be removed from a seed production plot. However, the procedure required both parents to be maintained through asexual means and has to be protected from viral contamination (Shifriss, 1997). In general, genetic male sterility systems for hybrid seed production in pepper have not been used to any significant level owing to the production of a high percentage of non-hybrid seed and because of the labour-intensive nature of the system (Daskalov and Mihailov, 1988).

### 2.2.1.2. *Cytoplasmic Male sterility*

Cytoplasmic male sterility (CMS) is another means by which hybrids may be produced. The advantage of a CMS system is that a population of sterile plants can be generated in which all the offspring are sterile. Sterility results from an interaction of nuclear and cytoplasmic factors. CMS is maternally inherited and is with a specific (mitochondrial) gene whose expression impairs the production of viable pollen (Budar and Pelletier, 2001). Since restorer of fertility (*Rf*) genes in the nucleus function to suppress the CMS phenotype, nuclear restoration allows commercial exploitation of the CMS system for the production of high yielding and heterotic hybrid seeds and avoids the need for intensive labour and extensive hand emasculation. Dominant restorer alleles have been identified in several hot pepper genotypes (Zhang *et al.*, 2000), to help breeders to differentiate restorer lines from maintainer lines to secure a completely sterile female parent, but fully fertile hybrids.

CMS in hot pepper plants (*Capsicum annuum*) was first documented by Peterson (1958) in the PI 164835 line introduced from India. Studies using Peterson's CMS material indicate that additional factors affect pollen sterility and stability (Novak *et al.*, 1971; Shifriss and Frankel, 1971; Shifriss and Guri, 1979). This trait was found to be controlled by a major recessive *rf* gene interacting with a specific S-cytoplasm. The S-cytoplasm of this line is usual source of CMS used in the production of hybrid seeds of chilli pepper. Other CMS systems reported later were proved to be identical to Peterson's (Shifriss, 1997). In India four stable CMS lines were identified in chilli along with their corresponding maintainers (Madhavi Reddy *et al.*, 2002) and using these male sterile lines three high yielding $F_1$ hybrids *viz.*, Arka Meghana, Arka Sweta and Arka Harita were released for commercial exploitation. Further, the technology was commercialized by selling the male sterile lines to the private industry on non-exclusive basis.

The correct and early identification of S-cytoplasm, N-cytoplasm and the restorer of fertility (*Rf*) genotype at the seedling stage would be useful in hot pepper breeding (Yu, 1990). To exploit CMS in hybrid seed production of hot pepper, new strategies must be used. Marker assisted selection is showing great promise in facilitating conventional breeding (Rafalski and Tingey 1993; Tanksley and McCouch, 1997). Zhang *et al.* (2000) identified two RAPD markers linked to major fertility restoration (*Rf*) gene in hot pepper. Both markers were repeatable and easy to score. These markers were validated at IIVR, Varanasi (Kumar *et al.*, 2002). Validity of two RAPD markers ($OPP13_{1400}$ and $OPW19_{800}$) associated with fertility restorer (*Rf*) gene were tested in a panel of 47 newly identified restorer and maintainer-inbred plants of hot pepper. Among the 37 restorer lines identified in this study, $OPP13_{1400}$ and $OPW19_{800}$ fragments were although repeatable and consistent, they were present only in 17 and 10 restorer lines, respectively. The case specific applications of both the repeatable and consistent RAPD markers in CMS heterosis breeding of pepper have been described (Kumar, 2004).

Further, Kim and Kim (2005) developed SCAR markers for early detection of sterile cytoplasm in chilli. To develop CMS specific sequence characterized amplified region (SCAR) markers, inverse PCR was performed to characterize the nucleotide sequences of the 5' and 3' flanking regions of mitochondrial *atp6* and *coxII* from the cytoplasms of male fertile (N-) and CMS (S-) hot pepper plants. Based on these data, two CMS-specific SCAR markers, 607 and 708 bp long, were developed to distinguish N-cytoplasm from S-cytoplasm by PCR. These CMS-specific PCR bands were verified for 20 cultivars containing either N- or S-cytoplasm. PCR amplification of CMS specific mitochondrial nucleotide sequences will allow quick and reliable identification of the cytoplasm types of individual plants at the seedling stage and assessment of the purity of $F_1$ seed lots. There two CMS-specific SCAR markers were used to screen the CMS lines developed at Indian Institute of Horticulture Research, Bangalore and are very clearly amplified in the male sterile lines confirming its validity. Moreover, the above-mentioned two markers have been validated in an array of genotypes possessing male sterile (S-) and normal/male fertile (N-) cytoplasms at Indian Institute of Vegetable Research, Varanasi. Briefly, a set of eight maintainer and restorer inbreds were crossed on four CMS lines possessing two independently isolated and commercially utilized S-cytoplasms. Based on fertility restoration/maintenance reaction of 32 resulted $F_1$s and on the presence of two SCARs ($atp6_{607}$ and $coxII_{708}$) in both the S-cytoplasms, it was concluded that although two S-cytoplasms were isolated and commercially utilized independently, they are genetically similar (Kumar *et al.*, 2009).

In chilli, stable cytoplasmic-nuclear male sterile lines (CMS) lines are available, which can be utilized as female parent in the hybrid seed production. Since male part (constricted anthers with little pollen) is sterile, use of such line precludes huge cost on manual emasculation. Thus, utilizing CMS line, the production cost of hybrid seeds can be drastically reduced. The CMS line (also called 'A'-line) is maintained by making crosses on it using pollen from maintainer line (also called 'B'-line). Through such crossing, 100 per cent male sterile seeds are obtained, unlike GMS line. The seedlings prepared from such crosses can be transplanted as female rows (4) along with the selected restorer line (called 'R'-line) as male row (1) in the hybrid seed production field. Since considerable amount of natural cross-pollination occur in chilli, no hand pollination is required. However, for economic seed yield of 300–350 kg/ha, hand pollination is recommended. Systematic study on honeybee activity in increasing natural seed yield in chilli using CMS lines is required.

## 3. REMOVAL OF NON-CROSSED FRUITS

After completion of hybridization programme, all the non-crossed (untagged) fruits developing/developed through natural cross-pollination (NCP) on female plants are removed, which facilitates vigorous development of crossed fruits and seeds. In case of CMS/GMS based hybrid seed production, this practice is not required, as all the fruits developed on the male sterile plants will be crossed fruits, provided recommended isolation distance is maintained.

### 3.1. Number of Hybrid Fruits

The number of crossed fruits per plant should be kept as many as possible. However, 50–70 crossed fruits per plant is optimum for getting a good seed yield of hybrids. It may also depend on the nature and genetic behaviour of the parental lines for fruit bearing habit, plant structure and other intercultural operations.

## 4. HARVESTING OF CROSSED FRUITS

Fruits of chilli mature at 35–50 days after pollination. The maturity index is red ripe fruit. Before harvesting of crossed fruits, open pollinated (non-hybrid) fruits are removed in order to eliminate chance contamination in hybrid fruits. Hence it should be secured that only tagged fruits are harvested. In case of CMS or GMS based hybrid seed production, all the fruits developed on CMS or GMS plants will be crossed fruits, provided recommended isolation distance is maintained.

## 5. EXTRACTION OF SEEDS AND PACKAGING

The harvested ripe fruits are dried and seeds are separated by maceration (commercial scale) or by longitudinal bifurcating (experimental scale) of the fruits. After extraction, seeds should be dried up to 8 percent moisture level. Before packing in appropriate moisture proof packing material, seeds should be cleaned on density gradient seed cleaner.

*Seed Yield*: On an average, 300–350 kg/ha—30–35 kg/10 guntas—10–15 g/plant.

**List of some chilli hybrids developed in India Public institutions**

| Chilli hybrid | Source |
|---|---|
| CH 1 (GMS based) | PAU, Ludhiana |
| CH 3 (GMS based) | PAU, Ludhiana |
| Arka Meghana (CGMS based) | IIHR, Bangalore |
| Arka Sweta (CGMS based) | IIHR, Bangalore |
| Arka Harita (CGMS based) | IIHR, Bangalore |
| CCH 2 (CGMS based) | IIVR, Varanasi |
| CCH 3 | IIVR, Varanasi |

Apart from public institutes, many private seed industries are extensively marketing different chilli varieties/hybrids. Some of the popularly grown private chilli hybrids in major chilli growing areas are:

**Private organizations**

| Chilli hybrid | Source |
|---|---|
| Wonder Hot/Ravindra | Seminis |
| Delhi Hot | Seminis |
| Indam 5 | IAHS |
| Indam 10 | IAHS |
| BSS 378 | Bejo Sheetal |
| BSS 275 | Bejo Sheetal |
| NS 1101 | Namdhari Seeds |
| NS 1701 | Namdhari Seeds |
| Roshni | Syngenta |
| HPH 232 | Syngenta |
| Tejeswini | MAHYCO Seeds |
| Devanur Delux | Nunhems Seeds |
| Sankranti | Nunhems Seeds |
| Soldier | Nunhems Seeds |
| ARCH 82 | Ankur Seeds |
| ARCH 228 | Ankur Seeds |

## REFERENCES

Anonymous (2008). Seasonal outlook on chilli. Karvy Comtrade Limited: 8th Sept 2008 http://www.karvycomtrade.com

Budar F and Pelletier G (2001). Male sterility in plants: occurrence, determinism, significance and use. *Life Sci.* **324**: 543–50.

Daskaloff S and Mihailov L (1988). A new method for hybrid seed production based on cytoplasmic male sterility combined with a lethal gene and a female sterile pollenizer in *Capsicum annuum* L. *Theor. Appl. Genet.* **76**: 530–2.

Kim DH and Kim BD (2005). Development of SCAR markers for early identification of cytoplasmic male sterile genotype in chilli pepper (*Capsicum annuum* L.) *Mol.Cells* **20(3)**: 416–22.

Kumar Sanjeet, Singh V, Kumar S, Singh M, Rai M and Kalloo G (2002). RAPD protocol for tagging of fertility restorer and male sterility genes in chilli (*Capsicum annuum* L.) *Veg. Sci.* **29(2)**: 101–5.

Kumar Sanjeet, Singh V, Singh M, Kumar Sanjeev, Kalloo G and Rai M (2004). Testing validity of fertility restorer (*Rf*) gene associated RAPD markers in newly identified restorer and maintainer lines of pepper (*Capsicum annuum* L). New directions for a diverse plant: Proceedings of the 4th International Crop Science Congress Brisbane, Australia, 26 Sept–1 Oct. 2004.

Kumar Rajesh, Kumar Sanjay, Dwivedi Neeraj, Kumar Sanjeet, Rai Ashutosh, Singh M, Yadav DS and Rai Mathura (2009). Validation of SCAR markers, diversity analysis of male sterile (S-) cytoplasms and isolation of an alloplasmic S-cytoplasm in *Capsicum*. *Sci. Hortic.* **120**: 167–72.

Madhavi Reddy K, Deshpande AA and AT Sadashiva (2002). Cytoplasmic genic male sterility in chilli (*Capsicum annuum* L.) *Indian J. Genet.* **62(4)**: 363–4.

Novak FJ Betlach and Dubovsky J (1971). Cytoplasmic male sterility in sweet pepper (*Capsicum annuum* L.) 1. Phenotype and inheritance of male sterile character. *Z. Pflanzenzucht* **65**: 129–40.

Peterson PA (1958). Cytoplasmically inherited male sterility in *Capsicum*. *Am. Nat.* **92**: 111–9.

Rafalski JA and Tingey SV (1993). Genetic diagnostics in plant breeding: RAPDs, microsatellites and machines. *Trends Genet.* **9**: 275–79.

Shifriss C (1997). Male sterility in pepper (*Capsicum annuum* L.). *Euphytica* **93**: 83–88.

Shifriss C and Frankel R (1971). New sources of cytoplasmic male sterility in cultivated peppers. *J. Hered.* **64**: 254–6.

Shifriss C and Guri A (1979). Variation in stability of cytoplasmic male sterility in *C. annuum* L. *J. Am. Soc. Hortic. Sci.* **104**: 94–96.

Shifriss C and Pilowsky M (1993). Digenic nature of male sterility in pepper (*Capsicum annuum* L.). *Euphytica* **67**: 111–2.

Shifriss C and Rylski I (1972). A male sterile (*ms-2*) gene in 'California Wonder' pepper (*C. annuum* L.). *HortScience* **7**: 36.

Shifriss C and Frankel R (1969). A new male sterility gene in *Capsicum annuum* L. *J. Am. Soc. Hort. Sci.* **94**: 385–7.

Tanksley SD and McCouch SR (1997). Seed banks and molecular maps: unlocking genetic potential from the wild. *Science* **277**: 1063–6.

Yu I (1990). The inheritance of male sterility and its utilization for breeding in pepper (*Capsicum spp.*) Ph. D. dissertation, Kyung Hee University, South Korea.

Zhang B, Huang S, Yang G and Guo J (2000). Two RAPD markers linked to a major fertility restorer gene in chilli pepper. *Euphytica* **113**: 155–61.

# 8

# Research on Bell Pepper at Indian Agricultural Research Institute, Regional Station, Katrain: An Overview

P.R. KUMAR AND S.R. SHARMA

## 1. INTRODUCTION

Bell pepper is one of the several names for bell-shaped fruits of *Capsicum annuum* plants. In India, bell pepper is commonly called as 'Shimla Mirch' probably due to its first cultivation at Shimla during the period of British rule. In British English, the fruit is simply referred to as a 'pepper', whereas in many Commonwealth of Nations countries, such as Australia, India, Malaysia and New Zealand, fruits are called 'capsicum'. In the United States and Canada, the fruit is often referred to simply as a 'pepper' or referred to by their colour (*e.g.* red pepper, green pepper, etc.), although the more specific term 'bell pepper' is understood in most regions. The colour can be green, red, yellow, orange and, more rarely, white, purple, blue, and brown, depending on genetically governed traits and/or the stage of harvesting. Green peppers are less sweet and slightly more bitter than red, yellow or orange peppers. The taste of ripe peppers can also vary with growing conditions and post-harvest storage treatment; the sweetest are fruits which are allowed to ripen fully on the plant in full sunshine, while fruit harvested green and ripened in storage are less sweet.

Indian Agricultural Research Institute, Regional Station, Katrain is located in Kullu valley in Himachal Pradesh, a Himalayan state in northwestern India. The elevation of the station and its three experimental farms ranges from 1560–1850 m above mean sea level. It possesses characteristic features of temperate monsoon or typical eastern China climate with two peaks of precipitation. It can be classified as warm temperate eastern margin climate, which is characterized by a warm moist summer, a cool dry winter and moderate rainfall. The rainfall distribution exhibits two distinct peaks; one in March and another in July-August, which

IARI Regional Station, Katrain, Kullu Valley (H.P.).

adequately caters the need for all agricultural purposes. There is no month without precipitation in the area. The cold alpine air streams bring considerable snow on peaks and in upper part of the Valley.

## 2. SOIL AND CLIMATE

Peppers thrive in a wide range of soil types with a pH range of 5.5–7.0, having good drainage facilities. A fertilizer dose of 150–200 kg N, 80 kg P and 80 kg K is applied in order to raise a good crop of bell pepper. Nitrogen is applied as one-third at transplanting, one-third when flowers form, and one-third a month after flowering. Additional nitrogen may be applied after the first harvest to improve fruit size and vigour. Excess nitrogen applied too early can cause flower drop. Bell pepper is a cool-climate crop (18–25°C). Production is low during summer months. When the temperature reaches above 32°C, the blossoms seldom set fruits.

## 3. PLANT GENETIC RESOURCES

The first two capsicum (sweet pepper) varieties released and recommended for cultivation in India were California Wonder and Yolo Wonder. These varieties were recommended by the institute in early sixties. Both of them are still widely cultivated in capsicum growing regions of the country and have been used extensively by public as well as private sector in hybridization and breeding programme. Later on the station was identified as centre for germplasm collection and maintenance for capsicum along with other crops. Consequently a vast collection came to the repository which gave fillip to further research.

Initial research was based on the few varieties and lines available at the station. Thakur *et al.* (1980) irradiated California Wonder seeds with five doses of X-rays *viz*., 5, 10, 15, 20 and 25 kR. They found a decrease in germination rate and increase in mortality after germination of irradiated seeds. Fruit shape was highly influenced resulting in tomato shaped as well as pointed fruits. Flower buds without sepals and protruded stigma were found at 10 and 15 kR dose levels. Moreover, male sterile mutants were detected at 5 and 10 kR dose levels. Joshi *et al.* (1987) reported a collection of 116 accessions both indigenous and exotic out of which 37 was vegetable paprika with round, blocky and conical shape fruits. A classification of 74 genotypes was done by Joshi *et al.* (1988) according to horticultural classification model of Smith *et al.* (1987).

## 4. INHERITANCE AND GENE ACTION

It has been observed that parents with high yield were usually good general combiners. The genetic diversity of parents and the extent of heterosis in the $F_1$ were directly proportional (Gill *et al.*, 1977). It has been suggested that selection for high yielding genotypes should be based on the number of fruits per plant because this trait alone accounted for 50 per cent variability in the yield. It is possible to isolate genotypes with high yield and early flowering resulting in high early yield. Thakur *et al.* (1980) found that additive gene action was responsible for control of fruit size, dominant gene action for days to flowering and over-dominance for plant height, number of fruits per plant and total yield. The main characters affecting early yield were days to flowering and days to first harvesting. After detail studies, Thakur *et al.* (1984) reported that these traits were governed by duplicate genes and suggested heterosis breeding programme for the improvement of bell pepper. Additive effects are less important than dominance and epitasis for early yield in sweet pepper. Fairly high degree of heterosis for early yield along with fruit size and weight of capsicum has been reported (Joshi, 1986; 1987).

In an attempt to develop a scheme for isolating high performing pure lines, Joshi and Singh (1987) selected superior hybrids and identified transgressive segregants in segregating generations of these crosses. They suggested that general combining ability estimates along with *per se* performance of a cultivar should be taken together for assessing its breeding value. Joshi (1987) observed significant inbreeding depression for all characters except days to flowering and first harvesting. He also found high heritability of traits in crosses showing high genotypic variance, and suggested that selection in advance generations would be more effective in the crosses. On the basis of performance of $F_1$ and $F_2$ generations, Thakur (1987) reported dominant × dominant interaction and high heritability for yield. Predominance of non-additive gene action indicates that heterosis breeding is of great use in improvement of capsicum.

Prior knowledge of heritability of different characters is useful in planning any scheme for genetic improvement and making choice of most appropriate breeding or selection method. Thakur *et al.* (1988) found that days to first harvest, fruit weight, fruit shape and plant height were traits of high heritability but early yield and total yield of low to medium heritability. Thakur (1993) established relationship of different trait with yield. Plant height was found positively correlated with number of branches per plant, number of fruits per plant and yield per plant. Number of branches also affected number of fruits per plant as well as yield. Joshi *et al.* (1995) also reported that yield is attributed by several contributing traits like number of fruits, fruit size, flesh thickness, and others.

## 5. DEVELOPMENT OF NEW VARIETIES AND HYBRIDS

Paprika genotype, Kt-Pl-19, was selected on the basis of yield, colour units and ideotype (Joshi *et al.*, 1993). This line is a non-pungent variety with high colour value suitable for processing. The plants are moderately branched, tall, indeterminate and fruits are pendant with two locules. The colour units of the powder have been found to be more than 85 thousand EOA units with a recovery of 68.7 per cent. The variety has been released for cultivation in 1994 from IARI Regional Station (RS), Katrain. California Wonder and Yolo Wonder released by IARI, RS, Katrain remained the predominantly cultivated varieties in larger parts of India until the first hybrid of capsicum, named Pusa Deepti was released for cultivation by the Central Sub Committee on Crop Standards and Release of Horticultural Varieties in 1997. It is a high yielding, profuse bearing and early maturing hybrid with light green conical fruits. Recently a new hybrid with dark green blocky fruits *viz.* KTCPH-3 has been identified for multiplication and release by the All India Coordinated Research Project on Vegetable Crops during 2005. This hybrid is also early maturing and profuse bearing with an average fruit weight of 85–100 g.

## 6. SEED PRODUCTION TECHNOLOGY

Inter-varietal crossing of bell and hot pepper has been suggested for augmenting the hybrid seed yield besides favourable dominant gene combination (Joshi *et al.,* 1991). However, due to dominance of pungency and pointed end of fruit, these $F_1$ hybrids can be used only as hot pepper.

Kumar and Thakur (2000) worked out the cost benefit ratio of hybrid seed production of bell pepper. They observed that cost of hybrid seed could be brought down using suitable male sterility system in capsicum. Kumar *et al.* (2007) found that spreading anthers on blotter paper overnight at a temperature of 20–25°C is the best method of pollen extraction while producing hybrid seed on large scale. They also reported that application of pollen on stigma through pollen pit is a quicker and efficient method of pollen transfer. In a separate study, Kumar *et al.* (2008) suggested that five fruits should be retained on each mother plant at a time for optimum yield and high quality of hybrid seed of capsicum under open field conditions. Seed production under polyhouse was found a good proposition on account of high setting percent and less disease infestation as compared to open field (Anon., 2008).

Thakur *et al.* (1988) in a study on the effect of aging and maximum safe storage period for sweet pepper seeds revealed that seed stored till third year under ambient conditions in temperate climate could be utilized safely as it retains the optimum germination. Thereafter, the process of aging considerably reduces and delays the germination.

## 7. TAKING CAPSICUM TO NON-CONVENTIONAL AREAS

Joshi and Munshi (2001) conducted a screening of germplasm lines available at Katrain centre to identify lines tolerant to high temperature. EG and Sel-2 among sweet pepper and Ancho-101, Jalapeno and Kt-Pl-8 among paprika type were found suitable to set fruits under temperatures as high as above 35°C. Varieties capable of bearing fruits under high temperature can be introduced into non-conventional areas creating a new opportunity to the farmers.

## 8. FUTURE THRUST

### 8.1. Male Sterility

At present entire hybrid seed of *Capsicum* is being produced by hand emasculation and pollination method. According to Kumar and Thakur (2004) more than 60 percent of cost of cultivation is consumed in the single process of emasculation and pollination. It is obvious that the cost of hybrid seed and consequently the cost of cultivation can be reduced drastically if male sterility system is available in this crop. Research in this direction is underway at IARI, RS Katrain. However, at Indian Institute of Vegetable Research, Varanasi male sterility is being extensively used for the production of $F_1$ hybrids on a large scale. These lines are also used for the development of new experimental cross combinations for evaluation and other trails. A hybrid, Kashi Surkh (CCH-2), suitable for green as well as dry chilli purpose has been released for cultivation by the farmers in different region.

### 8.2. Abiotic Stress Resistance

Very few of the available varieties and hybrids are able to set normal fruits when the temperature goes below 15°C or above 35°C. Therefore, it is an immense need to identify lines capable of setting fruits under low and high temperatures and use them for breeding for heat and cold tolerant varieties and hybrids. These will also help growers to cultivate bell pepper under non-traditional areas as well as during off-season. In the preliminary screening, two frost tolerant lines were identified at Katrain (Anon., 2004). More concerted efforts are required to make capsicum reach to the common people.

### 8.3. Biotic Stress Resistance

*Rhizoctonia* Root Rot—It is a soil borne disease, which causes mortality in nursery as well as after transplanting in the field. Some accessions at IARI, Regional Station Katrain have shown resistance under field conditions.

*Colletotrichum* (anthracnose)—It is an air borne disease which affects the marketability of fruits. So far no source of resistance could be identified. There is a need to screen vast number of germplasm lines and transfer resistance to ruling varieties and incorporate into new hybrids. Identification and earmarking of disease free zones will strengthen the concept of organic vegetable seeds.

### 8.4. Use of Molecular Markers

So far little application of modern biotechnological tools has been made in the field of capsicum breeding. Marker aided selection would enable faster and efficient selection of desired traits as well as backgrounds. At the same time it can also be applied for testing genetic purity of seed lots in a hybrid seed production programme.

## 9. CONCLUSION

More number of capsicum germplasm has to be collected from various sources for different traits in order to extend the cultivation of capsicum in non-conventional areas along with tolerance to high temperature. Plethora of work has been done on capsicum in other countries; it has to be intensified in our country to popularize the important vegetable.

## REFERENCES

Anonymous (2004). Annual Technical Report, IARI, R.S., Katrain.

Anonymous (2008). Annual Technical Report, IARI, R. S., Katrain.

Gill HS, Thakur PC, Asawa BM and Thakur TC (1982). Diversity in sweet pepper (*Capsicum annuum* L. var *grossum* Sendt.). *Indian J. Agric. Sci.* **52(3)**: 159–62.

Gill HS, Asawa BM, Thakur PC and Thakur TC (1977). Correlation, path coefficient and multiple regression analysis in sweet pepper. *Indian J. Agric. Sci.* **47(8)**: 408–10.

Gill HS, Thakur PC and Thakur TC (1977). Combining ability in sweet pepper (*Capsicum annuum* L. var *grossum* Sendt.). *Indian J. Agric. Sci.* **43(10)**: 918–21.

Joshi S, Thakur PC and Verma TS (1987). Germplasm resources of paprika (*Capsicum annuum* L.) from Katrain, India. *Capsicum & Eggplant Newslett.* **6**: 16.

Joshi S (1990). Genetics of six quantitative traits in sweet pepper (*Capsicum annuum* L.). *Capsicum & Eggplant Newslett* **8–9**: 26–27.

Joshi S and Singh B (1987). Results of the combining ability studies in sweet pepper. *Capsicum & Eggplant Newslett* **6**: 49–50.

Joshi S, Thakur PC and Verma TS (1995). Hybrid vigour in bell shaped paprika (*Capsicum annuum* L*). Veg. Sci.* **22(2)**: 105–8.

Joshi S, Thakur PC and Verma TS (1997). Kt-Pl 19—a paprika line for commercial cultivation. *Indian Hort.* **41(4)** Cov. II, 54.

Joshi S, Thakur PC, Verma TS and Verma HC (1990). Paprika germplasm with contrast qualitative traits from Katrain, India. *Capsicum & Eggplant Newslett.* **8–9**: 22–23.

Joshi S, Thakur PC, Verma TS and Verma HC (1988). Germplasm of paprika from India (Katrain). *Capsicum & Eggplant Newslett.* **7**: 27–28.

Joshi S, Thakur PC, Verma TS and Verma HC (1991). Inter-varietal crossing of bell and hot pepper augments the hybrid seed yield. *Capsicum & Eggplant Newslett.* **10**: 53–54.

Joshi S, Thakur PC, Verma TS and Verma HC (1993). Selection of spice paprika breeding lines. *Capsicum & Eggplant Newslett.* **12**: 50–52.

Kumar PR and Thakur PC (2004). Economics of seed production of Pusa Deepti—a Capsicum hybrid. *Seed Res.* **32(2)**: 163–5.

Kumar PR, Yadav Shiv K and Lal SK (2007). Seed production management of capsicum hybrid 'Pusa Deepti'. *Seed Res.* **35(2)**: 142–4.

Kumar PR, Yadav Shiv K and Lal SK (2008). Standardizing retention of fruits on plant for maximum seed yield and quality in sweet pepper (*Capsicum annuum* L.) *Haryana J. Hort. Res.* (*in press*).

Thakur PC, Joshi S, Verma TS and Verma HC (1988). 'Pusa Deepti' new capsicum hybrid. *Indian Hort.* **43(1)**: 6 Cov. III.

Thakur PC, Gill HS and Bhagchandani PM (1979). Diallel analysis of some quantitative traits in sweet pepper. *Indian J. Agric Sci.* **50(11)**: 811–7.

Thakur PC, Gill HS and Bhagchandani PM (1980). X-ray induced mutations in sweet pepper. *Veg. Sci.* **7(2)**: 118–21.

Thakur PC (1987). Gene acation an index for heterosis breeding in capsicum. *Capsicum & Eggplant Newslett.* **6**: 41–42.

Thakur PC (1990). Heritability in sweet pepper. *Capsicum & Eggplant Newslett.* **7**: 42–43.

Thakur PC (1993). Correlation studies in sweet pepper. *Capsicum & Eggplant Newslett.* **12**: 55–56.

Thakur PC, Gill HS and Choudhury B (1984). Genetics of characters related to earliness in sweet pepper (*Capsicum annuum* L. var *grossum* Sendt.). *Veg. Sci.* **111(1)**: 40–45.

Thakur PC, Joshi S, Verma TS and Kapoor KS (1988). Effect of storage period of germination of sweet pepper seeds. *Capsicum & Eggplant Newslett* **7**: 58–59.

# 9

# Current Trends in Production Technology

L.B. NAIK, M. PRABHAKAR, H.S. YOGEESHA, K. BHANUPRAKASH AND K. PADMINI

## 1. INTRODUCTION

Chilli is an important commercial vegetable crop grown throughout India. During the year 2007–08, it was grown on an area of 808.0 thousand hectare (ha) with a production of 1326.5 thousand tonnes (t) (Rao and Joseph, 2008). India is one of the largest producers of dry chilli in the world, accounting for over 46 per cent in area and 44 per cent of the total production. However, its per ha productivity is quite low at 1.55 t/ha compared to China (6.4 t/ha), USA (4.0 t ha), Turkey (2.2 t/ha) and Pakistan (2.0 t/ha).

The farm value of chilli production, was estimated at Rs. 3500 crores (US$ 867 million) per annuum while retail value of chilli and its products were estimated at Rs.7000 crores (US$ 1.7 billion) (Ali *et al.*, 2006). The importance of chilli has been gaining momentum in the last 2–3 years and during 2007–08 export of chilli from India (1.57 lakh tonnes) recorded an all time high of Rs. 848 crores (Salimath *et al.*, 2008). There is increasing demand for chilli powder, oleoresin, capsanthin and chilli oil, in the recent years. However, despite the increasing importance of chilli in the domestic and export market, adequate attention is not being paid for enhancing its productivity and quality in the country.

## 2. MAJOR BOTTLENECKS/GAPS

Even today many of the agronomic practices in chilli cultivation are old and need to be improved by adopting innovative and appropriate production technologies. In an exhaustive survey conducted by the World Vegetable Centre (Gowda *et al.*, 2006), it was observed that very few growers treat seeds and fields against infection, a great deal of area is still under local varieties, a great majority of chilli seedlings were prepared without any

Indian Institute of Horticultural Research, Bangalore.

protection against insects and diseases and raised mainly as bare root raised bed seedlings. Majority of chilli fields have flat beds without any shading, tunnel or staking, and very few farmers did mulching and fertilizer is mainly broadcasted. Despite high losses due to insects and diseases, about one fourth of the farmers do not apply any pesticide or any other control measure, especially in the rainfed production system. Weed control methods are very inefficient and herbicides are hardly used. While major area is under rainfed cultivation with all the attendant vagaries of nature, the irrigated growers still employ conventional methods of irrigation.

## 3. AVAILABLE TECHNOLOGIES

The low technology input in chilli cultivation in India resulted in high losses in yield due to several reasons listed above. Consequently, for substantial improvement in chilli production there is an urgent need for a paradigm shift in the various production techniques. Concerted efforts are, therefore, required to bring together all the improved practices for increasing chilli production. Some of such technologies/practices for enhancing productivity and quality of the produce are discussed in the following text.

### 3.1. High Yielding Varieties

A large number of high yielding varieties having better yields (up to 5.0 t/ha of dry chillies), resistance to major pests and diseases, seed content, colour and capsaicin content are evolved and recommended for different parts of the country. Some of these varieties are of dual type. The list of some of the varieties is listed in the Table 9.1. There is a need to select the variety suited to the region and the local requirements like chilli fruit type, colour, pungency, and resistance to pests and diseases etc.

**Table 9.1.** High yielding varieties of chilli evolved and recommended for different parts of India

| Sl.No. | Agency | Varieties |
|---|---|---|
| 1. | TNAU, Tamil Nadu | K 1, K 2, PKM 1, Co.1, Co 2, Co 3 (dual purpose) PMK 1, PLR 1 (green chilli), MDU 1 |
| 2. | IARI, New Delhi | NP 46A, Pusa Jwala, Pusa Sadabahar |
| 3. | UAS, Dharwad, Karnataka | KDC 1 |
| 4. | UAS, Bangalore | GPC 82 |
| 5. | GBPUA & T, Uttaranchal | Pant C 1, Pant C 2 |
| 6. | ANGRU, Hyderabad | G 1, G 2, G 3, G 4, G 5, Sindhur, Aparna, LCA 334, LCA 235, LCA 206 |
| 7. | MPKV, Rahuri, Maharashtra | Musalwadi (dual), Phule Jyoti, Phule Surya Mirchi, RHRC cluster erect, Sankeswhar-32 |

**Table 9.1.** *Contd.*

| Sl.No. | Agency | Varieties |
|---|---|---|
| 8. | JNKVV, Jabalpur, MP | JCA 283, Jawahar 218 |
| 9. | IIHR, Karnataka | Arka Lohit, A. Suphal, A. Abhir |
| 10. | Akola, Maharastra | AKC 86 39 |
| 11. | OUA&T Orissa | BC 142 |
| 12. | KAU, Kerala | Jwara Sanghi, Jwaramugh, Ujjhala, |

(*Source*: Johny *et al*., 2004).

Several varieties released at national level or at the local level are also very popular among the farmers and can be extended to other growing areas depending upon their suitability to the farmers and cultivation practices. Kashi Anmol (KA 2), developed at IIVR, Varanasi has become very popular amongst the farmers of northern India.

## 3.2. Hybrids

One of the major reasons for low productivity of the crop in India is lack of high yielding and high fruit quality hybrids. It is estimated that only about 2.6 per cent of area of chilli is under hybrids as against 90 per cent in USA and other developed countries. Hybrids in general are high in yield potential and are having wide adaptability with uniformity in the quality of the produce. Thus, there is an urgent need for developing hybrids suited to different chilli growing eco-regions and making them available to the growers at affordable seed prices. Some of the $F_1$ hybrids developed by public sector institutions (Table 9.2) and private organization (Table 9.3) are listed below:

**Table 9.2.** Chilli hybrids developed by public sector institutions

| Agency | Varieties |
|---|---|
| UAS, Dharwad | HCH 9646 |
| PAU, Ludhiana | CH 1, CH 3 |
| IIHR, Bangalore | Arka Meghana, Arka Swetha, Arka Harita |
| IIVR, Varanasi | CCH 2, CCH 3 |

In this context, it is worthwhile to note the story of hybrid chilli in Punjab. In the present scenario of agriculture in Punjab more and more farmers are now shifting towards hybrids for high productivity, improved quality, built-in resistance, environmental adaptation and earliness. With the evolution of chilli hybrid CH 1 by PAU, Ludhiana, the chilli cultivation in the state has been revolutionized. Area under the crop has increased three times and yield almost doubled during the last 4–5 years. Farmers are able to get 25 tonnes of red ripe fruit yield and 50 tonnes green fruit yield per hectare earning a profit between Rs. 1 to 1.25 lakh/ha. Progressive chilli growers on an average are producing an yield of 4 to 5 tonnes dry fruit yield/ha as compared to 1.0 t/ha average yield of the country. Its cultivation has spread to Haryana, Rajasthan, UP and MP. Hybrid seed of CH 1 is in abundance in Punjab. About 100 farmers are producing CH 1 on commercial scale after getting training at PAU on hybrid seed production technology (Dhall and Hundal, 2004).

**Table 9.3.** Chilli hybrids developed by private sector

| Company | Hybrid |
|---|---|
| Ankur | ARCH 006, ARCH 226, ARCH 236 |
| Hoechst | HOE 666, HOE 808, HOE 818 |
| Novartis | SHPHG 54, SHPH 35, SHPH 47, Picador |
| Namdhari | NS 1101, NS 1420, NS 1701 |
| Nath | Nath 70, Nath 120 |
| Sungro | Sungrow No. 16, Sungrow 86235 |
| Beejo Sheetal | BSS 138, BSS 141, BS 273 |
| Korean Hybrid | Kiran, Surya |
| Indo American | Indam 5, Indam 10 |
| Proagro | PROH 01, PR0H 02 |
| Century | Hybrid 3, |
| Seminis | Guntur Hope, Picabelto |
| Mahyco | MPH 5, MPH 58, MPH 59, MPH 1, Tejaswini |
| Nagarjuna | NARDI 712 |
| Nunhems | Dhyavanur Delux, Sankranti |

(*Source*: Johny *et al*., 2004).

## 3.3. Quality Seed

Quality seed determines the effectiveness of inputs and assume more significance when investment on inputs is increasing. There is a need for a paradigm shift in the perception from production (total quantity) to productivity (quantity/unit area) to profitability (quantity/unit area/unit time/man day). A good quality seed should be good looking, viable, and vigorous, genetically pure (true to type), bold and uniform in size, of the desired type, free from diseases, insect pests, weed seeds, foreign matter, fairly priced, has better longevity, good yielding ability and have wider adaptability.

## 3.4. Containerized Seedling Production

The quality of seedlings used in planting exerts great influence on ultimate crop performance. In India, the seedlings are raised in open field raised beds, which are lifted and planted as bare-root transplants. Although cheap, this method delays and affects adversely the initial growth of the plant due to root injury, transplanting shock, severe competition and the adverse effect of open air weather conditions. Recent years have seen many developments in the methods of raising transplants. One of these includes precision production of transplants in peat blocks/pro trays under protection especially in case of $F_1$ hybrids. More precision and control are possible when plants are raised under protected structures. Seedlings raised in pro trays offer an alternative to bare-root plants. It has been recognized that these new systems can provide: a means of controlling seedling growth, faster transplanting, improved crop establishment particularly under dry conditions and subsequent crop uniformity. Plants raised in containers are usually grown in specially prepared growing media. Many different media composts are available but there is an increasing trend of soil-less media based on peat, sand, grit, vermiculite, perlite, polystyrene or other materials and combination of these materials in different proportions.

### 3.5. Solarization for Healthy Seedlings

Soil solarization is a non-hazardous method of soil disinfection for the control of soil-borne pests including weeds. It is a method of heating the soil surface using plastic sheets placed on moist soil to trap solar radiation and thereby increasing the soil temperature. Use of 0.05 mm thick transparent polythene sheets for 20 days were found to enhance the soil temperature by 6 to 9°C and resulted in healthy seedlings of chilli (Haripriya *et al.*, 2004).

### 3.6. Raised Bed Cultivation

Chilli cultivation is mainly undertaken on flat lands under rainfed cultivation and on ridges and furrows or flat beds in the case of irrigated conditions. Raised bed cultivation with the size of beds ranging from 75 to 80 cm in breadth and 10 to 15 cm in height has shown great promise in recent years. Proper drainage, loose soil for proper root penetration and aeration, better microbial activity, enhanced nutrient availability etc. are perhaps the factors which provide a proper growing medium for the better growth and establishment of the crop. Bed-formers drawn either by tractors or bullocks can form uniformly sized, smooth raised seedbeds in a short time at a very low cost.

### 3.7. Nutrient Management

With the increasing area in high yielding varieties and hybrids, matching nutrient management is very essential. To support high biomass and economic yield, with improved produce quality, adoption of integrated nutrient management (INM) is a must in chilli production. In INM, it is to be seen that at least one-third of the nutrients are from organic sources.

Chilli being a long duration crop (6 to 8 months), responds very well to the application of fertilizers both in irrigated and rainfed conditions. N deficiency may cause stunted plant growth with pale green leaves. In P deficiency the plants produce smaller leaves with bluish green appearance. K deficient plants produce smaller leaves with crinkled surface. Hence, judicious application of fertilizers is very essential. About 20 to 25 tonnes of farm yard manure (FYM) is to be incorporated in the soil during last ploughing at the time of land preparation. Besides, 150 to 180 kg N, 60 kg $P_2O_5$ and 80–100 kg $K_2O$ per ha should be applied for getting high yield. High dose of N application may increase vegetative growth and delay in fruit maturity. For getting higher fruit yield of good quality, application of potassium to the tune of 50 to 65 per cent of recommended N is very important. N is to be applied in 3 splits *viz.*, before planting, 30 and 60 days after planting. But P and K should be applied in 2 splits *viz.*, before transplanting and 3 weeks after planting. Application of N at the commencement of flowering gives higher yield (Malawadi and Kulkarni, 2004). Under rainfed conditions full dose of P and K and half dose of N are usually applied 2 weeks after transplanting and the remaining N is to be applied as the top dressing one month later.

### 3.8. Use of Biofertilizers

Use of biofertilizers *viz.*, *Azotobacter*, *Azospirillum* has been recommended in chilli. *Azospirillum* is comparatively better and it can be applied as seed treatment, seedling treatment and as direct application in the soil. For seed treatment about 500 g seeds can be soaked in 200 g *Azospirillum* mixed with 200 ml of boiled and cooled rice water and dried in shade for 30 minutes before sowing of seeds. For seedling treatment, a solution can be prepared by dissolving 40 g of *Azospirillum* in 2 litres of water and roots of the seedlings may be dipped in the solution for 15 minutes before transplanting. *Azospirillum* at 2 kg/ha may be mixed with 20 kg of FYM or compost and the mixture may be applied directly in the soil. Similarly use of phosphorus solubilising bacteria also help in P availability.

### 3.9. Use of VAM

The Vesicular-Arbuscular Mycorrhizae (VAM) fungi are beneficial to plants which they colonize. They make more nutrients available to the plant, improve soil texture, water holding capacity, disease resistance and help in better plant growth. They also help in better plant growth and biological control of root pathogen and this need to be harnessed.

### 3.10. Vermiculture

The process of composting organic wastes through domesticated earthworms under controlled conditions is vermin-composting. During composting, the wastes are de-odorized, pathogenic microorganisms are destroyed and 40–60 per cent volume reduction in organic wastes takes place. It is estimated that the earthworms feed about 4–5 times their own weight of material daily. Thus 1 kg of worms decomposes approximately 4–5 kg organic wastes in 24 hours (Singh *et al.*, 2003).

#### 3.10.1. *Vermiwash*

It can be obtained by raising earthworms in containers made of plastic or in earthen or cement pots with holes at the bottom. The liquid, which trickle down can be collected and used as foliar spray after dilution or can be supplied with irrigation water.

### 3.11. Foliar Fertilization of Water Soluble Fertilizers

Water-soluble fertilizers containing fully soluble forms of NPK and micro nutrients in different concentrations are now available in the market. Because of their full water solubility these are highly suitable for foliar fertilization and have found to increase yield from 10 to 20 percent. Grades containing 19 percent each of N, $P_2O_5$ and $K_2O$ with micro nutrients such as Fe 1000 ppm, Mn 500 ppm, B 200 ppm, Zn 150 ppm, Cu 110 ppm and Mo 70 ppm are currently available in the market. Besides, potassium nitrate composed of 100 percent plant nutrients is one of the most effective and concentrated fertilizers. It contains 13 percent N and 46 percent $K_2O$. Three foliar sprays at 1 percent concentration starting from 30 days after planting at an interval of 15 days with a spray solution of 1000 l/ha are suitable.

### 3.12. Secondary Micronutrients

In addition to the major nutrients, secondary and micronutrients also play a vital role in increasing the photosynthetic efficiency and productivity of the crop. Among the second-generation problems of green revolution, one of the most important soil fertility constraints endangering the sustainability of high production agriculture is the emergence of secondary and micronutrient deficiencies. Increasing annual productivity in intensively cultivated areas particularly under irrigated ecosystems has increased removal of these nutrients from soils year after year. Use of concentrated high analysis NPK fertilizers further worsened the secondary and micronutrients turnover in soil plant system by increasing the removal of these nutrients on one hand and restricting their inadvertent supply in the form of impurities on the other. Decreased use of organic manures has also been one of the primary factors aggravating secondary and micronutrient disorders.

Application of calcium in the form of $CaCO_3$ (10 kg/ha) or gypsum (1 to 3 t/ha) increased total number of flowers per plant, number of fruits per plant and yield of chilli. Similarly, application of sulphates of Mg, B, Fe and Zn through soil or foliar applications have resulted in remarkable improvement of yield in chilli.

### 3.13. Organic Cultivation

Coupled with availability of disease resistant/tolerant varieties/hybrids and also biopesticides and bioagents and encouraged by the other benefits of IPM in chilli cultivation, some farmers/NGOs have started thinking on the lines of organic cultivation of chillies. Organic farming in chillies on a commercial scale has not yet been ventured by any farmers groups yet, however, if organic farming in chillies on commercial scale is done successfully and got certified, this will be a mile stone in the history of chilli cultivation. Organic farming enables several advantages. It is a holistic production management system, promotes and enhances agro-ecosystem health, including biodiversity, biological cycles and soil biological activity and lead to production of green food. The organic production system is designed to enhance biological activity within the whole production system, increasing soil biological activity, maintain long term soil fertility, duly relying on renewable resources in the locally organized agricultural systems. Without adequate organic matter content, soil gets poorer due to reduced nutrient and water holding capacity.

### 3.14. Micro Irrigation

It is an efficient method of providing irrigation water directly into soil at the root zone of plants, which permits the water to consumptive use of plants (increased water use efficiency, WUE) and facilitate utilization of water soluble fertilizers and chemicals. This system of irrigation saves water, increases the yield and improves the quality of the produce (Naik, 2003).

**Table 9.4.** Water saving and yield enhancement due to micro-irrigation (MI) in chilli

| Particulars | Surface irrigation | Drip irrigation |
|---|---|---|
| Yield (q/ha) | 42.30 | 60.90 |
| Irrigation (cm) | 109.00 | 41.70 |
| WUE (q/ha cm) | 0.39 | 1.50 |
| Advantage of MI (%) | 61.70 | 44.00 |

### 3.14.1. *Fertigation*

Fertigation offers the best solution for intensive and economical crop production, where both water and fertilizers are delivered to growing crops through drip irrigation system. Fertigation provides required fertilizer nutrients directly to active root zone, thus minimizing losses of expensive nutrients, which ultimately help in improving productivity and quality of farm produce besides savings in time and labour which makes fertigation economically profitable. The experiments have demonstrated that through fertigation 40 to 50 per cent of the nutrients could be saved which is otherwise wasted. Fertigation involves not only the efficient use of the two most precious inputs *i.e.* water and nutrients but also exploits the synergism of their simultaneous availability to plants. However, there is need to bring down cost of soluble fertilizers to reap this benefit.

## 3.15. Rainfed Production and Supplemental Irrigation

Productivity of rainfed chilli can be increased by better rainwater management which includes water harvesting to provide crop saving irrigations, supplemental irrigation during critical period and *in situ* rain water conservation and enhancing soil moisture holding capacity (Prabhakar and Hebbar, 1997). Dry chilli yield of cultivar Arka Lohit was significantly higher at 27.9 q/ha when two supplemental irrigations were given whenever the dry spell exceeded 10 days period during grand growth and fruiting period of the crop (Prabhakar and Naik, 1997). Mulching with plastic agrimulch substantially enhanced crop growth as well as yield and raised bed planting is well suited for this method of growing rainfed chili.

## 3.16. Plastic Mulching

Plastic mulching has proved to be of a great boon in conserving moisture, controlling weeds, keeping the soil warm during winter and cool in summer, enhancing yield and quality of produce. Large number of experiments conducted on mulching has established that mulching results in 40 to 50 per cent saving of water, 20 to 25 per cent increase in yield and suppressed weeds up to 90 per cent (Hosmani, 1982). The use of both drip irrigation and mulching has been found very useful and has resulted in further saving of water, increase in yield and weed suppression (Lee and Yoon, 1975).

However, the main difficulty in adopting this technology is the economic viability in view of the present day cost of agriculture, which is quite high. Besides, the availability of thinner mulching material (15–20 micron) and biodegradable mulch will encourage its use in commercial chilli production.

## 3.17. Protected Cultivation

Protected cultivation or green house production provides protection to crop and create an ideal environment for growing of the crops. This protection may be in the form of protection against adverse environmental conditions like high temperature, low temperature, high rainfall, high wind velocity, etc. and also protection against various pests and diseases and insects carrying diseases (vectors) and so on. In the case of chillies, protection of the crop against insects carrying diseases has given a big boost in chilli cultivation. Large number of viral diseases poses a major threat to chilli cultivation. Growing chillies under net houses has given a complete protection to the crop against vectors carrying viral diseases. Presently the hybrid seed production is entirely taken up in net houses in Karnataka. Besides, capsicum and Hungarian Yellow wax types are grown in green houses.

## 3.18. Integrated Pest Management (IPM)

Although chilli exports are gaining momentum for the past 2–3 years, there are several problems encountered in the exports. The presence of aflatoxin, pesticide residues and additives are posing major problems. Several consignments have been refused due to these problems. Therefore efforts are on to improve the product quality and post-harvest handling of chilli. As per the work carried out by the United Nations Development Programme (UNDP) in Guntur area of Andhra Pradesh, the farmers were motivated to use more of bio-pesticides and botanicals. They were also trained in combating pests using pheromone traps, bird perches, yellow stick traps, etc. In place of chemical fertilizers farmers were advised to use vermicompost, neem cake etc. Farmers were also advised to stop spraying pesticides at least a month prior to harvest. Major pesticides like DDT, BHC, Ethion, etc. were totally avoided. As a result the number of sprays was brought down from 20 to less than 8 in a crop period of 8 to 10 months. This has resulted in reduction of cost of cultivation, better productivity and improved product quality. Today this has been accepted by many growers in Andhra Pradesh. Some of the major chilli exporters from Kerala also have started backward and forward linkages in IPM cultivation to ensure sourcing of pesticide free chilli for export (Somasekharan and Devananda Shenoy, 2004).

## 3.19. Use of Plant Growth Hormones and Chemicals

The problem of flower and fruit drop is a severe cause of poor yield faced by many growers. This problem is due to low humidity, low light intensity and high temperature leading to excessive transpiration and water deficiency within the plant. It can be effectively controlled by foliar application of 20–50 ppm naphthalene acetic acid (NAA) at full bloom stage.

Application of tricontanol (vipul) at 0.5 ml/l of water was also found to be effective in reducing flower and fruit drop (Madalageri and Krishna Ukkund, 2004). Besides, water stagnation should be avoided during flowering and fruiting. Providing suitable plant and configuration and enhancing soil aeration will also help in controlling this problem.

### 3.20. Use of Herbicides

Good crop stand with early vigour of the seedlings goes a long way in obtaining higher yield. Control of weeds at the initial stages of crop growth is of paramount importance. This can be achieved through Good Agricultural Practices (GAP), micro irrigation, mulching, etc. At least 2–3 manual weeding is essential where no herbicide is used. However, for effective weed control it is advisable to use herbicides like pendimethalin @ 1.25 kg/ha or oxyflourfen @ 0.2 kg/ha in combination with one hoeing. However, 2–3 hoeings along with manual weeding also serves in controlling of weed growth (Mukhopadhyay, 2004).

### 3.21. Green Fruit Picking to Stimulate Growth

In chilli flowering starts 30–60 days after transplanting depending upon the cultivars and need another 30 days for harvesting of green chillies. However, first picking should be encouraged at green stage to stimulate further flush of flowering and fruit setting by encouraging further root and plant growth.

## 4. HARVESTING

Proper harvesting, packing and transporting is of great importance in view of the retail chains being established everywhere and quality being given the priority in the changed marketing scenario.

For dry chilli, well ripened, red coloured fruits are to be harvested. The harvested fruits should be kept in shade or room for 2 to 3 days for development of uniform red colour in a temperature of 22 to 25°C. The fruits should be dried in sun by spreading the fruits on clean mats or cemented surface and for uniform and quick drying, fruits are to be spread out in the drying yard in layers of 4 to 5 cm. Fruits should be frequently stirred for uniform drying and to avoid the growth of mould and discoloration. It takes around 5 to 10 days for proper drying of the chilies in order to use it for spice purposes.

## 5. CONCLUSION

Generally, 4 or 6 tonnes of dry chilli and 25 to 30 tonnes of green chilli can be obtained per ha. Productivity of chilli under rainfed condition, being very low at present, can be enhanced to 1.5 to 1.8 t/ha by adoption of high yielding varieties coupled with good crop management practices. Depending on the varieties 25–40 kg of dry chilli can be obtained from 100 kg of fresh ripe fruits. For export purpose the chilli should be harvested in proper maturity at right time and proper curing, grading and processing should be done. Utmost care should be taken for utilization of inputs for harnessing the potential yield of chillies.

## REFERENCES

Dhall RK and Hundal JS (2004). Successful cultivation of chilli hybrids. *Spice India* **17(4)**: 36–38.

Gowda MC, Mei-huey Wu and Mubarak Ali (2006). India. *In*: Ali, M (*ed.*) Chilli (*Capsicum spp.*) Food chain analysis: Setting research priorities in Asia.

Haripriya K, Manivannan K and Kamala Kannan S (2004). Nursery solarisation for early and healthy seedlings of chilli. *Spice India* **17(7)**: 37.

Hosmani MM (1982). Chilli. Publ. by Mrs.S.M.Hosmani, Narayanpur, Dharwad, pp. 109–10.

Johny A, Kallu Purackal and Ravindran PN (2004). Chilli varieties for higher yield. *Spice India* **17(4)**: 2–8.

Lee BX and Yoon JY (1975). The effect of polyethylene film mulches on soil temperature, growth and yield of red peppers. *J. Kor. Hort. Sci.* **16(2)**: 185–91.

Madalgeri MB and Krishnan Ukkund (2004). Management of flowering and fruiting in chilli (*Capsicum annuum*). *Spice India* **17(7)**: 20–23.

Malawadi MN and Kulkarni SS (2004). Nutrient Management in chilli (*Capsicum annuum*). *Spice India* **17(4)**: 26–30.

Mukhopadhyay TP (2004). Crop management of chilli. *Spice India* **17(4)**: 13–18.

Naik LB (2003). Irrigation aspects of *Capsicum*. *In*: The genus *Capsicum*. Amit Krishna De. (*ed.*) Taylor and Francis, London, p. 129–38.

Prabhakar M and Hebbar SS (1997). Water management in chilli under semiarid conditions. Proc. of the National seminar on water and nutrient management for sustainable production and quality of spices. Oct. 5–6, 1997. Calicut, Kerala.

Prabhakar M and Naik LB (1997). Effect of supplemental irrigation and nitrogen fertilization on growth, yield, nitrogen uptake and water use of green chilli. *Ann. Agric. Res.* **18(1)**: 34–39.

Rao Saideshwara and Joseph Rao P (2008). Chilli—The red star of Indian spices. *Spice India* **21(9)**: 25–27.

Salimath PM, Mohan Kumar HD and Basavaraj N (2008). Byadgi Chillies 'Dollar earning high quality chilli from Karnataka'. *Spice India* **21(7)**: 6–8.

Singh HP (2003). Hi-Tech horticulture and precision farming: Issues and approaches. *In*: Precision farming in horticulture, H.P. Singh, Gorakh Singh, Jose C. Samuel and R.K. Pathak (*eds.*). NCPAH, DAC, MOA, PFDC, CISH, p. 14.

Somasekharan KP and Devananda Shenoy (2004). Integrated pest management in chilli farming-A successful UNDP project intervention. *Spice India* **17(4)**: 11–12.

# 10

# Pest Management: Present Status and Future Thrust

A.B. RAI, S. SATPATHY AND R. GANDHI GRACY

## 1. INTRODUCTION

Chilli, *Capsicum annum* L., is an important condiments as well as vegetable known for its commercial and therapeutic value. The average national productivity is quite low (9.24 t/ha) compared to the world average productivity (14.43 t/ha). Among the different factors of low productivity, the biotic stress is the most important. Besides the fungal and viral pathogens, insect pests particularly the sucking pests *i.e.* thrips and mite adversely affect the qualitative and quantitative yield of chillies. The damage by sucking pests varies from 5 to 55 per cent (Table 10.1).

**Table 10.1.** Severity of important insect pests of chilli

| Important pests | Scientific name | Severity (%) |
|---|---|---|
| Thrips | *Scirtothrips dorsalis* (Hood) | 50–55 |
| Mites | *Ployphagotarsonemus latus* | 30–40 |
| White fly | *Bemisia tabaci* (Chilli leaf curl) | 30–50 |
| Aphid | *Aphis gossypii* (Chilli Mosaic) | 5–20 |
| Cut worms | *Spodoptera litura* | Negligible |
| Fruit borer | *Helicoverpa armigera* | Negligible |

Ahamed *et al*. (1987) reported the yield loss due to thrips, mite and *Spodoptera* to the tune of 76.68 per cent in chilli crop. Jagadeesh (2000) reported yield loss ranging from 40–70 per cent in chilli. This loss varies from season to season and from place to place in different states of India (Table 10.2).

Indian Institute of Vegetable Research, Varanasi.

**Table 10.2.** Yield loss due to thrips and mite in different states

| State | Yield loss (%) | References |
|---|---|---|
| **Yield loss due to thrips** | | |
| Karnataka | 88–92 | Kumar, 1995 |
| Tamil Nadu | 26.30 | Kavitha *et al.*, 2006 |
| Uttar Pradesh | 25.00 | Satpathy *et al.*, 2006 |
| Gujarat | 60.5–74.30 | Patel *et al.*, 1998 |
| **Yield loss due to mite** | | |
| Karnataka | 96.39 | Borah, 1987 |
| Tamil Nadu | 60.0 | Srinivasan *et al.*, 2003 |
| Gujarat | 27.78 | Desai *et al.*, 2007 |

Routine schedule application of insecticide at short intervals increases the cost of protection and chemical residues. Chilli receives approximately 20 rounds of pesticide application next to cotton and rice in the major growing areas of Andhra Pradesh, Maharashtra, Karnataka and Tamil Nadu, which contributes 75 per cent of total production in India. It ultimately causes the residual toxicity which also has the potential threat not only to our environment but also to economy by rejection in the export for meeting out the maximum residual limit (MRL) requirements. Intensive application results in pest resurgence and development of resistance in the target pest. Many commonly used pesticides like monocrotophos, thiometon, phosphamidon, methyl-o-demeton, fornothion, ethion, cypermethrin, deltamethrin, clocythain are known to cause the pest resurgence in chilli (David, 1991; Ashokan *et al.*, 1992). More recently report says that imidacloprid a valid pesticide for thrips control is known to cause resurgence in mite (Srinivasulu *et al.*, 2002a).

Thus, there is an urgent need for effective integrated pest management (IPM) with special importance to the biological control and host plant resistant aspect and in this paper we attempted to compile available information on effective IPM for thrips and mites and ultimately formulate the future strategies of present need.

## 2. PEST STATUS

Chilli thrips, *Scirtothrips dorsalis* (Hood) had a wider range of distribution in the world. It is an important pest in almost all the chilli growing countries including, Japan, China, India, Pakistan, Taiwan, Korea, Thailand, Africa, Australia and USA. The major attributes of thrips which made it an important pest on chilli are wider host range, fast mobility and excellent invaders, shorter generation time, and mode of reproduction varies from predisposition to parthenogenesis. It has become an economic and quarantine pests due to its role as vector of viral diseases of many vegetable crops. It causes mechanical injury by means of feeding and oviposition which ultimately damage the leaf tissue leading to disarrangement of cells and showing mostly upward leaf curling (Mound, 1997).

The broad mite or yellow mite, *Polyphagotarsonemus latus* (Banks) is another important sucking pest causing serious concern for the chilli growers in India. The broad mite got its name by means of enjoying the polyphagous status of more than 250 plants from 66 major families and has a cosmopolitan distribution (Waterhouse and Norris, 1987). Being microscopic in size with shorter life cycle of 2.8 to 5 days and fecundity of 40 eggs/female, showing resistance to most of the conventional insecticides, resurgence to some novel insecticides and causing by means of feeding leaf crinkling and mostly downward curling of leaves, all these contributed the mite to became a serious and threatening pest on chilli, demanding a real challenge in the management aspects. Often mite injury is confused with the leaf curl virus symptoms due to similarity in appearance of symptoms. The mite injury causes the physiological damage to the leaf tissues, dearrangement of the parenchyma cells of leaf and ultimately causes leaf crinkling and curling (Karmakar, 1995), but there was no significance of mite involvement in virus transmission.

### 2.1. Seasonal Occurrence

Regarding the seasonal occurrence of the thrips and mites in the country, it varies from states to state. Mote (1976) observed that mite population was in peak during February to May and again in October–November in Maharashtra during the chilli season. In South India and Gujarat, it is during November and February (Srinivasalu *et al.*, 2000; Solanki, 1999) and during October–November in U.P. (Anonymous, 2007). The mite infestation starts from second week of September and reaches its peak during first week of November, whereas the thrips population peaked during first week of October, though the occurrences were noticed from September to November. China has developed a model for pest forecasting of chilli mite and thrips with the correlation of weather parameters (Liu *et al.*, 1993). In India there is a need of successful pest forecasting model in chilli for reducing the use of pesticide and making the pest management system economically more viable.

## 3. MANAGEMENT OF THRIPS AND MITES

### 3.1. Cultural Control

Cultural practices include all the crop production and management techniques which are utilized by the farmers to maximize their crop productivity and/or farm income. It includes decision on crops/varieties to be grown, time and manner of planting, tilling of field and crops sanitation, application of fertilizers and irrigation, harvesting time and procedures and even off-season operations in fallow/cropped field.

In chilli, the protection should start from the seedling stage itself. Well protected pest free healthy seedling will always offer good tolerance limit in the main field. Studies showed that nursery raised in path under screen cover for 1–5 weeks gave better protection from thrips and mites in the seedling stages and which ultimately leads to good yield later in main field (Vos and Nurtika, 1995). The planting time also plays a major role in pest occurrence. Chilli planted during July showed less infestation of mite, whereas September planted chilli hosts more mite population (Anonymous, 2007). Mulches has also role in pest management, white and silvery plastic mulch in the main field reduced thrips incidence. Since nutrient of the plant has a direct role in pest management, the optimal dosage of fertilizer is crucial in any pest management system. Ram *et al.* (1996) reported that 125: 80: 100 kg NPK gave slight tolerance to thrips population and helped in reducing the population buildup in chilli. In case of mite, potassium and boron reduced the population buildup, whereas Mg and Zn enhanced the population (Raju Ram and Rammurthy, 2001; Karmakar, 1997). Some of the cultural practices like detopping of infested shoots reduced and delayed the mite and thrips population buildup (Chakraborti, 2004). Removal of weeds which are the hosts of mite and thrips offered certain role in reducing population buildup (Rao and Ahmed, 1986).

Intercropping is an important cultural practice with adverse effect on the pest in the main crop. Tomato as a main crop in chilli field reduced the thrips population (Manjunatha *et al.*, 2001a), whereas castor as a banker plant indirectly reduced the mite and thrips population by encouraging predatory mite activity. Report showed that some of the organic amendments like vermicompost, neem cake and neem seed kernel extract sprays reduced thrips and mite population and increased the chilli yield (George *et al.*, 2007).

### 3.2. Host Plant Resistance

Host plant resistance refers to the heritable qualities of cultivars to counteract the activities of pests so as to cause minimum percent reduction in yield as compared to other cultivars of the same species under similar conditions (Dhaliwal and Arora, 2001).

Chilli crop is severely attacked by *S. dorsalis* and *P. latus* causing leaf curl symptom called "murda". The potential role of host plant resistance (HPR) in the context of IPM has been well recognized. It acts as preventive measure against build up of pest population and is free from environmental pollution. Some resistant/tolerant sources of chilli reported so far against *P. latus* are presented in Table 10.3. However, these have not been commercially exploited or have not been used to intogress the resistance to popular cultivars. Researches have demonstrated various levels of responses to the infestation of chilli mite in terms of population growth and development of leaf curl symptoms, besides having differential reaction to *Capsicum* spp. and genotypes involved. Out of 101 pepper genotypes evaluated in greenhouse, 7.5 per cent of the genotypes of *Capsicum annuum,* 50 percent of *C. frutescens*, 57 per cent of *C. baccatum*, and *C. chinensis* have been found free from mite infestation (Lima *et al.*, 2003).

**Table 10.3.** Interspecific resistance against yellow mite

| Species | Percent resistant genotypes |
|---|---|
| *C. annuum* | 7.50 |
| *C. frutescens* | 50.00 |
| *C. baccatum* | 57.00 |
| *C. chinensis* | 57.00 |

Lot of work has been done on resistant sources of thrips and mites (Table 10.4). However, researchers have not paid much attention to this aspect in respect of chilli thrips and mite, in general. IAC Ubatuba derived from plants of Cambuci pepper (*C. baccatum*) has been observed to be resistant to *P. latus.* This line also showed resistance to mosaic virus Y and slender leaf (Anonymous, 1987). Selecting for resistance sources in *Capsicum* accessions at the Universidade Federal de Vicosa, Brazil, indicated the accessions BGH/UFV 1774 (*C. annuum*) and BGH/UFV 5086 (*C. frutescens*) to be resistant and highly resistant, respectively to *P. latus,* under severe testing conditions (Echer *et al.*, 2002). Sanap and Nawale (1985) reported differential reaction of chilli varieties to mite infestation and leaf curling. The variety19-1 and Ca (P) 247 had lowest mite infestations, while LIC 8 and LIC 30 had the greatest infestations. On the basis of leaf curling injury, however, LIC 8 was graded as resistant, and Pant C1 and LEC 7 as moderately resistant. Similarly, based on the reaction of 308 chilli (*C. annuum*) germplasm to mite leaf curl in Andhra Pradesh, 5 accessions *viz.*, EC 378630, EC 378633, EC 391082, IC 214991 and NIC 23897 were found resistant, with less than one per cent leaf curl. Some accessions reacted independently to the leaf curl caused by thrips and mites as was evident in EC 348630 and IC 214991 showing zero damage due to mites and heavy leaf curl due to thrips. However, one exotic entry (EC 391082, a paprika type) was found to be resistant to the leaf curl caused by both thrips and mites (Babu *et al.*, 2002). Initial screening work at National Bureau of Plant Genetic Resources (NBPGR), Hyderabad revealed that

PBC 613 and NIC 23906 were resistant to chilli thrips under field condition. The cultivar EC 391082, Pant C 1, LCA 304, LCA 312 were reported to be resistant against both mite and thrips (Tatagar *et al*., 2001). Initial studies conducted at IIVR, Varanasi, showed that PBC 473, Japani Longi, AVRDC 16, PBC 535 and 9852-173 have found to be resistant against both thrips and mites. Some of the cultivars such as Pant C 1, LCA 304 and Japani Longi are the known potential sources of resistant against both thrips and mite which has to be exploited under commercial breeding programme.

**Table 10.4.** Resistance/tolerance sources of chili against *P. latus*

| Resistant/tolerant genotypes | Screening | Place | References |
|---|---|---|---|
| S 7 | Field | Bangalore | Ningappa, 1972 |
| LIC 8, Pant C 1, LEC 7 19-1, Ca (P) 247 | Field | Maharastra | Sanap and Nawale, 1985 |
| IAC Ubatuba derived from *C baccatum* | Field | | Anonymous, 1987 |
| K 2, Lam X 235, Musalwadi, NP 46 A, Sel 118-2, Sindhur, Semiliguda local | Field | Orissa | Ram *et al*., 1997 |
| Pant C 1, LCA 304, LCA 312 | Field | South India | Tatagar *et al*., 2001 |
| LCA 235, LCA 330, EC 128946, Custer mutant, LIC 19, LCA 312, yellow anther mutant, LIC 13, LIC 15 | Field | AP | Khalid *et al*., 2001 |
| EC-378630, EC-378633, EC-391082, IC-214991 and NIC-23897 | Field | AP | Babu *et al*., 2002 |
| NEC, G 4, PDG 50, R Line and VNS 4 | Field | UP | Anonymous, 2007 |
| HD 60, HD 16, HD 12 (Doubled haploid) | Greenhouse | Cuba | Depestre and Gomez, 1995 |
| Accessions BGH/UFV 1774 (*C. annuum*) and BGH/UFV 5086 (*C. frutescens*) | Greenhouse | Brazil, America | Echer *et al*., 2002 |
| Jwala, RHRC Erect and AEG 77 | Field | Gujarat | Desai *et al*., 2007 |
| >50% genotypes of *C. frutescens*, 57% of *C. baccatum* and *C. chinensis* | Greenhouse | Outside India | Lima *et al*., 2003 |
| EC 378630, EC 378633, EC 391082 | Field | India | NBPGR, Hyderabad |

### 3.3. Biological Control

The commercial exploitation of biological control for chilli thrips and mite management is limited in India compared with other countries. Globally also the exploitation of biocontrol technique for chilli pest management is less compared to other crops. Chemical control of phytophagous mite is becoming increasingly difficult and pest resurgence is posing a serious problem. In view of this, more and more attention is focused throughout the world in utilizing the natural biocontrol agents. The biocontrol agents for thrips and mites are listed in Tables 10.5 and 10.6, respectively.

There are more than forty natural enemies reported against thrips in general. The natural enemies broadly belong to the category of predator, parasitoid and pathogens. Among predators, anthocorids are reported to be effective. Exploitation of anthocorids is more prevalent in European countries than India. *Orius tristicolor* (White) has shown promise in commercial vegetable greenhouses (Higgins, 1992). The use of *O. albidipennis* (Reuter), *O. niger* (Wolff) and *O. vicinus* (Ribaut) have also been considered for use in European glasshouses (Hejzlar and Kabicek, 2000). Other predators that have been recommended for the control of this pest include the common flower bug, *Anthocoris nemorum* (L.) and the lacewing, *Chrysoperla carnea* (Stephens) (Bennison *et al.*, 1998). Other species of the genera *Mallada* (Chrysopidae), *Macrolophus* (Miridae) and *Franklinothrips* (Aeolothripidae) have also been used commercially to control thrips (Quaries, 2001). In India, Ballal and Tripathi (2007) reported that anthocorids play a major role in the management of chilli thrips.

The predatory mite, *Hypoaspis miles* (Berlese) preys on soil-dwelling organisms, and is used to help control mature larvae of *F. occidentalis* which pupate in the soil (Linnamaki *et al.*, 1998). More recently, the predatory mites *Amblyseius (Iphiseius) degenerans* (Berlese), *A. umbraticus* (Chant), *A. andersoni* (Chant) and *A. limonicus* (Garman and McGregor) have been assessed under laboratory and glasshouse conditions for use against various thrips (Scott Brown *et al.*, 1999; van Rijn *et al.*, 1999; Sengonca and Drescher, 2001). Additional information about thrips predators and parasitoids has been summarised by Sabelis and van Rijn (1997) and Loomans *et al.* (1997), respectively. Information on the use of the predatory thrips genus *Franklinothrips* for pest control in glasshouses has been reviewed recently; this species appears to have some potential for use particularly against those thrips species with long developmental periods and where all development stages are present in or on the leaf (Loomans and Vierbergen, 1999).

The effectiveness of *Verticillium lecanii* (Ballal *et al.*, 2007) and *Metarhizium anisopliae* (Helyer *et al.*, 1995) against *S. dorsalis* has been reported in chilli. There are very few records of pathogens in thrips and most of these are fungi (Butt and Brownbridge, 1997). This may be because diseased thrips are difficult to find due to their small size and cryptic behaviour, and there appear to have been only a small number of searches conducted specifically for thrips pathogens. Insect fungal pathogens have the ability to penetrate their hosts directly which gives them an advantage over viruses and bacteria which have to be ingested. A range of nematodes have been shown to be active against pupae of

*F. occidentalis*, including *Heterorhabditis bacteriophora*, *Steinernema carpocapsae*, *S. feltiae* and *Thripinema nicklewoodii* (Greene and Parrella, 1993; Helyer *et al.*, 1995; Chyzik *et al.*, 1996).

**Table 10.5.** Natural enemies against chilli thrips

| Family | Species | References |
|---|---|---|
| **Predator** | | |
| Aeolothripidae | *Franklinothrips vespiformis* (Crawford) | Hirose *et al.*, 1993 |
| Anthocoridae | *Bilia* sp. | Hirose *et al.*, 1993 |
| | *Carayonocoris indicus* (Muraleedharan) | Sureshkumar and Ananthakrishnan, 1984 |
| | *Orius armatus* (Gross) | Young and Zhang, 2001 |
| | *Orius insidiosus* (Say) | Etienne *et al.*, 1990 |
| | *Orius maxidentex* (Ghauri) | Sureshkumar and Ananthakrishnan, 1984 |
| | *Orius minutus* (L.) | Ananthakrishnan and Sureshkumar, 1985 |
| | *Orius nagaii* (Yasunaga) | Nagai, 1993 |
| | *Orius sauteri* (Poppius) | Chang *et al.*, 1993 |
| | *Orius similis* (Zheng) | Kajita, 1986 |
| | *Orius strigicollis* (Poppius) | Yano, 1999 |
| | *Orius tantillus* (Motschulsky) | Nakashima and Hirose, 1997 |
| | *Wollastoniella parvicuneis* (Yasunaga) | Yasunaga, 1995 |
| | *Wollastoniella rotunda* (Yasunaga and Miyamoto) | Yasunaga and Miyamoto, 1993 |
| Berytidae | *Yemma exilis* (Horvath) | Kohno and Hirose, 1997 |
| Cecidomyiidae | *Arthrocnodax occidentalis* (Eelt.) | Chang *et al.*, 1993 |
| Chrysopidae | *Mallada basalis* (Walker) | Young and Zang, 2001 |
| Coccinellidae | *Coleomegilla maculata* (DeGeer) | Cooper, 1990 |
| | *Propylea japonica* (Thunberg) | Nagai, 1993 |
| Lygaeidae | *Geocoris lubra* (Kirkaldy) | Young and Zhang, 2001 |
| | *Geocoris ochropterus* (Fabr.) | Chang *et al.*, 1993 |
| | *Piocoris varius* (Uhler) | Hirose *et al.*, 1999 |
| | *Stigmatonotu minutum* (Malipatil) | Young and Zhang, 2001 |

**Table 10.5.** *Contd.*

| Family | Species | References |
|---|---|---|
| Miridae | *Campylomma austrina* (Malipatil) | Young and Zhang, 2001 |
| | *Campylomma chinensis* (Schuh) | Wang, 1995 |
| | *Campylomma livida* (Reuter) | Hirose *et al.*, 1993 |
| | *Deraeocoris* sp. | Young and Zang, 2001 |
| Phlaeothripidae | *Haplothrips victoriensis* (Bagnall) | Young and Zang, 2001 |
| Thripidae | *Neohydatothrips portoricensis* (Morgan) | Etienne *et al.* 1990 |
| Phytoseiidae | *Amblyseius cucumeris* (Oudemanns) | Wada, 1999 |
| | *Amblyseius longispinosus* (Evans) | Chang *et al.*, 1993 |
| | *Amblyseius mackenzie* (Schuster and Pritchard) | Kajita, 1986 |
| | *Amblyseius multidentatus* (Swirski and Shechter) | Chang *et al.*, 1993 |
| | *Amblyseius okinawanus* (Ehara) | Kajita, 1986 |
| | *Amblyseius tsugawai* (Ehara) | Nagai, 1993 |
| | *Phytoseius* spp. | Hirose *et al.*, 1993 |
| **Parasitoids** | | |
| **Egg parasitoids** | | |
| Trichogrammatidae | *Megaphragma* sp. | Hirose *et al.*, 1993 |
| **Larval parasitoids** | | |
| Eulophidae | *Ceranisus menes* (Walker) | Hirose, 1989 |
| | *Geotheana shakespearei* (Girault) | Young and Zhang, 2001 |
| **Parasites** | | |
| | *T. nicklewoodi* (North America) | |
| | *T. khrustalevi* (Asia, South America) | |
| | *T. fuscum* (North America) | |
| | *T. aptini* (Europe) | |
| | *T. reniroai* (Asia) | |
| | Undescribed species (New Zealand) | |
| **Fungal pathogens** | | |
| Hyphomycetes | *Beauvaria bassiana* | Saito, 1991 |
| | *Cladosporium cladosporioides* | Humber, 1992 |
| | *Hirsutella* sp. | Hall, 1992 |
| | *Verticillium lecanii* | Young and Zhang, 2001 |
| Zygomycetes | *Neozygites parvispora* | Saito *et al.*, 1989 |

The biggest impact of biological control of mite pests has been made by augmenting phytoseiid mite predators particularly in orchard and greenhouse systems abroad (Hussey and Scopes, 1977). In India, though several phytoseiid predators have been listed for their potential to control *P. latus* infesting chilli but they have not been exploited at field level. Pena (1992) investigated the predatory behaviour and feeding habits of phytoseiid *Typhlodromalus peregrinus* [*Amblyseius peregrinus*] in the laboratory and greenhouse in Florida, USA. It preferred *P. latus* and consumed 23–75 per cent of the prey population in 6 days. Castagnoli and Falchini (1993) in the laboratory at 25 ± 1°C, 90 ± 10 percent relative humidity and light duration 16:8 in Italy found P. *latus* as suitabile prey for *Amblyseius californicus.* The egg-to-egg time was 9.50 days, 2 days longer than when reared on *Tetranychus urticae*, while the mortality of juveniles was very low (0.81%). Its intrinsic rate of increase when fed on *P. latus* was 0.177 as against 0.194/day on *T. urticae*. Hariyapa and Kulkarni (1988) in laboratory evaluation at 23–27°C temperature and 65–70 percent relative humidity in Karnataka noted *Amblyseius longispinosus* to be a good predator of *P. latus*, on chilli, consuming 11.72 larvae, 9. 33 nymphs or 5.07 adults/ day. In another study with chilli leaves, the *Amblyseius ovalis* at predator-prey ratios of 1:25, 1:50 and 1:100 eliminated *P. latus* on the ninth, twelfth and seventeenth day, respectively (Hariyappa and Kulkarni, 1989). Karuppuchamy *et al.* (1994) found *A. ovalis* as an efficient predator of *P. latus* infesting chilli in South India consuming mite at an average of 5.76, 4.64, 3.20, 3.12 and 2.12 eggs, first instar, second instar nymphs and adults, respectively. Rodriguez and Ramos (2003) investigated the effects of *P. latus* reared on different substrates (*Solanum tuberosum*, *Capsicum annuum* and *Citrus latifolia*) on the biology of *A. largoensis*. The fecundity of *A. largoensis* was higher on *P. latus* fed with *Citrus latifolia* and comparable to that of chilli.

Manjunatha *et al.* (2001) studied the interaction between *A. ovalis* and the mite, *P. latus* at predator-prey ratios of 1:40, 1:20 and 1:100, in controlling *P. latus* infesting chillies in Karnataka, India. The *A. ovalis* successfully controlled the prey mite at a predator: prey ratio of 1:20. The rapid decline in adult prey population was observed by seventh day after start of the experiment and resulted in the lowest pest population at 45 days after release. Pena and Osborne (1996) studied the effect of predation by two phytoseiid species on *P. latus* on chilli in greenhouse and in field in Florida. *A. californicus* maintained mite density below economic damaging levels, whereas the effect of *Neoseiulus barkeri* was erratic. Among four release rates of *N. barkeri* tested under greenhouse, 10 or more predatory mites per plant of *Capsicum* cv. Hungarian Wax effectively reduced populations of *P. latus* from more than 100 mites/leaf to zero in a week. Influx experiments, in which there was continuous immigration of mite, single release of five *N. barkeri* adults per plant significantly reduced mite populations but failed to prevent all plants from mite injury, and that 3 weekly releases of five predatory mites per main stem provided adequate protection from mite injury for over 7 weeks (Fan and Petitt, 1994). The predatory mites were released twice (on day 1 and 5, or 15 days later) on each plant, every second plant or every fourth plant. Releasing *N. cucumeris* on each or every second plant was as efficacious in controlling broad mites as sulfur treatments. But when the predatory mites were released only on every fourth plant, the overall height and yield of the plants were adversely affected by broad mites (Weintraub *et al.*, 2003). Regarding the pathogens only under laboratory condition some of the pathogens *viz.*, *B. bassiana*, *H. thompsonii*, and *P. fumosoroseus* are effective against yellow mite, but their effectiveness under field condition are questionable.

**Table 10.6.** Natural enemies of *P. latus*

| Predatory mite | Experiment | Country | References |
|---|---|---|---|
| *N. barkeri* [*A. barkeri*], *N. californicus* | Green house and field condition | USA | Pena and Osborne, 1996 |
| *P. persimilis* | Glass house condition | Conthey | Granges and Leger, 1995 |
| *N. barkeri* [A *barkeri*] | 4 releases @ 10 mites/ plant 3 weekly releases @ mites/main stem | USA | Fan and Petitt, 1994 |
| *N. barkeri* [*A. barkeri*] | Green houses | USA | Petitt, 1992 |
| *P. persimilis* | Glass houses Spider mite control but *P. latus* outbreak | The Netherlands | Bravenboer, 1975 |
| **Pathogens** | | | |
| *B. bassiana* $1.16 \times 10^6$ conidia/ml | Laboratory | USA | Penna *et al.*, 1996 |
| *H. thompsonii* $2.39 \times 10^3$ conidia/ml | Laboratory | USA | Penna *et al.*, 1996 |
| *P. fumosoroseus* $1.29 \times 10^5$ conidia/ml | Laboratory | USA | Penna *et al.*, 1996 |

Though several natural enemies are identified for chilli thrips and mite, among these, relatively few phytoseiid predators have shown promise against thrips and tarsonemid mite on chilli but their successful utilization at field level are lacking. Another important setback in the biological control is lack of commercial scale production and failure of natural enemies under field condition. Large scale production of predatory mites, *Orius* sp. and any other anthocorids is facing the problem of cost effective large scale mass production technology. Apart from this, all these natural enemies are effective only under controlled condition like poly/green houses, whereas in field conditions they failed due to crops and hosts divergence.

### 3.4. Botanicals

The advantage of botanicals in pest management strategy is safer to environment and other non-target organisms. So far there was no successful report regarding botanicals against chilli thrips except for neem products. Among botanicals, neem, *Annona* and castor oil have shown promise against chilli mite. However, variable results have been obtained with neem products. Among different plant oils (neem, *Hibiscus cannabinus*, sesame, *Pongamia* spp. and castor (0.026%) tested for their synergistic property on dicofol (0.04%) against *P.latus* on *Kharif* chilli (cv. Byadagi) in Karnataka, the mite population was significantly lowest in dicofol + castor oil treatment with lowest leaf curl index and highest dry chilli yield (3.04 q/ha) (Smitha and Giraddi, 2003). Results on the effectiveness of neem-based treatments on the management of *P. latus*, including other sucking pests in red chilli revealed that all the integrated treatments, without or with the inclusion of one spray of phosphamidon at 45 days after transplanting (DAT) were highly effective and safe to natural enemies (Chakraborti, 2000). Among synthetic insecticides, neem preparations and chitin inhibitor evaluated for the control of pests of chilli in the field in Andhra Pradesh, neem oil and ready to use neem formulations (Neem guard, Repelin and Biosol) were less effective, while the chitin inhibitor (Duphar) was least effective (Rajasri *et al.*, 1991). Giraddi *et al.* (2001) found plant extracts (Bougainvillea, *Clerodendron* and sorghum leaf) to be ineffective in managing both thrips and mite on chilli (cv. Byadagi) during *Kharif* season, while the mixtures of acephate + dicofol and monocrotophos + dicofol were the most effective. In Thailand, among different plant extracts alone or in combination tested against *P. latus* on chilli, the treatment with 100 ppm *Annona* suspension was highly effective, killing 100 per cent eggs and larvae and 80 per cent adults in the laboratory and safer to predatory mite *A. longicaudatus* (Uraisakul and Kanok, 2003). In a field trial during summer in West Bengal, detopping of affected shoots at 16 DAT followed by application of neem cake 1 kg/m$^2$ at 20 days interval or foliar application of neem oil (10 ml/ litre) + azadiractin (4 ml/litre) at 7 days interval beginning at 17 DAT and need based application of profenofos (2 ml/litre) effectively controlled mites and thrips on chilli and were quite safe to natural enemies (Chakrabarti, 2004). Chlasson *et al.* (2004) reported that *Chenopodium* based oil has found to be effective against yellow mite. The main bottlenecks in botanicals are their low efficacy under field condition, faster biodegradation and lack of effective commercial formulations.

### 3.5. Chemical Control

Use of chemical pesticides is still the most prevalent practice to manage the pest especially for high value crops such as cotton and vegetables due to it quicker action. A number of acaricides/insecticides alone or in combination are labeled for the control of chilli thrips and tarsonemid mite on chilli (Tables 10.7 and 10.8).

Seal *et al*. (2005) reported six insecticides to be effective against thrips in vegetables and among them imidacloprid has the higher efficacy. Misra (2003) found NACLFMOA 1.9 percent EC a fermented metabolite to be promising insecticide against thrips under field condition.

**Table 10.7.** Effective chemical insecticides/acaricides against chilli thrips

| Pesticide | Effectiveness | References |
|---|---|---|
| Novaluron (Rimon 0.83 EC) | Effective under green house | Seal *et al*., 2005 |
| 1Chlorofenapyr (Pylon) | Effective under green/poly house condition | -do- |
| Imidacloprid (Provado) | Effective under field condition | |
| Spinosyn A + B (SpinTor 2 SC) | Effective under field condition | |
| Abamectin (Agrimek 0.15 EC) | Effective under field condition | |
| Cyfluthrin (Baythroid 2) | Effective under field condition | |
| Azadirachtin (Neemix 4.5) | Effective under field condition | |
| Monocrotophos (0.05%) + Mancozeb (0.25%) | Effective under field condition | Mandal and Beura, 2003; Reddy and Jagadish, 1980; Takre *et al*., 1979 |
| NACLFMOA 25 g/ha | Effective under field condition | Mishra, 2003 |
| Fenvalerate (0.05%) | Effective under field condition | Patnaik *et al*., 1985 |
| Imidacloprid (0.25 g/l) | Effective under field condition | Manjunatha *et al*., 2000 |
| Acephate (1 g/l) | Effective under field condition | Manjunatha *et al*., 2000 |
| Methyl-o-demeton (0.025%) | Effective under field condition | Kalaiyarasan *et al*., 2002 |

In Tamil Nadu, chilli seedling root dip in monocrotophos and carbosulfan have been found most effective against sucking pests including *P. latus* on chilli, giving protection for 28 days after transplantation (Dhandapani and Jayaraj, 1982). Seedling dip with imidacloprid + nursery treatment with monocrotophos and cypermethrin + dicofol at 7 weeks after planting resulted in the lowest pest population of mites (*P. latus*) infesting chilli during the *Kharif* season in Karnataka (Manjunatha *et al*., 2001b). Of several acaricides tested in the field in Maharashtra, monocrotophos, dicofol and dimethoate at 0.05, 0.05 and 0.3 per cent, respectively, gave fairly good control of chilli mite for 15 days and enhanced yields (Mote, 1976). In Punjab, 0.05 per cent carbophenothion, chlordimeform (chlorphenamidine) and monocrotophos eliminated both the mites and the curling symptoms; dicofol and binapacryl killed the mites but were phytotoxic to the chilli plants (Dhooria and Bindra, 1977). In a field trial in Singapore for the control of *P. latus*, the highest chilli yields were obtained with amitraz 0.02 per cent, propargite 0.03 percent, triazophos 0.04 per cent, chlordimeform 0.04 per cent and Cyanotox [cyanophos] 0.025 per cent. However, chlordimeform was mildly phytotoxic (Chuo and Ng, 1982). Dhandapani and Kumaraswami, (1985) in a field trial in Tamil Nadu found phosalone 0.07 percent and monocrotophos 0.1 per cent to be effective against chilli mite for up to 21 days after treatment. In a field experiment at Lam in Andhra Pradesh, among 13 insecticides tested for the control of insect and mite pests on chilli, phosalone and methamidophos at 0.5 kg a.i./ha proved effective against the mite (Rao and Ahmad, 1985). According to Khaire and Naik (1985), of nine pesticides tested against sucking pest complex on *Capsicum*, fluvalinate at 0.1 kg/ha gave the best control of *P. latus* (71.4%). The efficacy of Sevisulf (0.2%) [carbaryl with sulfur] evaluated against *P. latus* on chilli in Maharashtra and monocrotophos 0.05 per cent were also effective against *P. latus* and protected the crop up to 15 days after application. The list of chemicals which were found to be effective against chilli mite has been listed in Table 10.8.

**Table 10.8.** Effective chemicals against *P latus* infesting chilli in India and abroad

| Pesticide | References |
|---|---|
| Monocrotophos, dicofol, dimethoate | Mote (1976) |
| Carbophenothion, chlordimeform (chlorphenamidine), monocrotophos | Dhooria and Bindra (1977) |
| Seedling root dip in monocrotophos or carbosulfan | Dhandapani and Jayaraj (1982) |
| Amitraz, propargite, triazophos, chlordimeform, cyanotox | Chuo and Ng (1982) |
| Phosalone, monocrotophos | Dhandapani and Kumaraswami (1985) |
| Phosalone, methamidophos | Rao and Ahmad (1985) |
| Fluvalinate | Khaire and Naik (1985 |
| Sevisulf, monocrotophos | Sanap and Nawale (1986) |
| Phorate, phosalone | Nandihalli and Thontadarya (1986) |
| Dicofol, wettable sulfur, HCH dust | Karuppuchamy and Mohansundaram (1987) |
| Triazophos | Kandasamy *et al.* (1987) |
| Monocrotophos | Radke and Aherkar (1987) |
| Propargite | Dibyantoro (1988) |
| Monocrotophos, methyl-O-demeton, formothion, thiometon, ethion | David (1991) |

**Table 10.8.** *Contd.*

| Pesticide | References |
|---|---|
| Neem guard, repelin, biosol, neem oil, chitin inhibitor | Rajasri *et al*. (1991) |
| Triazophos, phosalone, amitraz | Rajasri *et al*. (1991) |
| Lliuyangmycin [preparation from *Streptomyces griseolus*] | Xie *et al*. (1992) |
| Dicofol | Jeyarajan *et al*. (1995) |
| Dicofol, chinomethionat, pyridaben, pyraclofos | Cho *et al*. (1996) |
| Fenazaquin | Walnuz and Pawar (2000) |
| Ethion | Mallapur *et al*. (2001) |
| Seedling dip in imidacloprid + nursery – monocrotophos, cypermethrin+ dicofol – after planting | Manjunath *et al*. (2001b) |
| Mixture of acephate + dicofol or monocrotophos + dicofol | Giraddi *et al*. (2001) |
| Abamectin, dicofol oxydemeton-methyl, imidacloprid | Srinivasulu *et al*. (2002) |
| Fenpyroximate | Srinivasan *et al*. (2003) |
| Neem oil + Azadirachtin, need based profenofos | Chakraborti (2004) |
| NACLFMOA, Abamectin | Mishra (2003), Sarkar *et al*. (2005) |
| Fenpropathrin | |
| Spiromesifen | Kavitha *et al*. (2006) |
| Lime sulfur, vicosa mixture (a nutrient amended with Bordeaux mixture) | Venzon *et al*. (2006) |

Though lots of chemical pesticides are said to be effective against thrips and mites, the major concern here is, some of the insecticides known to cause resurgence. Report shows that many conventional insecticides such as monocrotophos, thiometan, phosphamidon, methyl-o-demeton, formothion, ethion, cypermethrin, deltamethrin, clocythrin and more recently imidacloprid are known to cause resurgence of chilli mite (David, 1991; Ashokan *et al.*, 1992; Srinivasulu *et al.,* 2002). The complexity of the problem is the imidacloprid which is effective against chilli thrips known to cause resurgence in chilli mite.

This kind of problem has to be addressed clearly since in field it is necessary to manage both thrips and mites. The pest management has to be taken as an integrated approach not only to a particular pest but the different pests of that particular crop. Apart from these, the major focus has to be given to reduce or minimize the chemical spray due to its well known ill effects such as resistance, environment pollution and residual toxicity. The IPM here means to incorporate all the possible technique with minimum environment pollution and maximum benefit to the farmers: not to the particular pest but to the crop.

## 4. FUTURE STRATEGIES

The analysis of past research works on thrips and mite management in chilli indicates the focus on chemical control and field screening of germplasm for resistance which lack in depth basic information. The physiological adaptation in response to acaricide insecticide application in favour of the target pest makes this strategy less sustainable and encourages environmental pollution. Besides, in host plant resistance no systematic strategic attempt has been made through a multidisciplinary collaborative approach. The interspecific resistance needs to be exploited. The role of cultural and biological tactics with special reference to entomopathogens and bacterial toxins need to be exploited.

- To address the bioecology of chilli thrips and mite, development of effective forecasting method and model for crop loss due to thrips mite under different agroclimatic condition has to be made.
- Development of screening technique for thrips and mite resistant based on population grade, injury grade and fruit yield and exploitation of interspecific resistance.
- Screening has to be done for the new search for resistant cultivar against thrips and mites.
- Validation of resistant sources has to be done for different agronomical zone in India (region specific) against available resistant and newly found resistant cultivar.
- Research has to be carried out for the mechanism of resistance (biochemical basis or systemic induced resistance) in order to confirm the resistant source.
- Survey has to be made for the exploration of native natural enemies against thrips and mites.
- Research to be made on cost effective mass production technologies of effective natural enemies.
- Standardization and release of technology for various field conditions has to be made in order to exploit the bioagents under field condition.
- Search has to be made out for classical biological control if we failed to get native natural enemies.
- Evaluation has to be made for effective pathogens for microbial toxin.
- Search for novel insecticides/acaricides against thrips and mites has to be made.

- The urgent need of today is to develop resistant management strategy for thrips and mite.
- Research has to be made for utilization of synergists in the resistant management.
- Monitoring of insecticide residues is the present need.
- The detailed studies on integration of different pest management component under field condition for thrips and mite pests as a whole has to be made in order to develop a complete package for chilli thrips and mite management.

## REFERENCES

Ahamad K, Mohamed MG and Murthy NSR (1987). Yield losses due to various pests in hot pepper. *Capsicum News* **l.6**: 83–84.

Ahuja DB (1998). Influence of intercropping and mixed cropping sesame with pearl millet, green gram and moth bean on the incidence of insect and mite pests. *In*: Recent advance in management of arid ecosystem, Proc. of a symposium held in India.

Almaguel L, Perez R and Ramos M (1984). Life cycle and fecundity of the mite *Polyphagotarsonemus latus* on pepper. Cuba Ciencia y Tecnica en la Agricultura, *Proteccion de Plantas* **7(3)**: 93–114.

Ananthakrishnan J and Sureshkumar N (1985). Anthacolids as efficient biocontrol agents of thrips. *Curr. Sci.* **54**: 987–90.

Anonymous (1987). New IAC cultivars. (*Capsicum baccatum*). Novos Cultivars IAC. *Agronomico*. **39(2)**: 120–1.

Anonymous (2005). Insect and related pests of flowers and foliage plants. http://www.mrec.ifas.ufl.edu/lso/entomol/ncstate/mitel.htm

Anonymous (2007). Insect Pest Management, Annual Report 2006–07, Indian Institute of Vegetable Research, Varanasi.

Ashokan G, Venugopal MS and Mohansundaram M (1992). Resurgence of *Polyphagotarsonemus latus* (Banks) (Acari: Tarsonemidae) on chillies (*Capsicum annuum*) following insecticide application. *In*: *Proc*. IV Nat. Symp. Acarology, Oct 12–14, 1990. University of Calicut, Kerala, pp. 53–61.

Babu BS, Pandravada SR, Reddy KJ, Varaprasad KS and Sreekanth M (2002). Field screening of pepper germplasm for sources of resistance against leaf curl caused by thrips (*S. dorsalis* Hood) and mites (*P. latus* Banks). *Indian J. Plant Prot*. **30(1)**: 7–12.

Ballal CR and Gupta Tripti (2007). Anthocolids as potential predators. *In*: Biological and biotechnological approaches to insect pest and disease control. Ballal CE (*ed.*). Training manual, PDBC, India, pp. 162–85.

Ballal CR, Jalali SK, Preeseenth Mohanraj, Ramanujam B and Joshi Sunil (2007). Biological and biotechnological approaches to insect pest and disease control. Training manual 2007. Winter School, PDBC, Bangalore.

Bennison JA, Maulden KA, Wardlow LR, Pow EM and Wadhams LE (1998). Novel strategies for improving biological control of western flower thrips on protected ornamentals—potential new biological control agents. *In*: *Proc* of the Brighton crop protection conference: pests and diseases, 1: 193–8.

Borah DC (1988). Bioecology of *Polyphagotarsonemus latus* (Banks) (Acarina: Tarsonamidae) and *Scirtothrips dorsalis* Hood (Thysanoptera : Thripidae) infesting chilli and their natural enemies. Ph.D. Thesis, 1998, Hisar Agricultural University, pp. 1–180.

Bravenborer L (1975). Developments in biological control in glass house. *Zeitschrift tur Angewandte Entomologie* **7(4)**: 390–1.

Butt TM and Brownbridge M (1997). Fungal pathogen of thrips. *In*: Lewis T. (*ed.*) Thrips as crop pests. CAB International, Wallingford, UK, pp. 399–433.

Castagnoli M and Falchini L (1993). Suitability of *Polyphagotarsonemus latus* as prey for *Amblyseius californicus*. *Redia* **76(2)**: 273–9.

Chakraborti S (2000). Neem-based integrated schedule for the control of vectors causing apical leaf curling in chilli. *Pest Management and Econ. Zoology* **8(1)**: 79–84.

Chakraborti S (2004). Sustainable management of apical leaf curling in chilli. *J. Appl. Zoology Res.* **15(1)**: 34–36.

Chang NT, Hung CT, Hua T and Ho CC (1993). Notes on the predatory natural enemies of *Thrips palmi* on aubergine. *Plant Protection Bulletin* (Taiwan) **35**: 239–43.

Chen JL, Li LS, Jiang SQ and Chen DD (1985). Preliminary study on the spatial distribution patterns of pepper plants (*Capsicum frutescens*) infested by broad mite. *J. Southwest Agric. College*, China **3**: 175–87.

Chlasson H, Bostanian NJ and Vincent C (2004). Acaricidal properties of a *Chenopodium* based botanical. *J. Econ. Entomol.* **97(4)**: 1373–7.

Cho MyoungRae, Jeon HeungYong, La SeungYong, Kim Dong Soon, Yiem MyoungSoon, Cho MR, Jeon HY, La SY, Kim DS and Yiem MS (1996). Damage of broad mite, *P. latus*, on pepper growth and yield and its chemical control. *Kor. J. Appl. Entomol.* **35(4)**: 326–31.

Chuo SK and Ng BB (1982). Field evaluation of eleven acaricides for the control of the broad mite, *Hemitarsonemus latus* (Banks) on chilli plants, *Capsicum annuum* L. *Singapore J. Primary Industries*. **10(2)**: 64–70.

Cooper B (1990). Status of *Thrips palmi* in Trinidad. *FAO Plant Prot. Bull.* **39(1)**: 45–46.

David PMM (1991). Resurgence of yellow mite *Polyphagotarsonemus latus* (Acarina: Tarsonemidae) on chilli following application of insecticides. *Madras Agric. J.* 78 (1–4): 88–91.

Depestre T and Gomez O (1995). New sweet pepper cultivars for Cuban off-season production. *Capsicum & Eggplant Newslett.* **14**: 47–49.

Desai BK, Suresha BA, Allolli TB, Patil MG and Syed Abbas Hussain (2007).Yield and economics of chilli based intercropping system. *Karnataka J. Agric. Sci.* **20(4)**: 807–9.

Dhaliwal GS and Arora Ramesh (2001). Integrated pest management: concepts and approaches. Kalyani Publishers, New Delhi, India, pp. 427.

Dhandapani N and Jayaraj S (1982). Effect of chilli seedling root dip in insecticides for the control of sucking pests. *Pestology* **6(3)**: 5–10.

Dhandapani N and Kumaraswami T (1985). Persistence of toxicity in some foliar insecticides against sucking pests in chillies. *Indian J. Plant Prot.* **11(1–2)**: 20–23.

Dhooria MS (1984). Preliminary observation on the biology and host range of the tarsonemid mite, *Polyphagotarsonemus latus* (Acarina: Tarsonemidae), a mite pest of chilli and potato in Punjab. *Acar. Newslett.* **14**: 3–4.

Dhooria MS and Bindra OS (1977). *Polyphagotarsonemus latus* (Banks), a mite pest of chilli and potato in Punjab. *Acar. Newslett.* **4**: 7–9.

Dibyantoro ALH (1988). Field assessment of Omite 57 EC and Matador 25 EC against sucking insects on red chilli (*Capsicum annuum* L.). *Bull. Penelitian Hortikultura*. **17(1)**: 5–12.

Echer MM, Fernandes MCA, Ribeiro RLD and Peracchi AL (2002). Evaluation of *Capsicum* genotypes for resistance to the broad mite. *Horticultura Brasileira* **20(2)**: 217–21.

Etienne J, Guyot J and Waetermeulen X van (1990). Effect of insecticide, predation and precipitation on populations of *Thrips palmi* on aubergine (eggplant) in Guadeloupe. *Florida Entomologist* **73(2)**: 339–42.

Fan Y and Petitt FL (1994). Biological control of broad mite, *Polyphagotarsonemus latus* (Banks), by *Neoseiulus barkeri* Hughes on pepper. *Biol. Contr.* **4(4)**: 390–5.

Gerson U (1992). Biology and control of the broad mite, *Polyphagotarsonemus latus* (Banks) (Acari: Tarsonemidae). *Exp. Appl. Acar.* **13(3)**: 163–78.

Giraddi RS, Holihosur SN and Naik LK (2001). Bioefficacy of chemical and botanical toxicants against thrips and mite in chillies. *Karnataka J. Agric. Sci.* **14(3)**: 642–5.

Granges A and Leger A (1995). Seven years of experiments on biological integrated control on tomatoes and cucumbers under glass at the Rougeres centre at Conthey. *Rev. Science de Viticutture-d'-Arporicultuse-et-d'-Horticluture.* **27(3)**: 189–91.

Greene ID and Parella MP (1993). An entomophilic nematode, *Thripinema nicklewoodii* and an endoparasitic wasp, *Ceranisus* sp. parasitizing *Frankliniella occidentalis* in California. *IOBC/WPRS Bull.* **16**: 47–50.

Gupta SK (1985). Handbook: Plant Mites of India. Zoological survey of India, Calcutta, p. 520.

Hall RA (1992). New pathogen on *Thrips palmi* in Trinidad. *Florida Entomologist* **75**: 380–3.

Hanchinal SG and Kulkarni SV (2001). Effect of intercropping on incidence of mite and thrips in chilli. *Karnataka J. Agric. Sci.* **14(2)**: 493–5.

Hariyapa AS and Kulkarni KA (1988). Biology and feeding efficiency of the predatory mite *Amblyseius longispinosus* (Evans) on chilli mite, *Polyphagotarsonemus latus*. *J. Biol. Contr.* **2(2)**: 131–2.

Hariyappa AS and Kulkarni KA (1989). Interaction between the predatory mite, *Amblyseius ovalis* (Evans) and chilli mite, *Polyphagotarsonemus latus* (Banks). *Indian J. Biol. Contr.* **3(1)**: 31–32.

Hejzlar P and Kabicek J (2000). Predatory bugs of the genus *Orius* in biological control of pests. *Zahradnictvi Horticulture Science* (Prague) **27**: 141–9.

Helyer NL, Brobyn PJ, Richardson PN and Edmondson RN (1995). Control of western flower thrips (*Frankliniella occidentalis* [Pergande]) pupae in compost. *Ann. Appl. Biol.* **127**: 405–12.

Higgins CJ (1992). Western flower thrips in greenhouses: population dynamics, distribution on plants and associations with predators. *J. Econ. Entomol.* **85**: 1891–1903.

Hirose Y, Kajita H, Takagi M, Okajima S, Napompeth B and Buranapanichpan S (1993). Natural enemies of *Thrips palmi* and their effectiveness in the native habitat, Thailand. *Biol. Contr.* **3**: 1–5.

Hirose Y, Nakashima Y, Takagi M, Nagai K, Shima K, Yasuda K and Kohno K (1999). Survey of indigenous natural enemies of the adventive pest *Thrips palmi* on the Ryukyu Islans, Japan. *Appl. Entomol. Zoology* **34**: 489–96.

Hirose Y (1989). Exploration of natural enemies of *Thrips palmi* in Southeast Asia. Institute of Biological Control, Faculty of Agriculture, Kyushu University, Fukuoka, pp. 58.

Ho CC (1991). Life history of *Polyphagotarsonemus latus* feeding on lemon, tea and pepper. *J. Agric. Res.*, China. **40(4)**: 439–44.

Humber RA (1992). Collection of entomopathogenic fungal cultures : ARSEF Catalog of Strains, 1992. US Department of Agriculture, Agricultural Research Service, ARS-110, pp. 117.

Hussey NW and Scopes NEA (1977). Biological control by augmentation of natural enemies (Ridgway RL and Vinson SB, *eds*). Plennum Press, New York, pp. 349–77.

Jagadeesh RC (2000). Genetics of yield, yield components and fruit quality parameters to chilli (*Capsicum annuum* L.). Ph.D. Thesis, UAS, Dharwad, India.

Jeyarajan S, Natarajan S and Kandasamy OS (1995). Resurgence of yellow mite in semi-dry chilli. *South Indian Hort.* **43(3–4)**: 117–9.

Jones VP and Brown RD (1983). Reproductive responses of the broad mite, *Polyphagotarsonemus latus* to constant temperature-humidity regimes. *Ann. Entomol. Soci. America.* **76(3)**: 466–9.

Kalaiyarasan S, Sathiyanandan VK, Geetha C and Muthuswamy M (2002). Effect of cultural practices on the management of chilli thrips (*Scirtothrips dorsalis*). *South Indian Hort.* **50(4–6)**: 607–12.

Kandasamy C, Mohanasundaram M and Karuppuchamy P (1987). Evaluation of insecticides for the control of yellow mite, *Polyphagotarsonemus latus* (Banks) on chillies. *Madras Agric. J.* **74(8–9)**: 351–5.

Karmakar K (1995). Comparative symptomology of chilli leaf curl disease and biology of tarsonemid mite, *Polyphagotarsonemus latus* (Banks) (Acari : Tarsonemidae). *Ann. Entomol.* **13(2)**: 65–70.

Karmakar K (1997). Effect of micronutrients on the biology of *Polyphagotarsonemus latus* (Banks) (Acari: Tarsonemidae). *Environ. Ecology* **15(3)**: 699–701.

Karmakar R, Singh SB and Kulshrestha G (1995). Persistence and transformation of thiamethoxam, a neonicotinoid insecticide in soil of different agroclimatic zones of India. *Indian J. Public Health* **49(4)**: 227–30.

Karuppuchamy P and and Mohanasundaram M (1987). Bioecology and control of chilli muranai mite, *Polyphagotarsonemus latus* (Tarsonemidae: Acari). Coimbatore. *Indian J. Plant Prot.* **15(1)**: 1–4.

Karuppuchamy P, Balasubramanian G, Sundarababu PC and Gopalan M (1994). A potential predator of chilli mite *Polyphagotarsonemus latus* (Banks). *Madras Agric. J.* **81(10)**: 552–3.

Kavitha J, Kuttalam S and Chadrasekaran S (2006). Evaluation of spiromesifen 240 SC against chilli mite, *Polyphagotarsonemus latus* (Banks). *Ann. Plant Prot.* **14(1)**: 52–55.

Khaire VA and Naik RL (1985). Comparative efficacy of certain newer pesticides against the sucking pest complex on chilli. *South Indian Hort.* **33(6)**: 402–3.

Khalid S, Ahmad I, Shih SL, Tsai WS, Smith J, Green SK (2001). Molecular characterization of tomato and chili leaf curl begomoviruses from Pakistan. *Plant Protection Bulletin* [TW]. **43(4)**: 247–8.

Kohno K and Hirose Y (1997). The stilt bug *Yemma exilis* as a predator of *Aphis gossypii* and *Thrips palmi* on eggplant. *Appl. Entomol. Zoology* **32**: 406–9.

Kumar NKK (1995). Yield loss in chilli and sweet paper due to *Scirtothrips dorsalis* Hood (Thysanoptera: Thripidae). *Pest. Mngt. Hort. Ecosys.* **1(5)**: 61–69.

Lima ML, da P Melo, Filho P de A, CafeFilho AC, de A Melo Filho and Ciencia P (2003). Colonization by mites in hot and sweet pepper genotypes in greenhouse. *Rural.* **33(6)**: 1157–9.

Linnamaki M, Hulshof J and Vanninen I (1998). Biology and prospects for enhancing biocontrol of the western flower thrips *Frankiniella occidentalis* in cut roses. Proc. of the Brighton crop protection conference: Pests and diseases, **1**: 187–92.

Liu YQ, Lu SC and Wu WW (1993). Application of fuzzy function in forecasting the occurrence of the yellow mite (*Polyphagotarsonemus latus* Bank). *J. Southwest Agric. Univ.* **15(1)**: 24–26.

Loomans AJM and Vierbergen G (1999). *Franklinothrips*: perspectives for glasshouse pest control. *IOBC / WPRS working group on integrated control in glasshouse*. **22(1)**: 157–60.

Loomans AJM, Murai T and Green ID (1997). Interactions with hymenopterous parasitoids and parasitic nematodes. *In*: Lewis T. (*ed.*) Thrips as crops pests. CAB International, Wallingford, UK, pp. 355–97.

Mallapur CP, Kubsad VS and Hulihalli UK (2001). Effect of ethion on mites and thrips causing leaf curl in chilli. *Karnataka J. Agric. Sci.* **14(3)**: 668–70.

Mandal SMA and Beura SK (2003). Chemical control of leaf curl anthracnose and ripe fruit rot in chilli. *Indian J. Plant Prot.* **31(1)**: 137–8.

Manjunatha M (2001). Survey of yellow mite and thrips on chilli in North Karnataka. *Insect Environ.* **6(4)**: 178.

Manjunatha M, Hanchinal SG and Kulkarni SV (2001a). Interaction between *Amblyseius ovalis* and *P. latus* and efficacy of *A. ovalis* on chilli mite and thrips. *Karnataka J. Agric. Sci.* **14(2)**: 506–9.

Manjunatha M, Mallapur CP, Hanchinal SG and Kulkarni SV (2000). Evaluation of imidacloprid 75 WS with recommended chemical on chilli thrips and mite. *Karnataka J. Agric. Sci.* **13(4)**: 993–5.

Manjunatha M, Mallapur CP, Prabhu ST, Kulkarni SV and Hanchinal SC (2001b). Efficacy of few chemicals against important chilli pests. *Karnataka J. Agric. Sci.* **14(2)**: 474–8.

Martin NA (1991). Scanning electron micrographs and notes on broad mite *Polyphagotarsonemus latus* (Banks), (Acari: Tarsonemidae). *New Zealand J. Zool.* **18(3)**: 353–6.

Mishra HP (2003). Efficacy of newer insecticides in chilli leaf curl management. *Indian J. Agric. Sci.* **73(6)**: 358–60.

Moghe PG (1997). Investigations into the causes of churda murda (malformation) disease of chilli in Vidarbha. Akola, Maharashtra. *Curr. Sci.* **46(23)**: 831–2.

Mote UN (1976). Seasonal fluctuation in population and chemical control of chilli mites (*Polyohagotarsonemus latus* Banks). *Veg. Sci.* **3(1)**: 54–60.

Mound LA (1997). Biological diversity. *In.* Lewis T. (*ed.*) Thrips as crop pests. CAB International, Wallingfort, U.K., pp. 197–15.

Nagai K (1993). Studies on integrated pest management of *Thrips palmi*. The Special Bulletin of the Okayma Prefectural Agricultural Experiment Station **82**: 1–55.

Nakashima Y and Hirose Y (1997). Winter reproduction and photoperiodic effects on diapause induction of *Orius tantillus*, a predator of *Thrips palmi*. *Appl. Entomol. Zool.* **32**: 223–9.

Nandihalli BS and Thontadarya TS (1986). Efficacy of different insecticides-cum-acaricides in the control of chilli (*Capsicum annuum* L.) leaf-curl. *Mysore J. Agric. Sci.* **20(2)**: 122–6.

Natarajan K (1988). Transport of yellow mite *Polyphagotarsonemus latus* by cotton whitefly. *Curr. Sci.* **57(20)**: 1142–3.

Nawal Agatti CM, Chetti MB and Hiremath SM (1999). Biochemical basis of murda complex resistance in chilli (*C. annuum* L.) genotypes. *South Indian Hort.* **47(1–6)**: 310–2.

Ningappa MS (1972). Studies on the role of *Scirtothrips dorsalis* (Hood) and *Polyphagotarsonemus latus* (Banks) in causing chilli leaf-curl and their control. Univ. Agric. Sci. Bangalore.

Pandit NC (1996). Technique for estimation of yellow mite *Polyphagotarsonemus latus* Banks. *Environ. Ecol.* **14(4)**: 981–2.

Patel CB, Shah AH and Rai AB (1993). Problems of Mite Pests of Crop in Gujarat and their Management. Tech. Bull. No. 1: pp. 1–24. Gujarat Agric. Univ. Navsari.

Patnaik NC, Behra PK, Dash AN and Ghode MK (1985). Efficacy of insecticides against the chilli thrips, *Scirtothrips dorsalis* Hood. (Thysanoptrea : Thripidae). *Plant Protection Bulletin* **37(2)**: 1–2.

Pena JE (1992). Predator-prey interactions between *Typhlodromalus peregrinus* and *Polyphagotarsonemus latus*: effects of alternative prey and other food resources. *Entomologist*. **75(2)**: 241–8.

Pena JE and Osborne L (1996). Biological control of *Polyphagotarsonemus latus* in greenhouses and field trials using introductions of predacious mites (Phytoseiidae). *Entomophaga*. **41(2)**: 279–85.

Pena JE, Osborne LS and Duncan RE (1996). Potential of fungi as biocontrol agents of *Polyphagotarsonemus latus*. *Entomophaga*. **41(1)**: 27–36.

Petitt IL (1992). Biological control in the integrated pest management programme at the land, EPCOT center. *Bulletin OILB/SCOP* **16(2)**: 129–32.

Quaries W (2001). Directory of least-toxic pest control products. The IPM Practitioner 23 (11/12): 19.

Radke SG and Aherkar SK (1987). Studies on the efficacy of some pyrethroids for control of the pests of chilli. *PKV Res. J*. **11(2)**: 156–9.

Rai AB and Solanki VY (2002). Effectiveness of commercial of *Bacillus thuringiensis* and azadiractin based product, their integration with conventional chemical against sucking pests of chilli (*Capsicum annuum* L.). *Shashpa* **9(2)**: 169–73.

Rai AB and Solanki VY (2003). The rate of natural increase of tarsonemid mite *Polyphagotarsonemus latus* Banks (Acari: Tarsonemidae) when reared on chilli leaf (*Capsicum annuum* L.). *Shashpa* **10(1)**: 13–18.

Raisa Chyzik, Glazer I and Klein M (1996). Virulence and efficacy of different entomopathogenic nematode species against western flower thrips (*Frankliniella occidentalis*). *Phytoparasitica* **24(2)**: 103–10.

Rajasri M, Reddy GPV, Krishnamurthy MM and Prasad VD (1991). Bioefficacy of certain newer insecticides including neem products against chilli pest complex. *Indian Cocoa, Arecanut Spices J*. **15(2)**: 42–44.

Rajita H (1986). Predation by *Amblyseius* spp. and *Orius* spp. on *Thrips palmi*. *Appl. Entomol. Zool.* **21**: 482–4.

Raju Ram V and Ramamurthy R (2001). Effect of irrigation, nitrogen and potassium on mite incidence and yield of chilli. *Ann. Plant Prot. Sci*. **9(1)**: 127–9.

Raju Ram V and Ramamurthy R (2001). Influence of morphological characters on the incidence of chilli thrips *Scirtothrips dorsalis* and mite *Polyphagotarsonemus latus*. *Ann. Plant Prot. Sci*. **9(1)**: 54–57.

Ram S, Patnaik NC, Mohapatra AKB, Sahoo S and Samal KC (1997). Reaction of some chilli varieties to yellow mites *Polyphagotarsonemus latus* banks under eastern ghat high land zone of Orissa. *Environ. Ecol*. **15(1)**: 199–201.

Ram S, Patnaik NC, Mohapatra AKB, Shahoo S and Samel (1996). Effects of fertility levels and chilli varieties on the yield of green chillies and incidence of chilli thrips *Scirtithrip dorsalis* (Hood) in eastern ghat high land zone of Orissa. *Environ. Ecol*. **14(3)**: 642–5.

Rao DM and Ahmed K (1985). Evaluation of certain insecticides for the control of the pest complex on chilli (*Capsicum annuum* L.) in Andhra Pradesh. *Pesticides*. **19(2)**: 41–44.

Rao M and Ahmed K (1986). Control of chilli thrips. *Indian Farming* **36**: 7–25.

Reddly DNR and Jagadish A (1980). Insecticidal control of chilli thrips, *Scirtothrips dorsalis* Hood (Thysanoptera : Thripidae). *Pesticide* **4(10)**: 22–23.

Rijn PCJ van, Houten YM van and Sabelis MW (1999). Pollen improves thrips control with predatory mites. IOBC/WPRS working group on integrated control in glasshouses. **22(1)**: 209–12.

Rodriguez H and Ramos M (2003). Biology of *Amblyseius largoensis* (Muma) (Acari: Phytoseiidae) on *Polyphagotarsonemus latus* (Banks) reared in different substrates. *Revista de Proteccion Vegetal*. **18(1)**: 58–61.

Sabelis MW and Rijin PCJ van (1997). Predation by insects and mites. *In*. Lewis T. (*ed*.). Thrips as crop pests. CAB International, Wallingfort, U.K., pp. 259–353.

Saito T (1991). A field trial on an entomopathogenic fungus, *Beauvaria bassiana* for the control of *Thrips palmi*. *Japanese J. Appl. Entomol. Zool.* **35**: 80–81.

Saito T, Kubota S and Shimazu M (1989). A first record of the entomopathogenic fungus, *Neozygites parvispora* on *Thrips palmi* in Japan. *Appl. Entomol. Zool.* **24**: 233–5.

Sanap MM and Nawale RN (1985). Reaction of chilli cultivars to thrips and mites. *Maharashtra Agric. Univ. J.* **10(3)**: 352–3.

Sanap MM and Nawale RN (1986). Relative efficacy of modern synthetic pesticides for the control of mites (*Hemitarsonemus latus* Banks) on chilli (*C. annuum* Linn.). *Pesticides* **20(1)**: 31–33.

Sanap MN and Nawale RN (1987). Chemical control of chilli thrips, *Scirtothrips dorsalis* Hood (Thysanoptera: Thripidae). *Veg. Sci.* **14(2)**: 195–9.

Sarkar PK, Sarkar H, Sarkar MA and Somachoudhary AK (2005). Yellow mite, *Polyphagotarsonemus latus* (Banks): a menace in chilli cultivation and its management options using biorational acaricides. *Indian J. Plant Prot.* **33(2)**: 294–6.

Saumya George and Giraddi RS (2007). Management of chilli (*Capsicum annuum* L.) thrips and mites using organics. *Karnataka J. Agric. Sci.* **20(3)**: 537–40.

Scott Brown AS, Simmonds MSJ and Blaney WM (1999). Influence of species of host plants on the predation of thrips by *Neoselus cucumeris*, *Iphiseius degenerans* and *Orius laevigatus*. *Entomologia Experimentalis Applicata* **92**: 283–8.

Seal DR, Coimpeelik M, Richards ML and Klassen W (2005). Distribution of the chilli thrips, *Scirtothrips dorsalis* Hood. (Thysanoptera : Thripidae) in thin pepper field on St. Vincent. *Florida Entomol*. **89(3)**: 311–20.

Smitha MS and Giraddi RS (2003). Increased efficacy of dicofol in combination with different plant oils against yellow mite, *P. latus* (Banks) on chilli. Pest Management *Hort. Ecosyst*. **9(1)**: 67–70.

Solanki VY (1999). Ecology and control of sucking pest of chilli with special reference to *Polyphagotarsonemus latus* Banks (Acari: Torsonemidae). M.Sc. Thesis, Gujarat Agriculture University, Navsari, pp. 174.

Solanky VY and Rai AB (2006). Histological changes associated with chilli leaf curl. *Veg. Sci*. **33(2)**: 209–11.

Srinivasan MR, Natarajan N and Palaniswamy S (2003). Evaluation of buprofezin 25 SC and fenpyroximate 5 SC against chilli mite, *Polyphagotarsonemus latus* (Banks). *Indian J. Plant Prot*. **31(2)**: 116–7.

Srinivasulu P, Naidu VG and Rao NV (2000). Biology and bionomics of chilli mite, *P. latus* (Banks) on chilli. *J. Appl. Zool. Res*. **13(1)**: 19–21.

Srinivasulu P, Naidu VG and Rao NV (2002). Seasonal occurrence of chilli mite, *Polyphagotarsonemus latus* (Banks) with reference to biotic and abiotic factors. *J. Appl. Zool. Res*. **13(2/3)**: 142–4.

Srinivasulu P, Naidu VG and Rao NV (2002a). Evaluation of different pesticides for the control of yellow mite, *Polyphagotarsonemus latus* (Banks) on chilli. *J. Appl. Zool. Res*. **13(1)**: 71–72.

Sureshkumar N and Ananthakrishnan TN (1984). Predatory pre interactions with reference to *Orius maxidentex* and *Carayonocoris indicus*. Proc. Indian National Science Academy **50**: 139–45.

Tatagar MH, Prabhu ST and Jagadeesha RC (2001). Screening chilli genotypes for resistance to thrips, *Scirtothrips dorsalis* Hood and mite, *Polyphagotarsonemus latus* (Banks). *Pest Management Hort. Ecosystems* **7(2)**: 113–6.

Thakre SN, Tikar D and Borle MN (1979). Efficacy of some insecticides in the control of thrips on chilli. *Pesticides* **13(6)**: 31–32.

Ukey SP, Naitam NR and Patil MJ (1999). Determination of economic threshold level of mites on chilli crop. *J. Soil and Crop* **9(2)**: 268–70.

Uraisakul K and Kanok Uraisakul (2003). *Annona* seed extract and some herb extracts on chilli yield and control broad mite (*Polyphagotarsonemus latus* (Bank)) and some key pests in chilli. *In*: Proc. 41st Kasetsart Univ. Annual Conf., 3–7 Feb 2003, Kasetsart Univ. Bangkok, pp. 354–61.

Venzon M, Rosado M-da-C Pinto CMF, Duarte v-da-S, Euzebio DE and Pallini A (2006). Potential of alternative pesticides for control of broad mite on chilli pepper "Malagueta". *Horticultura Brasileira* **24(2)**: 224–7.

Vesterguard (1995). Pathogenicity of the Hympmycete fungai *Verticillum lecanii* and *Metarihizium amsopliae* to the western flower thrips *Franklinielle occidaelate*. *Biocontrol Sci. Tech*. **5(2)**: 185–92.

Vos JGM and Nurtika N (1995). Transplant production techniques in integrated crop management of hot pepper (*Capsicum* spp.) under tropical low land conditions. *Crop Prot*. **14(6)**: 453–9.

Wada T (1999). Development of *Cucumeris* and its future prospect. *Agrochemicals* Japan, **73**: 17–19.

Walnuz AR and Pawar SA (2000). Evaluation of fenazaquin, a new acaricide for the control of *Hemitarsonemus latus* on chilli. *Trade of Agrochemicals and Cultivars*. **21**: 1–2.

Wang CL (1995). Predatory capacity of *Campylomma chinensis* and *Orius sauteri* on *Thrips palmi*. *In*: Parker BL, Skinner M and Lewis T (*eds*.) Thrips biology and management. Plenum Press, New York, pp. 259–62.

Waterhouse DF and Norris KR (1987). *Polyphagotarsonemus latus* (Banks). *In*: Biological Control Pacific Prospects. Inkata Press: Melbourne. 454 pages.

Weintraub PG, Kleitman S, Mori R, Shapira N and Palevsky E (2003). Control of the broad mite (*P. latus*) on organic greenhouse sweet peppers with the predatory mite, *N. cucumeris*. *Biol. Contr*. **27(3)**: 300–9.

Xie M, Cheng HK and Zhao JH (1992). Use of liuyangmycin to control yellow mite, *Polyphagotarsonemus latus* (Acari: Tarsonemidae) infesting green pepper. *Chinese J. Biol. Contr*. **8(1)**: 29–32.

Yano E (1999). Recent advances in the study of biocontrol with indigenous natural enemies in Japan. Bulletin IOBC/WPRS working group on integrated control in glasshouse **22(1)**: 291–4.

Yasunaga T (1995). A new species of the genus *Wooastoniella*, predator of *Thrips palmi* in aubergine gardens of Thailand. *Appl. Entomol. Zool.* **30**: 203–5.

Yasunaga T and Miyamoto S (1999). Three anthocorid species, predators of *Thrips palmi* in aubergine gardens of Thailand. *Appl. Entomol. Zool*. **28**: 227–32.

Young GR and Zang L (2001). IPM of melon thrips, *Thrips palmi*, on eggplant in the top end of the Northern Territory. Proc. of the Sixth Workshop for Tropical Agricultural Entomology, Darwin, Australia, Technical Bulletin (288), pp. 101–11.

Zhi JunRui and Zhi JR (2002). The reproductive life table of *Polyphagotarsonemus latus*. *Entomological Knowledge*. **39(3)**: 199–202.

# 11

# Diseases of Chilli and their Management: Current Status and Future Directions

M. LOGANATHAN, S. SAHA, V. VENKATARAVANAPPA, R.K. SARITHA, T.K. BAG AND A.B. RAI

## 1. INTRODUCTION

Chilli belonging to genus *Capsicum* and family solanaceae is an important spice vegetable crop grown worldwide and consumed as fresh or after process. The genus *Capsicum* has been reported with five cultivated species *viz. C. annuum, C. chinense, C. baccatum, C. frutescens* and *C. pubescens*. These botanical classifications will always help the researchers to study the crop and its associated factors in detail as many species of phytopathogenic organisms are named based on the botanical name of the host (for instance in *Colletotrichum capsici*, a fungal pathogenic organism the species name *capsici* is derived from *Capsicum*). Chilli crop is known for its high nutritious values as rich source of vitamins A and C, potassium, folic acid and vitamin E, and also free from cholesterols (Osuna-Garcia *et al.*, 1998; Perez-Galvez *et al.*, 2003; Marin *et al.*, 2004). Production of such an important crop is highly affected by many biotic factors especially diseases (Table 11.1). Among them, the diseases of anthracnose, wilts and viruses are considered to be the important yield limiting factors of chilli cultivation in India.

**Table 11.1.** List of major diseases of chilli

| Name of the disease | Causal organism | References |
|---|---|---|
| Die back/anthracnose | *C. capsici* (Syd.) Butler and Bisby, *C. acutatum* (Simmonds), *C. gloeosporioides* (Penz.) Penz. and Sacc., *C. coccodes* (Wallr.) S. Hughes | von Arx and Die Arten der Gattung, 1957; Simmonds, 1965; Verma, 1973; Sutton, 1980; Hadden and Black, 1987 |
| *Phytophthora* root rot | *Phytophthora capsici* | Leonian, 1922 |

Division of Crop Protection, Indian Institute of Vegetable Research, Varanasi-221 305, Uttar Pradesh (India).

**Table 11.1.** *Contd.*

| Name of the disease | Causal organism | References |
|---|---|---|
| *Rhizoctonia* root rot | *Rhizoctonia solani* | Muhyi and Bosland, 1992 |
| *Fusarial* wilt | *Fusarium oxysporum* f.sp. *capsici* | Rivelli, 1989 |
| *Verticillium* wilt | *Verticillium dahliae* and *V. albo-atrum* | Sanogo, 2003 |
| Bacterial wilt | *Ralstonia solanacearum* | Yabnuchi *et al.*, 1995 |
| *Chilli veinal mottle virus* | ChVMV | Ong *et al.*, 1979 |
| *Cucumber mosaic virus* | CMV | Ong *et al.*, 1979 |
| *Pepper veinal mottle virus* | PVMV | Brunt and Kenten, 1971 |

## 2. FUNGAL DISEASES

### 2.1. Genus *Colletotrichum* and Crop Loss

Anthracnose pathogen infecting chilli has been reported with five different species: *C. capsici, C. acutatum, C. gloeosporioides, C. coccodes* and *C. dematium* (Kim *et al.*, 1989; Hong and Hwang, 1998; Gopinath *et al.*, 2006). Among the five species, *C. gloeosporioides* and *C. capsici* are the most predominant species infecting chilli crops especially the former which was reported from Taiwan and the latter one from India (Manandhar *et al.*, 1995; Gopinath *et al.*, 2006). *C. gloeosporioides* is reported with a wide host range. Apart from chilli it also infect fruit crops which comprises of avocado, guava, apple, almond, mango and strawberry, and plantation crop, arabica coffee (Agwana *et al.*, 1997; Freeman *et al.*, 1998; Martinez-Culebras *et al.*, 2000; Sanders and Korsten, 2003; Xiao *et al.*, 2004; Amusa *et al.*, 2005), whereas *C. capsici* is reported only in the crop genus *Capsicum* spp. (Taylor, 2007). This fungus can survive both internally and externally on the seeds as acervuli or micro-sclerotia or mycelia or stomata (Manandhar *et al.*, 1995; Pernezny *et al.*, 2003). It can over winter on plant debries and rotten fruits of other solanaceous or leguminous crops (Pring *et al.*, 1995).

In chilli cultivation, anthracnose is reported as one of the most destructive diseases worldwide including Asian countries (Sariah, 1989; Shin *et al.*, 2000; Oanh *et al.*, 2004; Sharma *et al.*, 2005; Taylor, 2007). Poonpolgul and Kumphai (2007) reported 80 per cent crop loss especially due to *Colletotrichum* infection in Thailand. Similarly, more than 30–50 per cent losses were noticed in United States of America (Haward *et al.*, 1992; Wilson *et al.*, 1992), and in tropical and subtropical regions

(Heggens 1930; Smith and Crasson 1959; Tindall, 1983). In severe cases about 95 per cent yield loss have been reported especially in Nigeria (Amusa *et al.*, 2004). Apart from this to post harvest losses due to fruit decay have also been reported due chilli anthracnose (Hadden and Black, 1989; Bosland and Votava, 2003). Damage symptoms noted on the infected plants are damping off, die-back of shoots, spots on leaves and fruit rots (Adikaram *et al.*, 1983; Mah, 1985).

### 2.1.1. *Epidemiology and Disease Development*

Disease development mainly depends on the interaction of factors, host, pathogen and environment. The environmental factors like duration and intensity of rainfall, crop geometry, pathogen inoculum load, leaf surface wetness and its duration, humidity, light, competitive microbiota and temperature influence the disease development (Royle and Butler, 1986; Dodd *et al.*, 1992; Roberts *et al.*, 2001). In general, optimum temperature around 27°C and 80 percent relative humidity (RH) are required for fast development of anthracnose disease (Roberts *et al.*, 2001). Many reports are available on hemibiotrophs or facultative biotrophs infection of *Colletotrichum* spp. (Kim *et al.*, 2004) but no clear cut studies have been made on *Colletotrichum* infecting chilli. However, infection process of *C. gloeosporioides* in susceptible chilli (*C. annuum* cv. Jejujaerae) showed that condensation of epidermal cytoplasm, increased number of small vacuoles, sub epidermal cell destruction, inter and intra cellular colonization which clearly indicates that infection process is by necrotrophism (Kim *et al.*, 2004).

### 2.1.2. *Detection and Diagnosis*

Effective management of anthracnose diseases required exact identification of pathogenic organisms and it can be done through morphological basis such as size and shape of conidia, appressoria, setae, teleomorph state, colony colour, growth rate and texture (von Arx and Die Arten der Gattung, 1957; Smith and Black, 1990). But these kind of morphological characterization may not be helpful always because there are several strains, which have a similar morphology infecting different host plants (Than *et al.*, 2008). But selection of appropriate management practice needs exact identification of pathogen at strain level which may not be possible in traditional method of identification. Alternatively, through advances in biotechnology, it is possible to identify anthracnose or its groups through molecular approaches like restriction fragment length polymorphisms (RFLP), random amplified polymorphic DNA (RAPD) and internal transcribed spacer (ITS) (Table 11.2). Many other molecular approaches have also been used to identify anthracnose of chilli.

**Table 11.2.** List of molecular techniques used to study the chilli anthracnose pathogen

| Pathogen | Technique/molecular tool | References |
|---|---|---|
| *Colletotrichum* species | ITS | Sreenivasaprasad *et al.*, 1994; 1996; Moriwaki *et al.*, 2002; Photita *et al.*, 2005 |
| *C. acutatum* | Protein coding genes of partial β-tubulin gene | Sreenivasaprasad and Talhinhas, 2005 |
| | Partial beta-tubulin 2 (exons 3-6) | Hong and Kim, 2007 |
| *C. acutatum* | Introns of glutamine synthase and glyceraldehyde-3-phosphate dehydrogenase genes | Guerber *et al.*, 2003 |
| *C. graminicola*, *C. gloeosporioides* and *C. acutatum* | MAT1-2 mating type | Du *et al.*, 2005 |
| *Colletotrichum* species | RFLP of ITS | Sheu *et al.*, 2007 |
| *C. capsici* and *C. gloeosporioides* | RAPD | Ratanacherdchai *et al.*, 2007 |

### 2.1.3. *Management of Anthracnose of Chilli*

#### 2.1.3.1. *Cultural Method*

To avoid or minimize the infection caused by anthracnose disease the cultural practices like use of pathogen-free seeds, elimination of weeds, crop rotation for 2–3 years with non-host crops, providing good drainage, removing infected plant debris, growing short duration crops, avoiding injury to fruits during intercultural operation because wounds pave the way for pathogen entry and deep ploughing need to be followed (Roberts *et al.*, 2001; Agrios, 2005).

#### 2.1.3.2. *Chemical Fungicides*

In practice, many chemical forms of fungicides are used to control the anthracnose of chilli. Fungicides, maneb, captafol and carbamate have been reported as effective chemicals against chilli anthracnose (Intanoo and Chamswarng, 2007) while manganese ethylenebisdithiocarbamate (Maneb) is a widely used (Smith, 2000). New generation fungicides, strobilurin, azoxystrobin, trifloxystrobin and pyraclostrobin have also been registered against the disease (Alexander and Waldenmaier, 2002; Lewis and Miller, 2003). Even though many fungicides are currently used against the dreadful disease for immediate effects, numerous negative effects on human beings and environment have been addressed (Voorrips *et al.*, 2004).

### 2.1.3.3. *Resistant Cultivars/Lines*

Among the management practices, use of resistant varieties was found to be the superior one since it is durable and eliminate adverse effects of chemicals. Though there is no strong evidence of high level resistance against anthracnose in pepper (Park, 2007), hitherto a wide range of studies have been undertaken to identify the genetic resistance against the disease from different parts of the world (Table 11.3) (Kim *et al*., 1987; Park *et al*., 1987; Hong and Hwang, 1998; AVRDC, 1999; Yoon and Park, 2001; Voorrips *et al*., 2004).

**Table 11.3.** Resistant cultivars/lines identified against anthracnose disease in chilli

| Cultivar/line | Reaction | References |
|---|---|---|
| Lines developed out of interspecific cross between Thai susceptible *C. annuum* cv. 'Bangchang' and anthracnose resistant *C. chinense* 'CM 021' | Resistant | Mongkolporn *et al*., 2004 |
| TC 6903, Taiwan 2, BS 35, Sikkim chilli and CM 334 | Resistant | IIVR 2007 |

### 2.1.3.4. *Plant Products*

Many plant based principles have been developed against viral diseases but only few citations were made against anthracnose of chilli. Different kinds of plant based products like stem and root bark extracts of *Azadirachta indica* and *Vernonia amygdalina* (Del.), stem and bark extract of *Cochlospermum planchonii* and sweetflag (*Acorus calamus* L.), *Ocimum sanctum* leaf extract, and oil obtained from palmorosa (*Cymbopogon martinii*) and neem (*Azadirachia indica*) were reported to have high level fungicidal effect against the *Colletotrichum* fungus (Jeyalakshmi and Seetharaman, 1998; Korpraditskul *et al*., 1999; Charigkapakorn, 2000; Nduagu *et al*., 2008).

### 2.1.3.5. *Biocontrol Agents*

Though the management of anthracnose of chilli through biocontrol agents has not been given much attention, different kinds of microbial agents were identified and used against anthracnose of chilli (Table 11.4).

**Table 11.4.** List of bio-fungicides used against anthracnose disease of chilli

| Biocontrol agent | References |
|---|---|
| *Pseudomonas fluorescens* | Jeger and Jeffries, 1988 |
| Bacterial formulation | Intanoo *et al*., 2007 |
| *Trichoderma* spp. | Boonratkwang *et al*., 2007 |
| *Bacillus subtilis* and *Candida oleophila* | Wharton and Diéguez-Uribeondo, 2004 |

## 3. BACTERIAL DISEASES

Three major bacterial diseases are wilt (*Ralstonia solanacearum*), leaf spot (*Xanthomonas axonopodis* pv. *vesicatoria*) and soft rot (*Erwinia carotovora* sub species *carotovora*).

### 3.1. Wilt (*R. solanacearum*)

Wilting begins with the youngest leaves during warm or hot weather conditions during the day time. The plants may recover temporarily in the evening under cooler temperatures and eventually permanent wilt occurs. The wilted leaves maintain their green color and do not fall as the disease develops. The roots and lower part of the stem have browning of the water conducting portion (*i.e.* vascular system) of the plant. Diseased roots or stems that are cut and placed in a small container of water shows steady yellowish or gray bacterial ooze coming from the cut end. Such oozing is not found with *Fusarium*-infected plants (Cerkauskas, 2004a). Wilt is more severe in high temperatures (30–35°C) and heavy soils with high moisture content in low-lying areas.

### 3.2. Leaf Spot (*X. axonopodis* pv. *vesicatoria*)

The symptoms appear on the leaves as the sunken lesions on the top surface and slightly raised on the lower surface. The lesions are brown, circular to irregular, water-soaked and when these coalesces cause necrotic areas, thereby giving the plants a blighted appearance. Diseased stems and petioles have elliptical, raised lesions. Flower infection results in severe blossom drop. On young fruit, lesions are small, circular, green spots that become brown and have a raised, coarse, wart-like surface (Bashan and Okon, 1983). The environmental factors *viz.*, temperatures (24–30°C), leaf wetness periods (more than 24 hrs), dew, fog, rain, and overhead irrigation influence the disease development (Cerkauskas, 2004b).

### 3.3. Soft Rot (*E. carotovora* sub sp. *carotovora*)

Initial symptoms often appear on leaves, which show dark veinal tissue followed by leaf chlorosis and necrosis. The pith and vascular system of stems may show internal dark brown discoloration. Later dry, dark brown or black stem cankers develop, often resulting in the breakage of branches. The affected plants wilt and die. The fleshy fruit peduncle is highly susceptible and is frequently the initial point of infection. Both ripe and green fruit may be affected and the lesions on the fruit are light to dark-colored, water-soaked, and somewhat sunken. In advanced stages, bacterial ooze may develop from affected areas, and secondary organisms follow, often invading the rotted tissue. The affected fruit hang from the plant like a water-filled bag (Cerkauskas, 2004c). Epidemiological factors like temperatures of 25°C to 30°C and RH of 95 percent are more conducive for the disease development.

## 3.4. Management of Bacterial Diseases

### 3.4.1. *Bleach Treatment*

Freshly harvested seeds are treated with acetic acid and clorox. Soak seeds in 13 ml of acetic acid per 1000 ml of water. Shake the seeds in the solution for four hours and rinse with water three times. Then soak seeds in 12.5 ml clorox bleach per 1000 ml of water for 5 minutes and finally rinse under running water for 15 minutes. This treatment may decrease the germination rate of some varieties (Berke, 2000).

### 3.4.2. *Hot Water Treatment*

Seed treatment with hot water is effective in reducing bacterial populations on the surface and inside the seeds. However, seed germination may be jeopardized by heat treatment more than by properly applied sodium hypochlorite treatment (Sanogo and Clary, 2004).

### 3.4.3. *Cultural Management*

Avoid cull piles near seedbeds or production fields. Remove chilli debris, burn it, or chop it and bury it immediately after harvest, and incorporate into the soil to assist in rapid decomposition of diseased pepper debris. After working in infested areas, tools should be decontaminated. Set transplants into soil with recommended rates of N and K, but at the high end of the scale. Losses from bacterial spot are greatest when chilli become deficient in N or K, and the disease can be minimized by maintaining high fertility, being careful not to get the plants into an overly vegetative state or fruit set will be seriously reduced. Strict water management practices by avoiding overhead watering in local transplant production to reduce spread of the disease (Goode and Sasser, 1980). Flood or furrow irrigation is used wherever possible. If overhead irrigation is necessary, it should begin early in the day so that the foliage can dry before the evening. Rain shelters to reduce water splash may reduce disease severity during periods of high rainfall. Use crop rotations of 2 to 3 years, excluding eggplant, tomatoes and tobacco from the total rotation. Control broadleaf weeds during the rotation and around the field borders, and volunteer chilli in affected fields.

### 3.4.4. *Resistance Breeding*

Resistant cultivars are becoming available but may not be resistant to all strains of the pathogen.(Pernezny and Collins, 1997). Cultivars ECW 10R,

ECW 20R and ECW 30R contain genes *Bs1*, *Bs2* and *Bs3*, respectively, which have been successfully used in resistance breeding programs (Hibberd *et al*., 1987; Kousik and Ritchie 1995, 1996). Follow other control options diligently to avoid bacterial spot from susceptible pepper varieties, which will continue to be released as well as to manage strains of the bacterium that may not be controlled by resistant varieties (Scott and Jones, 1989).

### 3.4.5. *Chemical Control*

Application of fixed copper fungicides copper oxychloride @ 4 gm/l or copper hydroxide @ 2.5 gm/l (Cox, 1982; Jones *et al*., 1991) manages the disease and reduce the losses. The relatively low cost and low toxicity to mammals of fixed copper compounds give them an advantage over other chemicals for control of foliar bacterial diseases. The efficacy of fixed copper compounds in control of bacterial diseases, however, has been variable (Egli and Sturm, 1980). Adding mancozeb to a fixed copper collected compound improves control (Conover and Averre, 1963). In a recent review, suggestions were made that copper hardens the plant surface, resulting in less penetration by bacterial pathogens (Egli and Sturm, 1980).

### 3.4.6. *Biological Control*

Products containing microorganisms can be used to enhance plant growth and reduce the negative effects of diseases. These products may contain plant-growth-promoting rhizobacteria (PGPR) or biological agents (Chen *et al*., 2000; Compant *et al*., 2005). PGPR are bacteria that live on plant roots or in soil around plant roots without causing any harm to plants. Instead, these bacteria enhance plant growth and also may induce resistance to plant pathogens. The resistance triggered in plants by PGPR bacteria is known as induced systemic resistance (ISR) (Marie *et al*., 1997; de Meyer and Hofte, 1997). Biological agents are microorganisms (fungi, bacteria or viruses) that are able to compete with or antagonize plant pathogens. As such, biological agents can reduce the negative effects of diseases on crops. Several products containing microorganisms are sold for biological control of diseases (Hofte *et al*., 1991; Jetiyanon *et al*., 2007). Examples are Serenade and Sonata, both of which contain beneficial bacteria (Polyanskaya *et al*., 2002). Another example is Agriphage, which contains bacteriophages that infect bacteria.

## 4. VIRAL DISEASES

About 45 viruses have been reported to infect chilli world wide (Bidari and Reddy, 1990; Green and Kim, 1991). Among these, 22 were found to occur naturally and remaining infect only through artificial inoculation. Out of these, *Potato virus* Y (PVY), *Tobacco etch virus* (TEV), *Pepper mottle virus* (PeMV), *Pepper mild mottle virus*, *Pepper veinal mottle virus* (PVMV), *Chilli veinal mottle virus* (ChVMV), *Ethiopian pepper mottle virus* (EpMV), *Pepper vein banding virus* (PVBV), *Pepper severe mosaic virus* (PeSMV) and *Peru tomato virus* (PTV) belong to potyvirus group and causes huge losses in terms of yield as well as in quality (Laird and Dickson, 1964; Zitter, 1972; Fernandez and Fulton, 1980; Nelson *et al.*, 1982; Ladera *et al.*, 1982; Gebre-Selassie *et al.*, 1983; Atiri and Dele, 1985; Agronovsky, 1993; Ravi *et al.*, 1997). CHVMV and PVBV of potyvirus group are known to causes economic losses upto 50 percent when infected at an early stage of the crop (Ong *et al.*, 1980; Ravi *et al.*, 1997). Among the above viruses, the most important and more constraints for production of chilli is posed by ChVMV, *Cucumber mosaic virus* (CMV), PVMV, *Potato virus* Y (PVY), *Capsicum chlorosis virus* (CaCV), *Peanut bud necrosis virus* (PBNV) and leaf curl begomovirus. These are discussed briefly in this chapter.

### 4.1. Non-Persistently Aphid-Transmitted Viruses

#### 4.1.1. ChVMV *genus Potyvirus*

ChVMV, flexuous particles 750 nm long and 12 nm wide was first reported on *Capsicum annuum* in Malaysia by Ong *et al.* (1979) and further it has been reported on chilli in different countries of Asia like Taiwan, Thailand, India, Korea, Indonesia, the Philippines and Tanzania (Anon., 1990a; Anon., 1990b; Anon., 1992; Green and Kim, 1991). The incidence of ChVMV was 42, 26 and 7.8 percent in Taiwan, Korea and Indonesia, respectively (Anon, 1998). Bidari and Reddy (1990) conducted survey during 1979–80 in the commercial chilli growing areas of Karnataka and reported that the average incidence of ChVMV was 53.0 percent. In pepper growing areas of Karnataka state reported with widespread occurrence of pepper mosaic diseases ranging from 15–31 percent and in rainfed conditions it was from 37–58 percent (Ravi *et al.*, 1997). In chilli growing areas of eastern U.P, the incidence ranged from 5–75 percent. The highest incidences upto 50 and 75 percent were recorded from Gorakhpur during the years 1997 and 1998, respectively (Sathya Prakash *et al.*, 2002).

ChVMV infected plant shows symptoms of dark green mottle, reduced leaf size and distortion with fewer and smaller fruits (Ong and Ting, 1977; Ong *et al.*, 1979). Yield reduction of more than 50 percent has been reported when the crop became infected at an early stage (Ong *et al.*, 1980). Leaf mottle and dark green vein banding are the most characteristic symptoms on the younger and smaller leaves. Plants infected when young become stunted and have dark-green streaks on their stems and branches. Most of their flowers drop before fruit formation. A few mottled, distorted fruits may be produced. Such symptoms contribute to significant yield losses.

ChVMV has broad host range which includes *C. annuum* (Delray Bell, Early Wonder, Florida VR-2, Yolo Wonder, Yolo Y), *C. chinense* (Miscucho), *C. frutescens* (Tobasco), *Datura stramonium, D. metel, Lycopersicon esculentum* (TK-70), *Nicandra physaloides, Nicotiana benthamiana, N. clavelandii, N. glutinosa, N. tabacum* (Samsun, Xanthi-NC, Xanthi, White Burley), *N. sylvestris, Petunia hybrida, Sesamum indicum* and *Physalis floridana*, African eggplant (*Solanum aethiopicum*) (Anon, 1990c; Green and Kim, 1991; Siriwong *et al.*, 1995; Krishna Reddy *et al.*, 2004). It is transmitted in a non-persistent manner by several aphid species namely, *Aphis gossypii, A. craccivora, Myzus persicae, A. spiraecola, Toxoptera citricidus, Hysteroneura setarieae, Rhopalosiphum maidis* and no seed transmission was observed (Ong *et al.*, 1979; Green and Kim, 1991; Satya Prakash *et al.*, 2002).

### 4.1.2. CMV genus *Cucumovirus*

Both hot and sweet peppers are highly susceptible to CMV and causes stunting of the plant, resulting in a bushy appearance. Leaves may show a mild, green mottling or mosaic, mottling, "shoestringing" on leaves and the infected fruit become wrinkled, bumpy appearance, mottling, uneven colour and ripening with dark spots. The maximum incidence of CMV in chilli ranged from 50–85 percent in the month of September, May and January (Sain *et al.*, 2005). Biswas *et al.* (2005) reported that the distribution, incidence and severity of chilli mosaic disease syndrome in *Capsicum annuum* in different location in Kalimpong hill region of West Bengal was up to 5–100 percent depending upon the altitude.

CMV is transmitted in a non-persistent manner by several aphid species namely, *Aphis gossypii, A. craccivora, Myzus persicae, A. spiraecola, Toxoptera citricidus, Hysteroneura setarieae* and *Rhopalosiphum maidis*. No seed transmission was observed (Ong *et al.*, 1979; Green and Kim, 1991; Satya Prakash *et al.*, 2002). CMV can also be transmitted mechanically. Weeds are hosts for the virus as well as for the aphid vectors. CMV has broad host range that includes tomato, pepper, cucumber, squash, spinach, melons, beets, petunia, chickweed, celery, mustard, sowthistle, and nightshade and other several crops, ornamentals, and weed species.

### 4.1.3. PVMV genus *Potyvirus*

PVMV occurs mainly in Africa. It was originally reported from Ghana (Brunt and Kenten, 1971). It consists of flexuous particles, mostly unaggregated and unbroken measured 760–800 x12 nm (Brunt and Kenten, 1971). The virus was well characterized by Joseph and Savithri (1999). The maximum incidence of PVMV on chilli crop ranged from 50–85 percent in the month of September, May and January (Sain *et al.*, 2005). In Nepal, the viral incidence on sweet pepper varied from 30–90 percent in different districts (Joshi *et al.*, 1997; Ong *et al.*, 1980). Ravi *et al.* (1997) reported that the virus is known to cause economic losses upto 50 percent when infected at an early stage of the crop. PVMV infected plant shows leaf distortion, vein-banding or vein-chlorosis in pepper and mosaic spots in chilli. The infected fruits are small, mottled, deformed with sunken or raised areas on skin.

The primary hosts of PVMV are hot and sweet pepper, tomato and eggplant (*Solanum melongena*).

In nature, PVMV is transmitted non-persistently by at least five aphid species: *A. gossypii*, *A. craccivora*, *A. spiraecola*, *M. persicae* and *T. citridus*. PVMV can be transmitted by grafting and mechanical inoculation but not by seeds. Handling and touching easily spread the virus from plant to plant.

### 4.1.4. PVY genus *Potyvirus*

PVY consists of non-enveloped flexuous rods with 694 nm length (Laird and Dickson, 1964) or 733 × 13 nm (Prasad Rao, 1976) or 694 × 12.5 nm (George *et al.*, 1993). PVY infected plant shows symptoms of mild to severe leaf mosaic, vein-banding or vein-clearing, interveinal chlorosis, mosaic patterns and chlorotic streaks on leaves, rough skin on fruits and general stunting of the plants. PVY has broad host range, which includes pepper (*Capsicum* spp.), potato and tomato. Jeyarajan and Ramakrishnan (1969) reported that PVY on pepper had a very limited host range. Among 18 hosts used, the virus passed on only to *Glyricidia maculata, N. tabacum* and *N. glutinosa*. It did not infect *C. chinensis, C. chacoense, C. pubescens* and *Chenopodium amaranticolor*.

PVY is transmitted by many aphid species in a non-persistent manner namely *M. persicae*, *A. fabae*, *A. gossypii* and *Macrosiphum euphorbiae* in efficient manner. A helper component is required for vector transmission. Virus contaminated seeds and infected plant debris are major sources of infection. The virus is easily spread from plant to plant by handling and touching.

### 4.1.5. *Epidemiology of Non-Persistently Aphid-Transmitted Viruses*

The incidence of non-persistently aphid-transmitted viruses like ChVMV, CMV, PVMV and PVY on pepper crop is mainly based on the population and activity of vector in a given area. The environmental factors like temperature and relative humidity also play an important role in spreading of disease (Jeyarajan and Ramakrishnan, 1969). Since the disease is transmitted by aphid vector in non-persistent manner the virus multiplication as well as builds-up of the vector population is affected by temperature. Berg (1984) reported natural increase in vector population from 10°C to maximum at 35°C then onwards the vector population declined.

## 4.2. Persistently Thrips-Transmitted Viruses

### 4.2.1. PBNV *and* CaCV

Both PBNV and CaCV belong to genusospovirus of the family *Bunyaviridae* and have more or less similar morphology but different genetic constitutions. Both consist of roughly spherical, enveloped particles, ranging in diameter from 70 to 110 nm. The genome of both viruses consists of three single-stranded RNA segments. CaCV was first identified in Queensland (Persley *et al.*, 2005). CaCV is distinct from TSWV and belongs to the Watermelon silver mottle virus serogroup or serogroup IV which are widespread and damaging in Asia (Thailand, Taiwan and China). The incidence of affected plants commonly exceeds 60 percent (Persley *et al.*, 2005). The detection of CaCV in chilli samples first time in India was reported by Ravi *et al.* (2005) and it is now the most serious disease on tomato and chilli in India.

The PBNV infected plant shows severe symptoms of yellowing (chlorosis), dead (necrotic) spots on leaves or terminal shoots. Fruits show chlorotic spots, red and/or green areas surrounded by yellow halos, concentric rings that may become necrotic. Although the symptoms of CaCV resemble those caused by PBNV, in *Capsicum*, chlorosis or yellowing on leaf margins and between the veins develops on young leaves, which often become narrow and curled, with a strap-like appearance. Older leaves become chlorotic with ring spots and line patterns developing. The fruit on infected plants is small, distorted and often marked with dark spots and scarring over the surface (McMichel *et al.*, 2002; Persley *et al.*, 2005).

At least 250 plant species are susceptible to PBNV (Cho *et al.*, 1989). The crop hosts of CaCV are *Capsicum* (including chilli types), tomato, peanut and several other weed species like, *Ageratum conyzoides*, *Sonchus oleraceus* and *Emilia sonchifolia* (Emilia or purple sowthistle). Both the viruses are transmitted by thrips in persistent manner namely *Frankliniella fusca*, *F. occidentalis*, *F. schultzei*, *Scirtothrips dorsalis* and *Thrips tabaci* (Ullman *et al.*, 1993; Wijkamp and Peters, 1993) and by mechanical inoculation, but not spread through seed or on cutting, pruning and cultivation equipment and do not survive in soil or decaying crop residues.

### 4.2.2. *Epidemiology of Persistently Thrips-Transmitted Viruses*

Persistently thrips-transmitted viruses in chillies like CaCV and PBNV are acquired from infected plants by second instar larvae of thrips during feeding periods of less than 30 minutes. Once acquired by immature thrips, the viruses circulate and multiply within body of the insect and are transmitted to healthy plants as the adult thrips pierce and suck the contents of plant cells. Few days are required to acquire virus by thrips from an infected plant until it is able to transmit the virus to another plant. This allows time for the virus to move and multiply in the insect gut and salivary glands. These viruses are not spread in seed or on cutting, pruning and cultivation equipment, or handling and do not survive in soil or decaying crop residues but it can be spreaded in infected plant parts used for plant propagation such as cuttings and bulbs (Persley *et al.*, 2007). The incidence on chilli crop is mainly based on the population and activity of thrips vector in a given area and environmental factors like temperature and relative humidity. Since the disease is transmitted by thrips vector in persistent manner, the virus multiplication as well as build-up of the vector population is affected by temperature, naturally increase in vector population from 10°C to maximum at 35°C then onwards the population get declined.

### 4.3. Management of Chilli Viral Diseases

#### 4.3.1. *Cultural Management*

The use of resistant or tolerant varieties, proper cultural practices that include intercropping with maize or the use of reflective mulches to reduce the vector population (Ong *et al.*, 1980) were found effective in controlling ChVMV in pepper. Grow the transplants in a net house or cover seedbeds with a 32-mesh or finer mesh net to prevent introduction of aphids. Use yellow sticky traps to monitor and to reduce aphid populations. Avoid touching or handling pepper plants prior to setting them in the field. Remove any diseased seedlings that show symptoms of the disease and place them in a refuse pile away from pepper production fields. Avoid handling other solanaceous plants prior to handling pepper plants. Spray weeds bordering the field with insecticides prior to seeding or planting the field. This will prevent the aphids from moving to other plants and infecting them when subsequent weed control is started. Destroy all annual weeds in the field. Avoid planting peppers close to tomato, tobacco, and pepper fields since these fields may harbor aphids. If possible, plant earlier to avoid high aphid populations that occur later in the season. Avoid possible sources of virus by removing volunteer pepper or tomato plants as soon as they appear in the field or nearby and keep fields free of weeds. Growing barrier crops in and around chilli crop at least 2 months before transplanting the chilli seedlings has been found beneficial. Spray barrier crop with suitable insecticide at least weekly to reduce population of insect vectors. The use of reflective mulches and yellow sticky insect traps can limit the virus spread by the aphid. Application of chemicals like furadon @ 1 kg a.i./ha at the time of transplanting and this should be followed by 3–4 foliar sprays of either monocrotophos (0.05%) or dimecron (0.05%) at 10 days interval for effective in controlling of this disease. Locate fields as far away as possible from very susceptible crops such as tomato, peanut and tobacco.

### 4.3.2. *Resistance Breeding against Chilli Viral Diseases*

The management of disease through host plant resistance has been an opt choice in all crop improvement programmes. Utilization of resistance is most simple, effective and economical method in the management of biotic stress. Increase in use of insecticide to control the sucking pest has led to their persistence and resurgences of the insect. To avoid this situation, identifying the resistant cultivars against viral disease is most significant one. Forty-eight chilli cultures, varieties and species of indigenous and exotic collections were screened against six mosaic viruses and their reactions were noted. Only crosses between Bydagi and Puri Orange were resistant to one or other viruses of PVBV, PVY and PeVMV, although none of these were resistant to TMV and CMV (Prasad Rao *et al.*, 1980). Out of 54 pepper lines selected for resistance to PVMV, PeMV, PVY and TEV in Nigeria, only Tca 14 was found resistant to PVMV (Alegbejo and Egwemi, 1999). AVRDC, Taiwan, reported that all 44 lines collected for resistance to various viruses, except VC 35, VC 36, VC 40, VC 41 and VC 71 were found 100 percent susceptible to both Japanese and Taiwan isolate of CVMV. VC 35 and VC 36 were 100 percent resistant to CVMV-AVRDC, but produced 68 percent and 73 percent resistant plants upon inoculation with CVMV-Japan (Anon., 1988). VC 37 produced 59 percent resistant plants upon inoculation with CVMV-Taiwan and 50 percent resistant plants upon inoculation with CVMV-Japan. Lines VC 40 and 41 contained 47 percent and 75 percent plants resistant to CVMV-Japan, respectively. AVRDC screened about 291 *C. annuum* lines for resistance to CMV from Taiwan, PVMV from England, CVMV from Taiwan and Japan by artificial inoculation on seedlings. It was reported that most resistant entries were VC 16, HDA 836 and Szechuan to PVMV; VC 35, VC 36, VC 37, VC 40 and VC 41 to CVMV and Kunja Kea Ryong San to CMV (Anon., 1990b).

## 4.4. Detection and Diagnostics of Chilli Viral Diseases

Though detection and diagnosis can be done on morphological basis, there is no scope to know the strain or related group which is only possible through molecular approaches. The DNA and protein based approaches used against chilli viral diseases are listed in Table 11.5.

**Table 11.5.** Detection/diagnostic techniques used against chilli viral diseases

| Viral disease | Techniques | References |
|---|---|---|
| ChVMV | DAC-ELISA | Kantharaju, 2003; Lakshiminarayanareddy, 2006 |
| | RT-PCR | Kantharaju, 2003; Krishna Reddy *et al.*, 2004, A non, 2002. Lakshiminarayanareddy, 2006; Cheiemsobat *et al.*, 1998 |
| CMV | RT-PCR | Hu *et al.*, 1995; Choijang *et al.*, 1998. |
| PVMV | DAC-ELISA | Brunt and Kenten, 1971; Siriwong *et al.*, 1995 |
| | RT-PCR | Gorsane *et al.*, 2001; Anindya *et al.*, 2004 |
| PBNV and CaCV | DAC-ELISA | Ravi *et al.*, 2005 |
| Chilli leaf curl virus | PCR | Rojas *et al.*, 1993; Deng *et al.*, 1994, Kumar *et al.*, 2006 |

### 4.4. Chilli Leaf Curl Begomovirus

The chilli leaf curl virus belongs to genus *Begomovirus* and family *Geminiviridae*. It has bipartite genomes with components referred to as DNA-A and DNA-B and is about 2.6–2.8 kb. The maximum incidence of leaf curl disease in chilli crop was up to 50 to 85 percent in the month of September, May and January in India (Sain *et al.*, 2005). The leaf curl diseases affecting peppers in India causes losses of up to 80 per cent in many parts of northern India (Singh, 1979). The geminivirus infected plant shows symptoms like abaxial and adaxial curling of the leaves accompanied by puckering and blistering of intervenial areas and thickening and swelling of the veins. Most of the geminiviruses have fairly restricted host ranges among economic crops, but most also infect weeds such as, *Datura* spp. and *Malva parvifolia* (cheeseweed).

The virus is transmitted by the whitefly, *Bemisia tabaci*, and is not known to be transmitted either mechanically or by seed. *B. tabaci,* the sweet potato whitefly, exists in several biotypes that are indistinguishable morphologically, but vary genetically and with respect to host preference. The 'B' biotype whitefly differs from the indigenous biotype in that it can survive on more hosts and possesses a degree of resistance to commonly used insecticides. These two characteristics have enabled the whitefly to build large populations in the field; both biotypes are efficient in transmitting geminivirus. Adult whiteflies acquire the virus after feeding on infected plants for 10 to 30 minutes, and can transmit the virus to pepper plants after 24 hours of incubation within the insect. A period of at least 15 minutes feeding on the new pepper host is subsequently required for transmission of the virus. The whitefly retains the virus for up to 20 days and does not transmit it to the progeny. Symptoms develop on young plants after 10 to 15 days. A hot and dry condition is most favourable for the whitefly and therefore helps to increase the spread of begomovirus infecting chilli.

### 4.4.1. *Management of Chilli Leaf Curl Begomovirus*

Control of geminiviruses is difficult once plants become infected. Management strategies include removing of diseased plants, rotating crop with non-host plants, seed treatment with imidacloprid @ 5 g/kg of seeds and spraying of imidacloprid @ 0.3 ml/lit of water. Additionally, controlling perennial weeds that may help the virus and the whitefly survive through the winter is important in reducing the potential for disease outbreaks.

### 4.5. Barrier Plants

These plants act as barrier for insect flight thereby preventing the movement of viruliferous and non-viruliferous insects in the field. Hence the insect infestation and incidence of viral disease reduced with increased yields of the crop. Different barrier crops used against chilli viruses are documented in Table 11.6.

**Table 11.6.** Barrier plants for control of plant viruses in chilli

| Virus targeted | Barrier crops | Factor involved | Country | References |
|---|---|---|---|---|
| CMV | Sunflower, sorghum, sesame, pearl-millet | Sunflower crop attract the sucking insects | India | Deol and Rataul, 1978 |
| | Maize, sorghum, sunflower | Barrier crops acted as sink for aphid colonization | India | Anandam and Doraiswamy, 2002 |
| ChVMV | Maize, brinjal | Brinjal is preferable host for sucking vector like whiteflies, there by prevent the spread of diseases in monocrop | Malaysia | Hussein and Samad, 1993 |
| CMV, PVY | Sorghum | Sorghum plants acted as a sink for both viruses | Spain | Avilla *et al.*, 1996 |
| CMV, PVY | Maize, vetch, sorghum | Barriers acted as a virus sink, but did not reduce aphid landing in crop | Spain | Fereres, 2000 |
| PVY | Sunflowers | Attract the sucking pests | USA | Simons, 1957 |

## 5. FUTURE LINE OF WORK

The following points are worth considering for a proper planning and development of techniques for better management of diseases of chillies:

- Hitherto no single cultivar is released commercially against the dreadful disease like anthracnose of chilli
- No specific bio-fungicide is developed and commercially utilized against chilli diseases
- More focuses are needed to develop transgenic chilli lines expressing multiple disease resistance
- Production of recombinant antibody against the chilli viral diseases for proper diagnosis and differentiation of viral diseases of chilli
- The generation of specific probes for monitoring of the current prevalence of chilli viral diseases in India
- Identification of resistance source in chilli germplasm to different viruses and strains of viruses
- Breeding of chilli lines to develop stable resistance to viral diseases
- Bio-formulation containing plant growth promoting endophytic bacteria to be developed and used in seed treatment so as to enhance growth and resistance in plants.

## REFERENCES

Adikaram NKB, Brown AE and Swinburne TR (1983). Observations on infection of *Capsicum annuum* fruits by *Glomerella cingulata and Colletotrichum capsici. Trnas. Br. Mycol. Soc.* **80**: 395–401.

Agrios GN (2005). Plant Pathology. 5th Ed. San Diego, Academic Press. p. 922.

Agronovsky AA (1993) Virus diseases of pepper (*Capsicum annuum* L.). *Ethiopia. J. Phytopathol.* **138**: 89–97.

Agwana CO, Lashermes P, Trouslot P, Combes M and Charrier A (1997). Identification of RAPD markers for resistance to coffee berry disease, *Colletotrichum kahawae*, in arabica coffee. *Euphytica* **97**: 241–8.

Alegbejo MD and Egwemi J (1999). Performance of some pepper selections screened for resistance to the potyviruses that commonly infect pepper. *Capsicum & Eggplant Newsl.* **18**: 56–59.

Alexander SA and Waldenmaier CM (2002). Management of anthracnose in bell pepper. Fungicide and Nematicide Tests. Vol. 58. New Fungicide and Nematicide data committee of the American Phytopathological Society; p. 49. (http://apsjournals.apsnet.org/doi/abs/10.1094/PDIS.2004.88.11.1198).

Amusa NA, Kehinde IA and Adegbite AA (2004). Pepper fruit anthracnose in the humid forest region of south-western Nigeria. *Nutrition and Food Science* **34(3)**: 130–4.

Amusa NA, Ashaye OA, Oladapo MO and Oni MO (2005). Guava fruit anthracnose and the effects on its nutritional and market values in Ibadan, Nigeria. *World Journal of Agricultural Sciences* **1(2)**: 169–72.

Anandam RJ and Doraiswamy S (2002). Role of barrier crops in reducing the incidence of mosaic disease in chilli. *J. Plant Dis. Prot.* **109**: 109–12.

Anindya R, Joseph J, Gowri TDS and Savithri HS (2004). Complete genomic sequence of *Pepper vein banding virus* (PVBV): a distinct member of the genus *Potyvirus*. *Arch. Virol.* **149**: 625–32.

Anonymous (1988). Annual progress report of Asian Vegetable Research and Development Center for the year 1986, Shanhua, Tainan, Taiwan, p. 414.

Anonymous (1990a). Annual progress report of Asian Vegetable Research and Development Center for the year 1989, Shanhua, Tainan, Taiwan, pp. 351.

Anonymous (1990b). Annual progress report of Asian Vegetable Research and Development Center for the year 1988. Shanhua, Tainan, Taiwan, pp. 414.

Anonymous (1990c). Annual progress report of Asian Vegetable Research and Development Center for the year 1987, Shanhua, Tainan, Taiwan, p. 411.

Anonymous (1992). Annual progress report of Asian Vegetable Research and Development Center for the year 1991, Shanhua, Tainan, Taiwan, pp. 413.

Anonymous (1998). Annual progress report of Asian Vegetable Research and Development Center for the year 1998, Shanhua, Tainan, Taiwan, pp. 415.

Anonymous (2002). Annual progress report of Asian Vegetable Research and Development Center for the year 1994, Shanhua, Tainan, Taiwan, Summaries, pp. 75–78.

Atiri GI and Dele HW (1985). *Pepper veinal mottle virus* infection, host reaction, yield and aphid transmission in pepper plants. *Tropic. Agric.* **62**: 190–2.

Avilla C, Collar JL, Duque, M, Hernaiz P, Martýn B and Fereres A (1996). Cultivos barrera como m´etodo de control de virus no persistentes en pimiento. *Bol. San. Veg. Plagas* **22**: 301–7.

AVRDC (1999). Off-Season Tomato, Pepper, and Eggplant. Taiwan, China: Asian Vegetable Research and Development Centre. Progress Report , 1998.

AVRDC (2004). Asian Vegetable Research and Development Fact sheet 2004, 04-589, pp. 1–2.

Bashan Y and Okon Y (1983). Internal and external infection of fruits and seeds of peppers by *Xanthomonas campestris* pv. *vesicatoria*. *Canadian J. Botany* **64**: 2865–71.

Berg GN (1984). Non-persistant viruses: epidemiology and control. *In*: Advances in virus research (Raccah B, *ed.*) **31**: 387–429.

Berke TG (2000). Multiplying seed of pepper lines, AVRDC Pub.-XXX: 1–8.

Bidari VB and Reddy HR (1990). Identification of naturally occurring viruses on commercial cultivars of chilli. *Mysore J. Agric. Sci.* **24**: 45–51.

Biswas KK, Pant RP, Chouwdhury Anil, Pun KB, Das DR and Ahlawat YS (2005). Incidence, detection and management of chilli mosaic disease syndrome in *Capsicum annuum* cultivar in Kalimpong local in the hills of West Bengal. *Indian J Virol.* **16(1& 2)**: 132–6.

Boonratkwang C, Chamswarng C and Intanoo W (2007). Proceeding of the 8th National plant protection conference. Phisanulok, Thailand: Naresuan University: Effect of secondary metabolites from *Trichoderma harzianum* strain *Pm9* on growth inhibition of *Colletotrichum gloeosporioides* and chilli anthracnose control; pp. 323–36.

Bosland PW and Votava EJ (2003). Peppers: vegetable and spice capsicums. England: CAB International; p. 233.

Brunt AA and Kenten RH (1971). *Pepper veinal mottle virus*—a new member of potato virus Y group *from pepper* (*Capsicum annuum L.*) *and C. frutescens* (*L.*) *in Ghana. Ann*. Appl. Biol. **69**: 235–43.

Cerkauskas R (2004a). Bacterial wilt. AVRDC Fact sheet 04–573: 1–2.

Cerkauskas R (2004b). Bacterial spot. AVRDC Fact sheet 04–572: 1–2.

Cerkauskas R (2004c). Bacterial soft rot. AVRDC Fact sheet 04–571: 1–2.

Charigkapakorn N (2000). Control of chilli anthracnose by different biofungicides. Thailand. http://www.arc-avrdc.org/pdf_files/029-Charigkapakorn_18th.pdf.

Chen C, Belanger R, Benhamou N and Paulitz TC (2000). Defense enzymes induced in cucumber roots by treatment with by plant growth promoting rhizobacteria (PGPR) and *Pythium aphanidermatum*, *Physiol. Mol. Plant Pathol.* **56**: 13–23.

Chiemsobat P, Sae-Ung N, Attathom S, Patarapuwadol S and Siriwong P (1998). Molecular taxonomy of a new potyvirus isolated from chilli pepper in Thailand. *Arch. Virol*. **143**: 1855–63.

Cho JJ, Mau RFL, German TL, Hartmann RV, Yudin LS, Gonsalves D and Provvidenti R (1989). A multidisciplinary approach to management of tomato spotted wilt virus in Hawaii. Plant Dis. **73**: 375–83.

Choijang K, Kim H, Yoon J, Park S, Kim D and Lee S (1998). Detection of virus in fruit and seed of vegetables using RT-PCR. *Korean J. Pl. Pathol*. **14**: 630–5.

Compant B, Duffy B, Nowak J, Clement C and Barka EA (2005). Use of plant growth promoting bacteria for biocontrol of plant diseases: Principles mechanisms of action, and future prospects. *Applied Environmental Microbiol.* **71**: 4951–9.

Conover RA and Averre CW (1963). Tomato bacterial spot, *Xanthomonas vesicatoria*. Pages 82–83, *In*: Fungic. Nematic. Tests. Vol. 19. F.H. Lewis, (*ed.*) Am. Phytopathol. Soc. St. Paul, MN.

Cox RS (1982). Control of bacterial spot of tomato in southern Florida. *Plant Dis.* **66**: 870.

de Meyer G and Hofte M (1997). Salicylic acid produced by the *Rhizobacterium, Pseudomonas aeruginosa* 7NSK2 induces resistance to leaf infection by *Botrytis cinerea* on bean. *Phytopathol.* **87**: 588–93.

Deng D, McGrath PF, Robinson DJ and Harrison BD (1994). Detection and differentiation of whitefly transmitted geminiviruses in plants and vector insects by the polymerase reaction with degenerate primers. *Ann. Appl. Biol.* **125**: 327–36.

Deol GS and Rataul HS (1978). Role of various barrier crops in reducing the incidence of cucumber mosaic virus in chilli, *Capsicum annuum* L. *Indian J. Entomol*. **40**: 261–4.

Dodd JC, Estrada A and Jeger MJ (1992). Epidemiology of *Colletotrichum gloeosporioides* in the Tropics. *In*: Bailey JA, Jeger MJ (*eds.*). *Colletotrichum*: Biology, Pathology and Control. Wallingford: CAB International; pp. 308–25.

Du M, Schardl CL and Vaillancourt LJ (2005). Using mating-type gene sequences for improved phylogenetic resolution of *Colletotrichum* species complexes. *Mycologia*. **97(3)**: 641–58.

Egli T and Sturm E (1980). Bacterial plant diseases. *In*: Chemie der Pflanzenschutz and Schadlingsbekampfungs-mittel. Band 6. R. Wegler, (*ed.*) Springer-Verlag, Berlin. pp. 345–88.

Fereres A (2000). Barrier crops as a cultural control measure of nonpersistently transmitted aphid-borne viruses. *Virus Res*. **71**: 221–31.

Fernandez NEN and Fulton WR (1980). Detection and characterization of Peru tomato virus strains infecting pepper and tomato in Peru. *Phytopathol.* **70**: 315–50.

Freeman S, Katan T and Shabi E (1998). Characterization of *Colletotrichum* species responsible for anthracnose diseases of various fruits. *Plant Dis.* **82**: 596–605.

Fujisava I, Hanada T and Saharan A (1986). Virus diseases occurring on some vegetable crops in west Malaysia. *Japan. Agri. Res.* **20(1)**: 78–84.

Gebre-Selassie K, Dumas de Vaulx R and Pochard E (1983). Biological and serological characterization of PVY strains affecting peppers and other related strains. *Capsicum Newslett.* **3**: 134–6.

George TE, Anand N and Singh SJ (1993). Isolation and identification of two major viruses infecting bell pepper in Karnataka. *Indian Phytopath.* **46**: 24–28.

Goode M J and Sasser M (1980). Prevention. *Plant Dis.* **64**: 831–4.

Gopinath K, Radhakrishnan NV and Jayaral J (2006). Effect of propiconazole and difenoconazole on the control of anthracnose of chili fruit caused by *Colletotrichum capsici*. *Crop Protection* **25**: 1024–31.

Gorsane F, Fakhfakh H, Tourneur C, Marrakchi and Makni M (2001). Nucleotide sequence comparison of 3′ terminal region of the genome of Pepper vein mottle virus isolated from Tunisia and Ivory Coast. *Arch. Virol.* **146**: 611–8.

Green SK and Kim JS (1991). Characterization and control of viruses infecting peppers: a literature review. AVRDC Tech. Bull No. **18**, pp. 78.

Guerber JC, Liu B, Correll JC and Johnston PR (2003). Characterization of diversity in *Colletotrichum acutatum sensu lato* by sequence analysis of two gene introns, mtDNA and intron RFLPs and mating compatibility. *Mycologia* **95(5)**: 872–95.

Hadden JF and Black LL (1989). Anthracnose of pepper caused by *Colletotrichum* spp. Proc. of the International symposium on integrated management practices: tomato and pepper production in the tropics; Taiwan: Asian Vegetable Research and Development Centre, pp. 189–99.

Hadden JF and Black LL (1987). Comparison of virulence of tomato and pepper isolates of *Colletotrrichum* spp. (Abstr). *Phytopathol.* **77**: 641.

Haward CM, Maas JL, Chawder CK and Albregts EC (1992). Anthracnose of strawberry caused by the *Colletotrichum* complex in Florida. *Plant Dis.* **76**: 976–81.

Heggens BB (1930). A pepper fruits rot. Georgia experiment station bull. Georgia, USA, pp. 145–9.

Hibberd AM, Stall RE and Basset MJ (1987). Different phenotypes associated with incompatible races and resistant genes in bacterial spot disease of pepper. *Plant Dis.* **71**: 1075–8.

Hofte KY, Seong E, Jurkrvitch and Verstraete W (1991). Pyoverdin production by plant growth beneficial *Pseudomonas* strain 7NSK2: Ecological significance in soil. *Plant and Soil* **130**: 249–58.

Hong JK and Hwang BK (1998). Influence of inoculum density, wetness duration, plant age, inoculation method, and cultivar resistance on infection of pepper plants by *Colletotrichum coccodes*. *Plant Dis.* **82**: 1079–83.

Hong JK and Kim DH (2007). Taxonomic characteristics of *Colletotrichum* spp. and their teleomorphs causing anthracnose of chilli cepper. *In*: Oh DG and Kim KT (*eds.*) Abstracts of the first International symposium on chilli anthracnose. Republic of Korea: National Horticultural Research Institute, Rural Development of Administration, pp. 31.

Hu JS, Li HP, Barry K, Wang M and Jordan R (1995). Comparison of dot blot, ELISA and RT-PCR assays for detection of two cucumber mosaic virus isolates infecting banana in Hawaii. *Plant Dis.* **79**: 902–6.

Hussein MY and Samad NA (1993). Intercropping chilli with maize or brinjal to suppress populations of *Aphis gossypii* and transmission of chilli viruses. *Int. J. Pest Manage.* **39**: 216–22.

IIVR (2007). Indian Institute of Vegetable Research. Annual Report 2007–2008.

Intanoo W and Chamswarng C (2007). Proceeding of the 8th National plant protection conference. Phisanulok, Thailand: Naresuan University. Effect of antagonistic bacterial formulations for control of anthracnose on chilli fruits; pp. 309–22.

Jeger MJ and Jeffries P (1988). Alternative to chemical usage for disease management in the post-harvest environment. *Aspects Appl. Biol.* **17**: 47–57.

Jetiyanon K (2007). Defensive-related enzyme response in plants treated with a mixture of *Bacillus* strains (IN937a and IN937b) against different pathogens *Biological Control* **42**: 178–85.

Jeyalakshmi C and Seetharaman K (1998). Biological control of fruit rot and die-back of chilli with plant products and antagonistic microorganisms. *Plant Dis. Res.* **13**: 46–48.

Jeyarajan R and Ramakrishnan K (1969). Potato virus Y on chilli (*Capsicum annuum* L.) in Tamilnadu. *Madras Agric. J.* **56**: 761–6.

Jones LB, Woltz SS, Jones JP and Portier KL (1991). Population dynamics of *Xanthomonas campestris* pv. *vesicatoria* on tomato leaflets treated with copper bactericides. *Phytopathol.* **81**: 714–9.

Joseph J and Savithri HS (1999). Determination of 3′ terminal nucleotide sequence of *Pepper vein banding virus* RNA and expression of its coat protein in *Eschirichia coli*. *Arch. Virol.* **144**: 1679–87.

Joshi S, Shrestha K, Timila RD and Shrestham SK (1997). Leaf curl virus of tomato and chili crops in Nepal. Collaborarive vegetable research in South Asia: Proc. of the phase I final workshop of the South Asian Vegetable Research Network held in Kathmandu, Nepal, 23–28 January 1996. Asian Vegetable Research and Development Center, Taiwan. Pub. No. 97–458: 156–67.

Kantharaju HM (2003). Identification and diagnosis of *Chilli veinal mottle potyvirus* (CVMV) infecting hot pepper (*Capsicum annuum* L.). M.Sc (Agri) Thesis, Univ. Agri. Sci., Bangalore, India, pp. 80.

Kim BS, Park HK and Lee WS (1989). Resistance to anthracnose (*Colletotrichum* spp.) in pepper. *In*: Tomato and pepper production in the tropics. AVRDC, Shanhua, Taiwan, China, pp. 184–8.

Kim BS, Park KS and Lee WS (1987). Search for resistance to two *Colletotrichum* species in pepper (*Capsicum* spp.). *J. Korean Soc. Hort. Sci.* **28**: 207–13.

Kim KK, Yoon JB, Park HG, Park EW and Kim YH (2004). Structural modifications and programmed cell death of chilli pepper fruits related to resistance responses to *Colletotrichum gloeosporioides* infection. *Genetics and Resistance*. **94**: 1295–1304.

Korpraditskul V, Rattanakreetakul C, Korpraditskul R and Pasabutra T (1999). Proceeding of Kasetsart University Annual Conference. Bangkok: Kasetsart University; Development of plant active substances from Sweet flag to control fruit rot of mango for export, p. 34.

Kousik CS and Ritchie DF (1995). Characteristics of pepper bacterial spot pathogen races that overcome the *BS2* gene for resistance. *Phytopathol.* **85**: 1163.

Kousik CS and Ritchie DF (1996). Race shift in *Xanthomonas campestris* pv *vesicatoria* within a season in field grown pepper. *Phytopathol.* **86**: 952–8.

Krishna Reddy M, Madhavi Reddy K, Lakshminarayana Reddy CN, Smitha R and Jalali S (2004). Molecular characterization and genetic variability of *Chilli veinal mottle virus* and its reaction on chilli pepper genotypes. *In*: 12th Eucarpia meeting on Capsicum and Eggplant, 1719 May, 2004, Noordwijk, Netherlands, pp. 171–7.

Kumar S, Kumar S, Singh M, Singh AK and Rai M (2006). Identification of host plant resistance to *Pepper leaf curl virus* in chilli (*Capsicum* species). *Sci. Hortic.* **110**: 359–61.

Ladera P, Lastra R and Debrot EA (1982). Purification and partial characterization of a potyvirus infecting peppers in Venezuela. *J. Phytopathol.* **104**: 97–103.

Laird EF and Dickson RC (1964). *Tobacco etch virus* and *Potato virus* Y from pepper in Southern California. *Plant Dis. Rep.* **48**: 772–4.

Lakshminarayana Reddy CN (2006). Molecular characterization of *Chilli veinal mottle virus* (ChVMV) infecting chilli (*Capsicum annuum* L.) Ph.D Thesis, Univ. Agri. Sci., Bangalore, India, p. 123.

Leonian LH (1922). Stem and fruit blight of chillies caused by *Phytophthora capsici*. *Phytopathol.* **12**: 401–8.

Lewis IML and Miller SA (2003). Evaluation of fungicides and a biocontrol agents for the control of anthracnose on green pepper fruit, 2002. Nematicide test report. Vol. 58. New Fungicide and Nematicide Data Committee of the American Phytopathological Society, p. 62.

Mah SY (1985). Anthracnose fruit rot (*Colletotrichum capsici*) of chilli (*Capsicum annuum*): causal pathogen, symptom expression and infection studies. *Teknol. Sayur-sayuranEl* **1**: 35–40.

Manandhar JB, Hartman GL and Wang TC (19950). Anthracnose development on pepper fruits inoculated with *Colletotrichum gloeosporioides*. *Plant Dis.* **79**: 380–3.

Marie SY, Sundin PB and Waechter-Kristensen Jensen P (1997). Induction of phenolic compounds in tomato by rhizosphere bacteria. *In*: Ogoshi A, Kobayashi K, Homma Y, Kodama F, Kondo N and Akino S (*eds.*) Plant Growth-Promoting Rhizobacteria-Present Status and Future Prospects. Proc. Fourth International workshop on plant growth promoting rhizobacteria Japan-OECD joint workshop, Sapporo, Japan, pp. 340–4.

Marin A, Ferreres F, Tomas Barberan FA and Gil M (2004). Characterization and quantization of antioxidant constituents of sweet pepper (*Capsicum annuum* L.). *J. Agric. Food Chem.* **52(12)**: 3861–9.

Martinez-Culebras PV, Barrio E, Garcia MD and Querol A (2000). Identification of *Colletotrichum* species responsible for anthracnose of strawberry based on the internal transcribed spacers of the ribosomal region. *FEMS Microbiol. Lett* **189(1)**: 97–101.

McMichael LA, Persley DM and Thomas JE (2002). A new tospovirus serogroup IV species infecting capsicum and tomato in Queensland. *Australasian Plant Pathol.* **31**: 231–9.

Mongkolporn O, Dokmaihom Y and Kanchana-udomkarn C (2004). Genetic purity test of $F_1$ hybrid *Capsicum* using molecular analysis. *J. Hortic. Sci. Biotech.* **79**: 44–451.

Moriwaki J, Tsukiboshi T and Sato T (2002). Grouping of *Colletotrichum* species in Japan based on rDNA sequences. *J. Gen. Plant Pathol.* **68(4)**: 307–20.

Muhyi R and Bosland P W (1992). Evaluation of *Capsicum* germplasm for sources of resistance to *Rhizoctonia solani*. *HortScience* **30**: 341–2.

Nduagu C, Ekefan EJ and Nwankiti AO (2008). Effect of some crude plant extracts on growth of *Colletotrichum capsici* (Synd & Bisby), causal agent of pepper anthracnose. *J. Appl. Biosci.* **6(2)**: 184–90.

Nelson MR, Wheeler RE and Zitter TA (1982). Pepper mottle virus. CMI/AAB Descriptions of plant viruses No. 253. Commonwealth Mycological Institute, Kew, England, p. 4.

Oanh LTK, Korpraditskul V and Rattanakreetakul C (2004). A pathogenicity of anthracnose fungus, *Colletotrichum capsici* on various Thai chilli varieties. *Kasetsart J.* (Natural Science). **38(6)**: 103–8.

Ong CA and Ting WP (1977). A review of plant virus diseases in peninsular Malaysia. *In*: Virus disease of tropical crops, September 1976, TARC Trop. Agr. Ser. Tsukuba, Japan, **10**: 155–64.

Ong CA, Varghese G and Poh TW (1979). Aetiological investigation on a veinal mottle virus of chilli (*Capsicum annuum* L.), newly recorded from peninsular Malaysia. Malaysian Agri. Res. Dev. Inst. (MARDI) Res. Bull. **7**: 78–88.

Ong CA, Varghese G and Poh TW (1980). The effect of chilli veinal mottle virus on yield of chilli (*Capsicum annuum* L.). Malaysian Agri. Res. Dev. Inst. (MARDI) Res. Bull. **8**: 74–79.

Osuna-Garcia JA, Wall MW and Waddell CA (1998). Endogenous levels of tocopherols and ascorbic acid during fruits ripening of New Mexican-type chilli (*Capsicum annuum* L.) cultivars. *J. Agric. Food Chem.* **46(12)**: 5093–6.

Park HG (2007). Problems of anthracnose in pepper and prospects for its management. *In*: Oh DG and Kim KT (*eds.*) Abstracts of the first international symposium on chilli anthracnose. Republic of Korea: National Horticultural Research Institute, Rural Development of Administration, p. 19.

Park WM, Park SH, Lee YS and Ko YH (1987). Differentiation of *Colletotrichum* spp. causing anthracnose on *Capsicum annuum* L. by electrophoretic method. *Korean J. Plant Pathol.* **3**: 85–92.

Perez-Galvez A, Martin HD, Sies H and Stahl W (2003). Incorporation of carotenoids from paprika oleoresin into human chylomicron. *British J. Nutr.* **89(6)**: 787–93.

Pernezny K and Collins J (1997). Epiphytic populations of *Xanthomonas campestris* pv. *vesicatoria* on pepper: relationships to host-plant resistance and exposure to copper sprays. *Plant Dis.* **81**: 791–4.

Pernezny K, Roberts PD and Murphy JF (2003). Compendium of pepper diseases. St. Paul, Minnedota: The American Phytopathological Society; p. 73.

Persley D, Sharman M, Thomas J, Kay I, Heisswolf S and Michael L (2007). Thrips and tospovirus: A management guide; department of primary industries and fisheries pp. 1–17.

Persley DM, Thomas JE and Sharman M (2005). Tospoviruses: an Australian perspective. *Australasian Plant Pathol.* pp. 35.

Photita W, Taylor PWJ, Ford R, Lumyong P, McKenzie HC and Hyde KD (2005). Morphological and molecular characterization of *Colletotrichum* species from herbaceous plants in Thailand. *Fungal Diversity*. **18**: 117–33.

Polyanskaya LM, Vedina OT, Lysak LV and Zvyagintsev DG (2002). The growth-promoting effect of *Beijerinckia mobilis* and *Clostridium* sp. cultures on some agricultural crops. *Microbiol.* **71:** 109–15.

Poonpolgul S and Kumphai S (2007). Chilli pepper anthracnose in Thailand. *In*: First international symposium on chilli anthracnose, Seoul National University, Korea, p. 23.

Prasad Rao RDVJ (1976). Characterization and identification of some chilli mosaic viruses. Ph.D. Thesis, Univ. Agric. Sci., Bangalore, India, p. 198.

Prasad Rao RDVJ, Yaraguntaiah RC and Govindu HC (1980). Reaction of chilli (*Capsicum* spp.) cultures, varieties and species to six chilli mosaic viruses. *Mysore J. Agric. Sci.* **14**: 11–14.

Pring RJ, Nash C, Zakaria M and Bailey JA (1995). Infection process and host range of *Colletotrichum capsici*. *Physiol. Mol. Plant Pathol.* **46(2)**: 137–52.

Ratanacherdchai K, Wang HK, Lin FC and Soytong K (2007). RAPD analysis of *Colletotrichum* species causing chilli anthracnose disease in Thailand. *J. Agric. Tech.* **3(2)**: 211–9.

Ravi KS, Joseph J, Nagaraju N, Krishna Prasad S, Reddy HR and Savitri HS (1997). Characterization of *Pepper vein banding virus* from chilli pepper. *Plant Dis.* **81**: 673–6.

Ravi KS, Suresh Kunkalikar M, Bhanupriya P, Sudarsana PR, Usha B Zehr and. Naidu RA (2005). Current status of tospoviruses infecting vegetable crops in India, International symposium on emerging and re-emerging viruses in vegetable crops AVRDC-Regional Center for South Asia, ICRISAT campus, Patancheru, 502 324, Andhra Pradesh. pp. 67.

Rivelli VC (1989). A wilt of pepper incited by *Fusarium oxysporum* f. sp. *capsici* f. sp. nov. M.S. Thesis, Louisiana State University, Baton Rouge.

Roberts PD, Pernezny K and Kucharek TA (2001). Anthracnose caused by *Colletotrichum* sp. on pepper. J. Univ. of Florida/Institute of Food and Agric. Sci. p. 104.

Rojas MR, Gilbertson RL, Russell DR and Maxwell DP (1993). Use of degenerate primers in the polymerase chain reaction to detect whitefly transmitted geminiviruses. *Plant Dis.* **77**: 340–7.

Royle DJ and Butler DR (1986). Epidemiological significance of liquid water in crop canopies and its role in disease forecasting. *In*: Ayres PG and Boddy L (*eds.*) Water, Fungi and Plants. Cambridge: Cambridge University Press, pp. 139–56.

Sain SK and Chadha ML (2005). Major viral diseases incidence on important vegetable crops in Hyderabad, India, International symposium on emerging and re-emerging viruses in vegetable crops, AVRDC-Regional Center for South Asia, ICRISAT Campus, Patancheru, 502 324, Andhra Pradesh, p. 68.

Sanders GM and Korsten L (2003). Comparison of cross inoculation potential of South African avocado and mango isolates of *Colletotrichum gloeosporioides*. *Microbiological Research* **158**: 143–50.

Sanogo S (2003). Chile pepper and the threat of wilt diseases. Online. Plant Health Progress doi: 10.1094/PHP-2003-0430-01-RV.

Sanogo S and Clary M (2004). Bacterial leaf spot of chile pepper: A short guide for growers. New Mexico Chile Association Report 30, pp. 1–7.

Sariah M (1989). Detection of benomyl resistance in the anthracnose pathogen, *Colletotrichum capsici*. *J. Islamic Acad. Sci.* **2(3)**: 168–71.

Satya Prakash, Singh SJ, Singh RK and Upadhyaya PP (2002). Distribution, incidence and detection of a potyvirus on chilli from eastern Uttar Pradesh. *Indian Phytopath.* **55**: 294–8.

Scott JW and Jones JB (1989). Inheritance of resistance to foliar bacterial spot of tomato incited by *Xanthomonas campestris* pv *vesicatoria*. *J. Am. Soc. Hort. Sci.* **14**: 111–4.

Sharma PN, Kaur M, Sharma OP, Sharma P and Pathania A (2005). Morphological, pathological and molecular variability in *Colletotrichum capsici*, the cause of fruit rot of chillies in the subtropical region of North-western India. *J. Phytopathol.* **153**: 232–7.

Sheu Z, Chenand J and Wang T (2007). Application of ITS-RFLP analysis for identifying *Colletotrichum* species associated with pepper anthracnose in Taiwan. *In*: Oh DG and Kim KT (*eds.*) Abstracts of the first international symposium on chilli anthracnose. Republic of Korea: National Horticultural Research Institute, Rural Development of Administration, p. 32.

Shin HJ, Xu T, Zhang CL and Chen ZJ (2000). The comparative study of capsicum anthracnose pathogens from Korea with that of China. *J. Zhejiang University* (Agric. and Life Sci.). **26**: 629–34.

Simons J (1957). Effects of insecticides and physical barriers on field spread of pepper vein banding mosaic virus. *Phytopathol*. **47**: 139–45.

Singh SJ (1979). Reaction of chilli varieties (*Capsicum* spp.) to mosaic and leaf curl viruses under field conditions. *Indian J. Hort*. **30**: 444–7.

Siriwong P, Kittipakam K and Ikegaru M (1995). Characterization of *Chilli vein banding mottle virus* isolated from pepper in Thailand. *Plant Pathol*. **49**: 710–27.

Smith KL (2000). Peppers. *In*: Precheur RJ (*ed.*) Ohio Vegetable Production Guide. Columbus, Ohio: Ohio State University Extension; pp. 166–73.

Smith BJ and Black LL (1990). Morphological, cultural, and pathogenic variation among *Colletotrichum* species isolated from strawberry. *Plant Dis*. **74(1)**: 69–76.

Smith KW and Crasson DF (1959). The taxonomy, etiology and control of *Colletotrichum piperatum* (E and E) and *C. capsici* (Synd) (BB). *Plant Dis. Rep*. **42**: 1099.

Soum Sanogo and Marvin Clary (2005). Bacterial leaf spot of chile pepper: A short guide for growers, New Mexico Chile Association, Report 30, pp. 1–7.

Sreenivasaprasad S and Talhinhas P (2005). Genotypic and phenotypic diversity in *Colletotrichum acutatum*, a cosmopolitan pathogen causing anthracnose on a wide range of hosts. *Mol. Plant Pathol*. **6(4)**: 361–78.

Sreenivasaprasad S, Mills P and Brown A (1994). Nucleotide sequence of the rDNA spacer 1 enables identification of isolates of *Colletotrichum* as *C. acutatum*. *Mycological Res*. **98**: 186–8.

Sreenivasaprasad S, Mills P, Meehan BM and Brown A (1996). Phylogeny and systematics of 18 *Colletotrichum* species based on ribosomal DNA spacer sequences. Genome **39(3)**: 499–512.

Sutton BC (1980). The Coelomycetes. Commonwealth Mycological Institute, G.B. pp. 523–37.

Taylor PWJ (2007). Anthracnose disease of chilli pepper in Thailand. Proc. international conference on integration of science and technology for sustainable development (ICIST) Biological Diversity, Food and Agricultural Technology, Bangkok, Thailand. 26–27 April 2007, pp. 134–8.

Than PP, Prihastuti H, Phoulivong S, Taylor PWJ and Hyde KD (2008). Chilli anthracnose disease caused by *Colletotrichum* species. *Zhejiang Univ Sci*. **9(10)**: 764–78.

Tindall HD (1983). Vegetable in the Tropics. Macmillan, London, pp. 533.

Ullman DE, German TL, Sherwood JL, Westcott DM and Cantone FA (1993). Tospovirus replication in insect vector cells: Immunocyto-chemical evidence that the non-structural protein encoded by the S RNA of tomato spotted wilt tospovirus is present in thrips vector cells. *Phytopathol*. **83**: 456–63.

Verma ML (1973). Comparative studies on virulence of isolates of four species of *Colletotrichum* parasitic on chillies. *Indian Phytopathol*. **26**: 28–31.

von Arx JA and Die Arten der Gattung (1957). *Colletotrichum* Cda. *Phytopathologische Zeitschrift*. **29**: 414–68.

Voorrips RE, Finkers R, Sanjaya L and Groenwold R (2004). QTL mapping of anthracnose (*Colletotrichum* spp.) resistance in a cross between *Capsicum annuum* and *C. chinense*. *Theor. Appl. Genet*. **109(6)**: 1275–82.

Wharton PS and Dieguez-Uribeondo J (2004). The biology of *Colletotrichum acutatum*. *Anales del Jardin Botanico de Madrid*. **61**: 3–22.

Wijkamp I and Peters D (1993). Determination of the median latent period of two tospoviruses in *Frankliniella occidentalis*, using a novel leaf disk assay. *Phytopathol*. **83**: 986–91.

Wilson LL, Maddan LV and Ellis MA (1992). Over winter survival of *Colletotrichum acutatum* in infected strawberry fruit in Ohio. *Plant Dis*. **76**: 948–58.

Xiao CL, MacKenzie SJ and Legard DE (2004). Genetic and pathogenic analyses of *Colletotrichum gloeosporioides* from strawberry and non-cultivated hosts. *Phytopathol*. **94**: 446–53.

Yabuuchi E, Kosako Y, Yano I, Hotta H and Nishiuchi Y (1995). Transfer of two *Burkholderia* and *Alcaligenes* species to *Ralstonia* gen. nov.proposal of *Ralstonia pickettii* (Ralston, Palleroni & Doudororoff, 1973) comb. nov., *Ralstonia solanacearum* (Smith, 1896) comb. nov. and *Ralstonia eutropha* (Davis, 1969) comb. nov. *Microbiol Immunol*. **39**: 897–904.

Yoon JB and Park HG (2001). Screening method for resistance to pepper fruits anthracnose: Pathogen sporulation, inoculation methods related to inoculums concentrations, post-inoculation environments. *J. Korean Soc. Hortic. Sci*. **42**: 382–93.

Zitter TA (1972). Naturally occurring pepper virus strains in South Florida. *Plant Dis. Rep*. **56**: 586–90.

# 12

# Post Harvest Handling, Storage and Value Addition in Chilli

K. Uma Jyothi, T. Vijaya Lakshmi, S. Surya Kumari, P. Venkata Reddy and K.V. Siva Reddy

## 1. INTRODUCTION

Chilli, an important spice and vegetable crop cultivated in all the states and Union territories of India. No country in the world has so much area and production as much as India. It has two important commercial qualities; some varieties are famous for typical red colour and others are known for pungency. India is rich in several types of varieties with different quality factors. Breeding works on chillies are being carried out at different centres and improved varieties have been released for commercial cultivation. However, the post-harvest handling and value addition of chilli has not been given much importance. Although Indian chilli is exported to over 90 countries, the percentage of chilli export in the international trade from India is very low. Pesticide residues and aflatoxins are the issues affecting chilli imports from India. Moreover, preservation of attractive typical red colour and retention of the pungency during storage have become problems. The major operations with respect to handling chillies during after harvest has been reviewed and discussed in this chapter.

## 2. HARVESTING

Various operations followed during post-harvest handling of chillies has been depicted in Fig. 12.1. The time of harvesting chillies depends on the purpose for which it is grown. Chilli fruits can be harvested either at green immature or red mature stage. For vegetable purpose, green fruits are harvested. The quality of green chilli is decided by the size, colour and plumpness of fruit. Long, green, smooth, shiny and moderately plumpy fruits are preferred, however, the qualities may vary from region to region. Pickles are made from either green or red chillies. For the preparation of

Regional Agricultural Research Station, Lam, Guntur (A.P).

dry chilli, the chilli fruits should be harvested at fully red-ripe condition. Under well managed situation, 10–12 pickings can be harvested as green chilli or 6–8 pickings as ripe chilli under irrigated conditions and 3–4 pickings as ripe chilli under rainfed conditions. Harvested red matured fruits should be kept at room temperature for 1–2 days for uniform ripening of immature fruits.

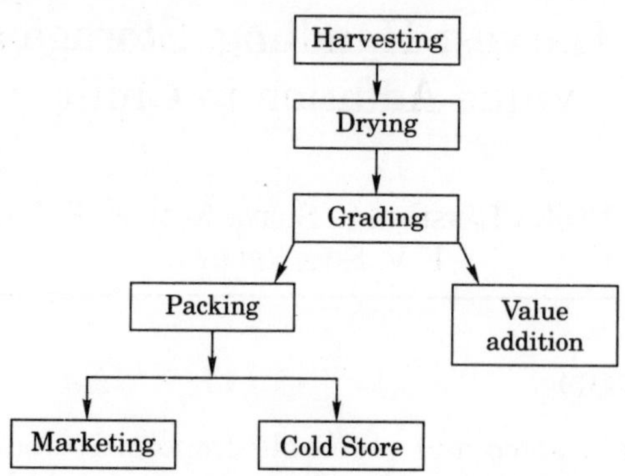


**Fig. 12.1.** Post-harvest handling operations in chilli

Harvesting of chilli is labour intensive operation and is carried out manually. The suggested alternative is mechanical harvesting. For cost effective machine harvesting the accurate plant spacing and determinate cultivar with large pendulous fruits need to be selected.

## 3. POST-HARVEST OPERATIONS

### 3.1. Drying of Chillies

Drying constitutes the most important processing step in post-harvest technology. Traditionally in the states of Andhra Pradesh, Karnataka and Maharashtra, after harvest fruits are spread uniformly on mats, rooftops, and tarpaulins or on floor and sun dried. Frequent stirring is done to avoid any mould growth and for obtaining uniform drying. The post-harvest loss of chillies by traditional method of drying is estimated at around 35 to 50 per cent. Chillies have a thick waxy skin that prevents rapid drying. The initial moisture content of fresh red ripe chilli ranged from 70–80 per cent (wet basis) which is too high for further processing and storage. The decrease of moisture content to a safe level of 8–9 per cent on wet basis is necessary before its processing and storage (Mangaraj *et al.*, 2001). The better retention of colour with good quality of pungency of dried chillies with considerable decrease of breakage of pods and loss of seeds are achieved by adopting improved simpler techniques over traditional method of sun drying (Sumathi Kutty and Mathew, 1987).

Depending upon the weather, variety and order of picking *i.e.* first or last picking takes 15 to 20 days to dry chillies to a final moisture content of 10–12 per cent. Generally fruits of first and second picking take more time, as the moisture is relatively high. Also the hybrids which have thick pericarp, generally grown as irrigated crop take more time. The final dried product is generally 1/4 to 1/3 of the initial weight of the ripe pod depending upon the variety and order of picking. High temperatures cannot be used to accelerate drying as the spice colour becomes brown. During sun drying, light energy is absorbed by the free water, which utilizes this energy for vaporization. When the water content decreases to about 10 per cent or less, light energy promotes lipid oxidation through which free radicals are generated. Generation of short-lived radicals is likely to be involved in the mechanism of rapid photo degradation of carotenoids. In the case of ground paprika (powder), because of the high surface area, sensitivity of carotenoids to light is many folds and exposure to sunlight for a few hours is enough to completely decolorize red pigments. Energy efficient drying systems should be standardized with optimum moisture content of 8 per cent. Moisture content above 11 per cent allows mould growth and below 4 per cent causes excessive colour loss. During cloudy weather and intense rains, damages as high as 50 per cent are reported. Such unfavourable conditions also lead to discolouration with white spots over the surface of final product. In view of its direct exposure to environment, dirt and dust also get deposited on the chillies. Better retention of colour and glossiness can be obtained by improved drying technologies.

### 3.1.1. *Driers*

#### 3.1.1.1. *Energy Efficient Heat Pump Driers*

Energy efficient heat pump driers are well suited for drying fruit, as they operate at low temperatures. This is essential for chilli drying as the spice otherwise becomes brown instead of bright red colour. The drying temperature should be below 60°C for 6 hrs followed by 40°C for hot air driers. The initially high temperature does not cause browning as the moisture content of the fruit is still too high at the early stage of the drying. To accelerate drying, fruit should be cut into small, regular segments. This allows water to move out of the tissues without interference from the waxy skin. Using a drying oil also increased drying rates. However, this was not effective as cutting fruit and high rates of oil were needed due to tough nature of the waxes.

#### 3.1.1.2. *Solar Drier*

Garg and Krishnan (1974) reported that by the use of solar drier, the drying time for chillies can be reduced to nearly half of the open drying method.

The quality of the chillies dried by the solar drier was also superior. Another study conducted at CAZRI, Jodhpur revealed that the solar drier took 9 days to reduce the moisture content of chilli from 83.6 to 8.5 per cent in comparison to 21 days required for dehydration under traditional open yard drying.

3.1.1.3 *Solar Tunnel Drier*

This drier has been developed by Post Harvest Technology Centre of Tamil Nadu Agricultural University, which ensures quality drying. It contains stainless steel trays with 40 per cent perforations that helps uniform stacking of chillies to achieve the dry processing with final moisture content at four percent within two to three days as compared to seven to ten days in traditional sun drying. The optimum drying condition created by the drier also drastically brings down the high post-harvest wastage percentage of chillies normally associated with the traditional ground spread drying of chillies (Gurumurthy, 2007).

3.1.1.4. *Modified Tobacco Barn*

This technology was developed by Acharya N.G. Ranga Agricultural University. About 10–12 q of ripe chillies can be loaded in the barns by using galvanised iron (GI) wire mesh trays. Drying time varied with pericarp thickness and size of the fruit and takes about 40–50 hrs only as against 12–15 days in open yard sun drying. The cost of barn drying is approximately Rs. 1.5 to 2.00 per kg of dry chillies.

3.1.1.5. *Polyhouse Solar Drier*

A polyhouse solar drier of size 40 × 250 × 8′ has been developed by Acharya N.G. Ranga Agricultural University in collaboration with ITC-ILTD, Guntur to dry about 2 tons of ripe chilli. The drier essentially consists of an arch type polyhouse to hold chillies on two different tiers made of wire mesh fixed to a frame assembled by nuts and bolts. The whole frame structure is covered with a UV stabilized 120 gsm cross laminated semi transparent polyethylene sheet with ventilators at bottom and top to facilitate movement of air. The drying time is 5–8 days to reduce moisture from 75 to 10 per cent in comparison to 15–20 days required to dry chilli in open yard sun drying. Based on the success of this unit, the Government of Andhra Pradesh has announced subsidy for promotion of polyhouse driers.

## 4. GRADING

In general, grading is done after drying. Colour, glossiness, size, moisture content, pest and disease infected fruits and whitened fruits are some of the important factors that traders consider when they buy the produce from farmers. But, it is advisable to do grading at various stages starting from harvesting to prevent microbial contamination and for quality produce.

## 5. PACKING AND STORAGE

The dried fruits are collected, heaped and packed picking-wise in gunny bags (35–40 kg capacity) generally during morning hours when the temperatures are relatively lower as fully dried chillies become brittle causing considerable breaking at higher temperatures. Use of natural colours *viz.* carmocin, erythrocin, indigocaramin, brilliant blue, etc. are suggested instead of synthetic colours for marking the gunny bags. The dried chillies are stored in gunny bags for marketing. During this period, the deterioration in quality is more due to improper packing and storage. To slow colour loss during storage, it is important to keep the spices in cool and out of light. The temperature inside the cold store is maintained from 5–8°C and the relative humidity is usually about 55–65 per cent. The use of commercial cold store for chilli storage has now become almost a general practice among farmers in some states.

## 6. MARKETING

The chillies have well established spot markets at Guntur, Warangal, Khammam in Andhra Pradesh; and Raichur, Bellary in Karnataka. The trade channel involves several members, *viz.* a village level trader, commission agent, wholesaler, retailer and agents for exporters. The commodity changes hands several times, exposing all these members to price risk. The commodity displays high volatility, with the prices heavily dependent on season, production in different producing tracts spread across the country, demand from exporters and the stock available at the cold storages. Guntur is Asia's largest market for chillies. The marketing season begins in the first week of February and peaks during the month of April, and closes by the middle of May. In addition to whole chilli, the marketing is also done in the form of stem-less chilli and crushed chilli.

## 7. GRINDING

Different grinding techniques do not alter powder quality as long as they produce the same particle size. Seeds dilute the colour of the spice, but their exclusion reduces yield. To minimize colour loss, cultivars with high seed antioxidant levels should be selected. However, seed oil that is added to increase gloss of the spice produces off odours.

## 8. VALUE ADDITION

Chilli is valued for its diverse commercial uses. It is gaining importance in the global market because of its by-products like powder, oleoresin, capsanthin, capsaicin, chilli paste and chilli oil. Important products made from chillies are discussed hereunder. A flow chart showing steps followed in value addition of chillies has been depicted in Fig. 12.2.

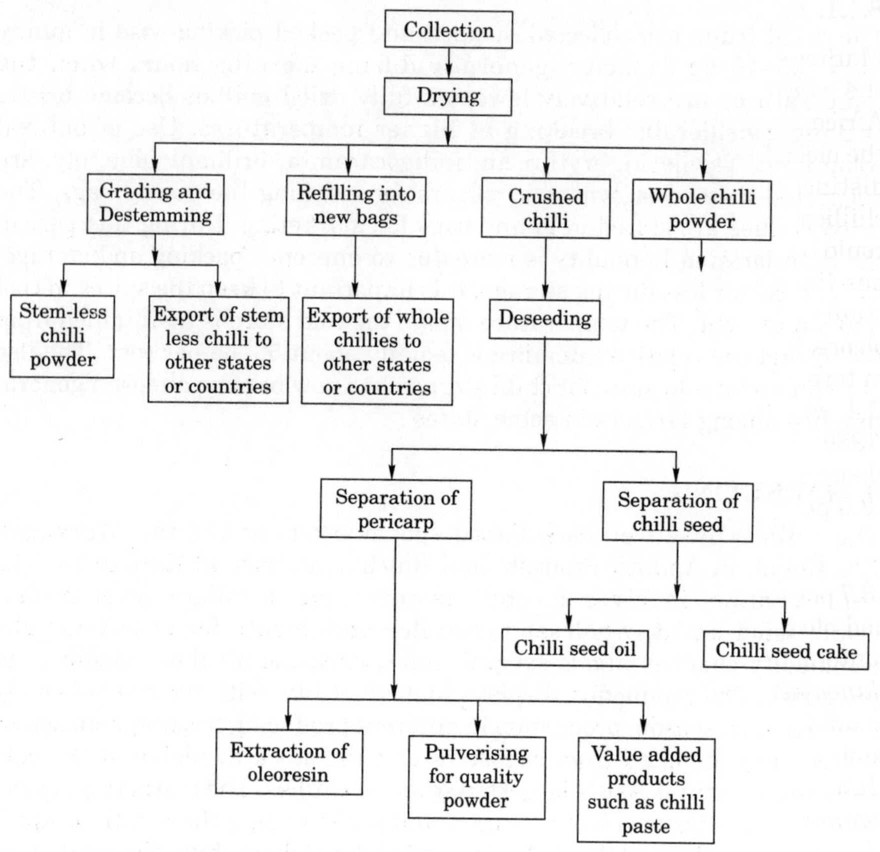


**Fig. 12.2.** Flow chart showing value addition in chilli

## 8.1. Oleoresins

In food and beverage industry, chilli has acquired a great importance in the form of oleoresin that permits better distribution of colour and flavour in food as compared to chilli powder. The food industry generally prefers to use large, highly coloured and less pungent chillies for the preparation of oleoresins.

Oleoresin, in general contains capsaicinoids, carotenoids, colouring principles, fatty components, proteins, vitamins, etc., the relative proportion of which depends on the kind of capsicum, the anatomical parts used and method of extraction (Verghese *et al.*, 1992). Chilli oleoresin, which is prepared from dried chilli powder by solvent extraction method, represents the complete flavour or the true essence. Considerable variability exists among the chilli cultivars with respect to oleoresin recovery.

### 8.1.1. *Influence of Cultivar on Oleoresin Content*

Mathew *et al*. (1971) analyzed the oleoresin yield of some major chillies in the world, which included three varieties from India, four varieties from Africa, two varieties from Japan and one from Bahamas and reported that the oleoresin yield varied from 8.7 to 16.5 per cent. In 1972, Lewis reported distinct differences in quality and yield of oleoresin among varieties of chillies. Mathew and Sankaracharya (1974) found that the Indian chilli could be used to produce a product equivalent to oleoresin red pepper, which has limited demand in food industries. Bajaj and Kaur (1980) conducted studies on the evaluation of 24 entries of chillies and reported that the oleoresin content among the entries varied from 29.52 to 111.52 expressed in terms of ASTA units. The highest (111.52 ASTA units) was recorded by G 4, whereas the lowest (29.52) by Kalyanpur Red No. 1. Raina and Teotia (1986) analyzed chillies of Jammu & Kashmir and reported variability in oleoresin recovery. Teotia and Raina (1987) observed a variation of 9.6 to 18.0 per cent in oleoresin recovery of chillies grown in Himachal Pradesh.

Pradeepkumar (1990) reported that the oleoresin content varied from 18.7 per cent in *C. annuum* to 31.7 per cent in *C. chinense*, while *C. fruitesens* had oleoresin content of 27.3 per cent. The $F_1$ hybrids with high oleoresin content were *C. frutescens* × *C. chinense* (35.37%) and *C. annuum* × *C. chinense* (34.4 per cent). Pruthi (1993) analyzed 15 chilli cultivars grown in different regions of India and reported that the oleoresin content varied from 6.2 to 12.4 per cent on moisture free basis. Tirupathireddy and Mahendranathreddy (1994) reported that the chilli cultivar LCA 235 contained 13.5 per cent oleoresin. Indira *et al.* (1994) reported a range of 14 to 28 per cent oleoresin in chilli accessions belonging to *C. annuum, C. baccatum* and *C. chinense*. Lakshmanachar (1993) found that the varieties with low seed content and free of stalk and calyx were suited for chilli oleoresin production. Gbolade *et al.* (1997) indicated that the Nigerian chillies could be used to produce oleoresin capsicum for use in pharmaceutical preparations. Govinda Reddy *et al.* (1999) evaluated 12 chilli varieties and reported that the highest oleoresin content (14.4%) was recorded by Sel-1, followed by LCA 235 (13.5%). Uma Jyothi *et al.* (2003) analyzed 8 varieties of chillies developed at Regional Agricultural Research Station, Lam and reported that the highest oleoresin content (10.7%) was recorded by LCA 235 followed by LCA 334 with 10.3 per cent. LCA 424 recorded the lowest oleoresin content of 7.1 per cent. In another study, 300 chilli germplasm were evaluated and reported a variation of 4.0 to 16.72 per cent in oleoresin content. The highest oleoresin content was recorded in GP 125 and the lowest in GP 152 (Uma Jyothi *et al.*, 2007). Twenty three cultivars supplied by the coordinated centres of All India Co-coordinated Vegetable Improvement Project (AICVIP) were evaluated for two consecutive years *viz*; 2005–06 and 2006-07. The highest oleoresin content was recorded in DCL 352 (13.82%) and the lowest was recorded in PC 7 (6.91%) (Uma Jyothi *et al.*, 2008).

### 8.1.2. *Influence of Storage Condition on Oleoresin Content*

With good storage methods, the quality of oleoresin was maintained for fairly long periods (Govindarajan *et al.*, 1986). Gopalakrishna and Babylatha (2000) conducted storage studies on raw and roasted samples of chillies; whole and deseeded at 5°C in the refrigerator and at 37°C in an incubator. They reported a loss of 8 per cent in oleoresin in whole chillies of Guntur variety. In case of raw and roasted chilli pericarp powder, the loss of oleoresin was more in samples stored at 37°C than the samples stored at 5°C. Irrespective of storage conditions (*i.e.* either cold store or ambient) all the cultivars showed a declining trend in storage for oleoresin content. The storage of dried chilli under cold store proved better compared to ambient condition. After 9 months of storage in cold store, the chilli cultivars lost 18.10 per cent of oleoresin as against 44.20 per cent of oleoresin in chilli stored at ambient conditions. Among the cultivars, LCA 206 was proved better in retention of oleoresin (Uma Jyothi *et al.*, 2008).

### 8.1.3. *Effect of Chemical Treatment on Oleoresin Content*

The chemical emulsions had significant effects on oleoresin content in chilli variety LCA 235 with maximum oleoresin content (11.66%) by potassium sulphate emulsion treatment (Papakumari, 2001). Significant differences were reported among the chemicals employed for pre-treatment of fresh ripe chilli after their drying. The chilli treated with sodium carbonate (2%) and kept in cold store recorded minimum loss of 26.05 per cent of oleoresin after 9 months of storage. However, the maximum oleoresin content of 8.95 per cent even after 9 months of storage in cold store was recorded by chilli treated with calcium chloride (2%) with a loss of 31.54 per cent. Therefore, the treatment of chilli with calcium chloride (2%) and placement in cold store was found to be more effective (Uma Jyothi and Ravi Shankar, 2007).

## 8.2. Capsanthin (EOA Colour Value)

There is a great demand for the natural chilli fruit colour, which is used in processed foods in place of synthetic colours. Colour is one of the important quality attributes in the spice trade. Basically the colouring matter of chillies is a mixture of carotenoids, yellow and red pigments, which encompass carotenes and xanthophylls, the former structured with carbon and hydrogen and the latter with carbon, hydrogen and oxygen. Capsanthin and capsorubin are the red pigments and the yellow include beta carotene, cryptoxanthin and zeaxanthin (Zechmeister and Cholnoky, 1950). Capsanthin is the major pigment in chillies constituting about 35 per cent of the total pigments (Curl, 1962). Considerable variability exists among the genotypes with respect to colour values.

### 8.2.1. *Influence of Cultivar on Capsanthin Content*

Lease and Lease (1956) reported that the green capsicums do not have capsanthin. Brown capsicums such as chocolate coloured Mexican Pasilla and Mulato, simultaneously contain chlorophyll and capsanthin, a mixture of red and green that produces a brown colour. Davies *et al.* (1970) indicated wide variation in the total carotenoid content as well as the ratio of carotenoid components. The varieties with higher chlorophyll compounds in their fruits had also more carotenoid pigments (Laszlo, 1970).

Narayanan *et al.* (1980) reported that in Mundu, pericarp contained Essential Oil Association of America (EOA) colour value of 39650 and dissepiments contain EOA colour value of 5978. In Jwala, pericarp contained EOA colour value of 41480 and dissepiments contained EOA colour value of 6100. Bajaj and Kaur (1979) informed wide variation in extractable colour among the 24 chilli genotypes. It varied from 22.30 (Kalyanpur Red No. 1) to 118.63 (76-196-2-1-1) expressed in ASTA units. The cultivars/genotypes possess a dominant role in determining the quality attributes like colour (Purseglove *et al.,* 1981). Teotia and Raina (1987) showed a variation of 73.3 to 190.0 mg percentage in total extractable colour among the seventeen samples collected from commercial chilli growing areas of Punjab. Srivastava *et al.* (1990) indicated that the cultivar A 8 had brightest red colour and recorded (0.3 OD). Cultivar A 12 was found to have the lowest total colour and recorded only 0.14 OD among the 33 chilli cultivars evaluated. Joshi *et al.* (1993) reported that paprika types have generally high colour value.

Govindareddy *et al.* (1999) reported significant difference in colour value of different cultivars of chillies. The highest colour value was recorded by LCA 235 (94.0 ASTA units) whereas Sel-1 recorded the lowest colour value of 76.0 ASTA units. Uma Jyothi *et al.* (2003) reported considerable variability in capsanthin content of 8 varieties developed at Regional Agricultural Research Station, Lam. It was in the range of 8570 to 43645 EOA units. The highest EOA units was recorded by LCA 353, followed by LCA 424 with 36844 EOA. Twenty three cultivars supplied by AICVIP were studied and among the them, PC 7 and PC 6 from Pantnagar recorded the highest mean capsanthin content of 50782 and 49456 EOA colour value, respectively, and the lowest mean capsanthin content was recorded by DCL 352 (19932 EOA colour value) (Uma Jyothi *et al.*, 2008).

### 8.2.2. *Influence of Storage Condition on Capsanthin Content*

Storage has a marked influence on the colour of the dried chillies, though it has little effect on their pungency. Since colour is one of the main determinants of the price, which a producer receives and it may be months before the dried ground product reaches the consumer, the problems of loss of colour during the necessary storage period is one the needs to be cared of. The temperature less than 18.8°C is the best for quality retention of red bell peppers compared to 30°C. The loss in aroma and flavour at 30°C was due to deteriorative chemical changes and enzymatic reactions such as the formation of brown water soluble pigments, the formation of cysteic acid and taurine from cysteine and the loss of amino acids and carotenoids (Daoud and Luh, 1967). Chen and Gujmanis (1968) showed that the deterioration of extractable colour pigments of dehydrated ground chilli peppers during storage was due to an auto oxidative process having the kinetics of second order reaction. Philip and Francis (1971) gave a scheme of oxidation of capsanthin by oxygen. Oxidation of capsanthin involves primarily the oxidation of hydroxyl groups followed by scission of the chain at the carbon-carbon alpha to inchain carbonyl group. Once the keto compounds are formed, they are rapidly decomposed into colourless compounds or react with other compounds to give dark coloured compounds.

Oleoresin stored in dark was higher in colour value compared to samples kept in open condition (Krishnamurthy and Natarajan, 1973). Sunlight exhibited profound effect in bleaching the colour and brings about discolouration of red pigments. They further reported that the drying and storage at higher temperature accelerated the rate of colour destruction. The blackening of whole chillies was found to be accelerated by the higher storage temperatures. The loss of colour was markedly slower at low temperature storage. Kanner *et al.* (1978) reported poor stability of carotenoids in stored chilli oleoresins as a major economic problem. Discolouration and blackening in chillies stored at higher temperature and higher moisture contents have also been reported by Sankaranarayana (1987). Rani (1996) determined capsanthin content of 21 and 29 *C. annuum* genotypes in 1980 and 1981, respectively, after one year of storage. Capsanthin contents differed significantly in both years and decreased during storage. Storage loss of capsanthin was lowest in *cv*. Duncale. Gomez *et al.* (1997) reported that the rate of pigment degradation increased for all paprika types at higher storage temperatures and decreased as relative humidity increased. Significant differences with respect to degradation rate and pigment content were found between the cultivars in most of the storage conditions assayed.

Gopalakrishna and Babylatha (2000) conducted storage studies on raw and roasted samples of chillies (whole and deseeded) at 5°C in a refrigerator and in an incubator at 37°C. The deterioration of chillies in colour was faster compared to other spices. They reported a loss of 70 per cent in colour in whole chillies of Guntur variety in comparison to 85 per cent in colour in roasted whole chillies. In case of raw and roasted chilli pericarp powder, the loss of colour was more in samples stored at 37°C than the samples stored at 5°C. Puranaik *et al.* (2001) carried out packaging and storage studies in commercial varieties of Indian chillies *viz*; Guntur and Byadgi whole chillies and reported that the colour retention was good in both MPP (Metallised Polyester Polyethylene) and HDPE (High Density Polyethylene) packing stored at 4–5°C (refrigerator) than at ambient conditions. They also reported that the retention of colour in MPP pouches was better than HDPE pouches at 4–5°C and at ambient conditions irrespective of varietal differences.

The storage of dried chilli under cold store proved better compared to ambient condition with respect to longevity of the produce. After 9 months of storage in cold store, the chilli cultivars lost 32.70 per cent of capsanthin (EOA colour value) as against 67.70 per cent of capsanthin (EOA colour value) in chilli stored at ambient conditions. Among the cultivars, LCA 357 was proved better in retention of capsanthin content (Uma Jyothi *et al.*, 2008).

### 8.2.3. *Effect of chemical treatments on capsanthin content*

It is well established fact that any stored biological product like spices is subjected to deterioration depending upon number of dependent variables. It is also influenced by storage techniques and treatment given to the produce. Lease and Lease (1956) reported that antioxidants such as santoquin, L-tocopherol, butylated hydroxyl anisole, propyl gallate and ascorbic acid were effective for the retention of red colour in chilli powder during storage. When butylated hydroxyl anisole, an antioxidant was added, it markedly improved the colour retention of cured pods. Further, they reported that the changes of colour in paprika during processing and storage with subsequent browning could be attributed to oxidative attack catalyzed by light. Chen and Gujmanis (1968) observed that chilli powder samples stored with moisture contents of 9 to 10 per cent retained better colour than did the samples stored with lower moisture contents of below 7 per cent in oxygen atmosphere. Treatment of chilli pods with ethoxyquin (santoquin) afforded both substantial protection against colour deterioration and an improvement in surface colour during storage.

Natarajan *et al.* (1969) found that the treatment given to whole chillies with butylated hydroxyl anisole, propyl gallate or ascorbic acid did not show a very significant effect on the preservation of colour during storage. Laul *et al.* (1970) informed that the chillies treated with potassium carbonate alone showed poor colour retention. The product was dull and unpleasant in appearance whereas the chillies treated with potassium carbonate and olive oil showed sharp colour and glossy appearance. They mentioned that the presence of oil brings out the colour. While Krishnamurthy and Natarajan (1973) reported that capsanthin which is keto-carotenoid of chillies have strong adsorption affinity with adsorbants like $CaCO_3$, $Ca(OH)_2$ or MgO. Pruthi *et al.* (1980) informed that the simplest method of fresh vegetable preservation was treatment of vegetables in brine solution (12–14% salt + 0.5–1.2% acetic acid, 350–750 ppm $SO_2$) at room temperature for 6–9 months. Harold Egar *et al.* (1981) reported that chillies, Cayenne pepper and paprika when treated with mineral oil, it improved the luster of fruits. Pawar *et al.* (1985) reported that blanching and sulphitation were the important pre-treatment for drying and dehydration of fruits and vegetables. These treatments not only helped in accelerating the drying process but also in maintaining the quality of produce during processing and storage. Sulphitation retarded browning, retained the colour and ascorbic acid. Nagargoji and Singh (1986) reported that the use of pre- and post-harvest additive treatments helped in extending shelf life of fresh vegetables. Shivhare *et al.* (1987) reported that when chillies smeared with oil of mahuwa (*Madhuca longifolia*) imparted glossiness. Lindsay (1985) reported that the bicarbonate salts are widely used in the food industry at levels up to 2 per cent for leavening, pH control, and taste and texture development. Bicarbonate salts have efficacy against aerobic and anaerobic bacteria and yeasts that are food borne (Corral *et al.*, 1986) and against oral pathogens (Miyasaki *et al.,* 1986). Freshly harvested green chillies cleaned thoroughly in water and steeped in saturated brine solution after destalking showed that the enzymatic action was inhibited and the brown discolouration was prevented (Kachru and Srivastava, 1990).

Mohammad *et al.* (1991) dipped the red and yellow fruits of two local cultivars in sodium hypochlorite (95 or 190 mg/l) or in benomyl, calcium chloride, alum, milton (sodium hypo chlorite and sodium chloride solution 2%) or vitamin-C, each at 500 mg/l. After dipping, the fruits were packed in perforated low-density polythene bags and stored at 10°C and 90–95 per cent relative humidity for up to 15 days. Red fruits appeared to be more susceptible to decay than yellow fruits. After 15 days storage, percentage decay was lowest with 10 and 2 per cent, respectively in red and yellow fruits dipped in Milton compared with 76 and 68 per cent decay, respectively, in un-dipped control red and yellow fruits.

Significant differences were reported among the chemicals employed for pre-treatment of fresh ripe chilli after their drying. The capsanthin content was found to be decreasing as the period of storage increased.

Considering the EOA colour value at 9 months after storage, maximum EOA colour value of 8695 was recorded by sodium benzoate 0.1 per cent followed by ascorbic acid 0.5 per cent with EOA colour value of 8540. Although the EOA colour value was the lowest (6354), the minimum loss of EOA colour value of 30.1 per cent was recorded by calcium chloride 2 per cent (Uma Jyothi and Ravi Shankar, 2007).

## 8.3. Capsaicin

The hot flavour of chilli is caused by capsaicin and allied constituents, which have good counter irritant function. It is used in pharmaceuticals for this property and perhaps helps in absorption and movement of bowels. The pungent principle of chillies, capsaicin, is an acid amide of vanillylamine. Capsaicin is considered as double-edged sword providing nutritional and toxic effects. Physiological and pharmacological effects of capsaicin have been studied. The actions on digestive, cardiovascular, respiratory and nervous system are fairly well known. Because of its specific excitatory and neuro toxic properties on fibres, capsaicin has been extensively used as a main tool for neuropharmacological studies concerning pain and thermoregulation. Hence, this molecule is emerging to target diseases associated with deep pain. The molecular action of capsaicin is mediated by neuropeptides such as neurokinins and glutamate (Kal Caulibaly *et al.*, 1998). Considerable variability exists among the different chilli cultivars for capsaicin content is reviewed here under.

### 8.3.1. *Influence of Cultivar on Capsaicin Content*

Pungency is one of the most important characteristic attributes of the genus *Capsicum*. A consistent pungency level is important for processors and consumers. Tirumalachar (1967) reported that the capsaicin is an alkaloid present in chillies responsible for the pungency of chilli and observed that a considerable variability in capsaicin content of sun dried fruits of some varieties of chilli. The range being 0.2723 to 1.1267 mg per 100 mg of the powdered sun dried fruits and was attributed to genotypic differences. According to Balbaa *et al.* (1968) capsaicinoid content increased with fruit maturation in relation to increase in dry matter content. Fruits in summer also reported to posses higher pungency. Mathew *et al.* (1971) compared some Indian varieties of chillies with African varieties with regard to quality and reported that the African varieties were found rich in capsaicin content.

Pankaj and Magar (1978) evaluated ten commercially important varieties of chillies and reported that the capsaicin content varied from 0.053 to 0.912 per cent in whole fruit, 0.082 to 1.53 per cent in pericarp and 0.012 to 0.098 per cent in seeds. Among the varieties, NP 46-A and Pusa

Jwala contain the highest capsaicin of 0.87 and 0.91 per cent, respectively. The two varieties, Byadgi and Kashmiri were found to be less pungent containing only 0.053 and 0.083 per cent capsaicin, respectively. Dry fruits of genus *Capsicum* containing capsaicinoids of 0.4 per cent and above should be called chillies and those below capsicums (Thain *et al.*, 1980). Narayanan *et al.* (1980) reported that the pericarp contained almost all the pungency whereas the chilli seeds contained only traces of pungency with a capsaicin of 0.2723 to 0.1267 mg per 100 g and 0.005 per cent, respectively. The pungency of pericarp was mostly concentrated in the dissepiments. They analyzed the cvs. Mundu and Jwala. In cv. Mundu, pericarp contained capsaicin of 0.17 per cent and dissepiments contain capsaicin of 6.6 per cent. In cv. Jwala, pericarp contained capsaicin of 0.58 per cent and dissepiments contained capsaicin of 7.7 per cent. Bajaj *et al.* (1980) reported a wide variation in capsaicin content among 24 different genotypes studied. Paprika type chilli contained comparatively less quantity of capsaicin than Cayenne and other hot peppers. The capsaicin content between different genotypes varied from 0.15 (Kalyanpur Red No. 2 to 0.925 per cent (K2) on dry weight basis. Purseglove *et al.* (1981) reported that the cultivars/ genotypes possess a dominant role in determining the quality attribute like pungency. Balasubramanian *et al.* (1982) evaluated 53 chilli cultivars for their capsaicin content. The capsaicin content of cultivars varied within the wide limits of 0.09 to 0.59 per cent, the highest was present in the culture No. 516. They further reported a positive correlation between capsaicin; long length of leaves, breadth of leaves, surface area of leaves, number of fruits per plant and weight of fruits per plant. Govindarajan *et al.* (1985) reported that Indian chillies have moderate pungency with capsaicin 0.2–0.3 per cent and are not suitable for the manufacture of high capsaicin oleoresin for pharmaceuticals and export. However, the capsaicin content of Pusa Jwala is 0.7 per cent, thus is a prized raw material for manufacture of high capsaicin oleoresin in the world market. A variation of 0.24 to 0.54 per cent was found in capsaicin content among 17 samples collected from commercial chilli growing areas of Punjab (Teotia and Raina, 1987). According to Sankaranarayana (1987), the pungency in chillies was directly related to the size of the fruit and smaller the fruit, the more pungency. Capsaicinoid content increased with fruit maturation in relation to increase in dry matter content. Tewari (1988) reported that the cv. Pusa Sadabahar was a superior quality chilli with 12 per cent capsaicin content in oleoresin compared to 8 per cent of capsaicin in oleoresin of cv. Pusa Jwala. Ideal extraction material for the industry should contain about 1 per cent capsaicin (Tewari, 1990). De (1992) reported that the maximum amount of capsaicin was found in the inner walls (placenta and dissepiments) of capsicum fruits and showed that the pericarp and seeds contained practically no capsaicin and also reported that capsaicin crystals were observed in the secretion organs of the dried fruit.

Gbolade *et al.* (1997) determined the capsaicin content of fresh and dried chillies (*Capsicum frutesens* cultivars Ata Were and Sombo and *C. annuum* cultivars Rodo and Tatase), obtained from a local market in Nigeria. The most pungent chilli was *C frutesens* cv. Ata Were with 0.49 and 0.37 per cent capsaicin in fresh and dried fruits, respectively. Govinda Reddy *et al.* (1999) evaluated 12 chilli varieties for capsaicin and reported that the highest capsaicin content was recorded by LCA 235 (0.525) and LCA 206 (0.465) and the lowest amount of capsaicin by TC 2 (0.141). Uma Jyothi *et al.* (2003) reported that the capsaicin content of 8 chilli varieties developed at RARS, Lam was in the range of 0.226 to 0.658 per cent. The highest percentage was recorded by LCA 235 and the lowest by LCA 424. In another study, 300 germplasm lines of chilli were evaluated for capsaicin content and it was in the range of 0.094 (GP 158) to 0.99 per cent (GP186) (Uma Jyothi *et al.*, 2007).

### 8.3.2. *Influence of Storage Condition on Capsaicin Content*

The loss of capsaicin in chilli cultivar Pant C1 after one year of storage was more in chilli samples dried and stored with stalk than the samples without stalk. The loss of capsaicin from open chilli samples was up to 82.75 per cent whereas minimum loss of capsaicin was seen in sealed chilli samples (39.69%) (Laxmichand *et al.*, 1990). They also indicated that the loss in capsaicin content in a local cultivar of chilli increased during storage. The loss of capsaicin in opened samples was 37.6 and 41.4 per cent and in sealed samples 33.6 and 37.9 per cent when dried at 52°C and 65°C, respectively. The retained moisture levels (6–10%) in the dried chillies might be responsible for the loss of capsaicin though at a slower rate. Rani (1996) determined the capsaicin contents of 21 and 29 *C annuum* genotypes in 1980 and 1981, respectively, after one year of storage. Capsaicin contents differed significantly in both years and decreased during storage. Storage loss of capsaicin was the lowest in IHR 309-4.

The analysis of samples at different periods in varying containers showed that the capsaicin content of oleoresin samples was reduced to about 45 to 67 per cent within one month (Mini, 1997). After that a gradual increase was noticed in capsaicin content. But plastic bottles were superior in preserving a higher pungency value, followed by pet jar after 8 months of storage. Reduction in capsaicin content of oleoresin was higher in open condition, irrespective of bottles up to one month. Gopalakrishna and Babylatha (2000) conducted storage studies on raw and roasted samples of chillies (whole and deseeded) at 5°C in the refrigerator and in an incubator at 37°C. The deterioration of chillies in pungency was faster compared to other spices. They reported a loss of 20 per cent in capsaicinoids in whole chillies of Guntur variety in comparison to 27 per cent in capsaicinoids in roasted whole chillies. In case of raw and roasted chilli pericarp powder, the loss of capsaicinoids was more in samples stored at 37°C than the samples stored at 5°C. Puranaik *et al.* (2001) carried out packaging and storage studies in commercial varieties of Indian chillies *viz.*, Guntur and Byadgi whole chillies and reported that the both MPP (metallised polyester polyethylene) and HDPE (high density polyethylene) packaging and storage conditions showed a decreasing trend irrespective of storage conditions with respect to capsaicin content. In other words, the capsaicin content gradually decreased in chillies in both the packaging under control, ambient and accelerated conditions. The storage of dried chilli under cold store proved better compared to ambient condition in retention of capsaicin content. The cold store showed a loss of 47.8 per cent capsaicin after 9 months of storage as compared to 63 per cent loss in the chilli stored at ambient condition (Uma Jyothi *et al.*, 2008).

### 8.3.3. *Effect of Chemical Treatments on Capsaicin Content*

Papakumari (2001) reported that the chemical emulsions showed significant effects on capsaicin content of chilli variety LCA 235 recording maximum capsaicin content by potassium sulphate emulsion and less percent of damaged pods by potassium nitrate emulsion. Uma Jyothi and Ravi Shankar (2007) reported that the ripe chilli treated with sodium benzoate 0.1 per cent followed by sodium hypochlorite 2 per cent were found to be effective in recording the maximum capsaicin of 0.863 and 0.825 per cent, respectively immediately after drying. It is also observed that during storage, the capsaicin content was decreasing with an increase in the age of storage. However, after storage for a period of 9 months, the chilli treated with ascorbic acid 0.5 per cent was found to be effective in recording a loss of 39.9 per cent only. In context of total capsaicin content after 9 months of storage, among the chemicals employed, the treatment sodium benzoate 0.1 per cent was found to be superior.

### 8.3.4. *Effect of Drying Methods and Chemical Treatment on Oleoresin, Capsanthin and Capsaicin Content*

Studies on drying methods and chemical treatments on quality aspects of chilli cv. 334 revealed that tobacco barn drying gave higher oleoresin content of 10.5 per cent and was found to be significantly superior to other methods of drying *i.e.* ground tarpaulin and mechanical drying. None of the chemical (*viz.*, BHA, $K_2CO_3$, dipsol) exerts any significant influence on oleoresin content. However, chilli treated with dipsol and dried in tobacco barn recorded the highest oleoresin content of 10.7 per cent followed by chilli treated with BHA and dried in tobacco barn (10.5%). With regard to capsanthin content, drying methods, chemical treatments and their interactions exert significant influence among the drying methods, the chemical drying gave the highest capsanthin content (14503 EOA colour value) followed by tobacco barn drying with 13893 EOA colour value. Among the chemicals, dipsol recorded the highest value (13923 EOA) and was superior to others chemicals. Among the combinations, chilli treated with dipsol and dried in mechanical driers recorded the highest capsanthin content (15616 EOA colour value). Chilli dried in tobacco barn recorded the highest capsaicin content. Among the chemicals, BHA recorded the highest capsanthin content (0.508%). Among the interactions, chilli treated with BHA and dried in tobacco barn recorded the highest capsaicin content of 0.576 per cent (Uma Jyothi *et al.,* 2008).

## 9. CONCLUSION

Value added products such as dehydrated chilli, pickle powder, paste, sauce etc., can be prepared for higher returns. Almost all chillies grown are sold directly, through value addition, we can increase its market price and shelf life in the quality. Efforts of scientific extension agencies and the state government should be strengthened for exploitation of value added products in chillies. Chilli is highly perishable in nature. It requires more attention during harvest, storage and transportation. Most of the growers sell their fresh crop in the market and it is consumed as green chilli as well as in powder form. Through value addition we can increase its market price, shelf life and the quality. For this care has to be taken in picking, cleaning, sorting, grading, drying, packing, storage and transportation.

## REFERENCES

Bajaj KL and Kaur G (1979). Colorimetric determination of capsaicin in *Capsicum* fruits with the folin-ciocalteu reagent. *Mikrochimica Acta* **1**: 81–86.

Balasubramanian T, Raja D, Kashrus R and Runaway P (1982). Capsaicin and plant characters in chillies. *Indian J. Hort.* **39(3&4)**: 239–42.

Chen SL and Gujmanis F (1968). Automation of extractable colour pigments in chilli pepper with special reference to ethoxyquin treatment. *J. Food Sci.* **33**: 274.

Corral LG, Post LS and Montiville TJ (1988). Antimicrobial activity of sodium bicarbonate J. *Food Sci.* **53**: 981–2.

Curl AL (1962). The carotenoids of red bell pepper. *J. Agric. Food Chem.* **10**: 504.

Daoud HN and Luh BS (1967). Packaging of foods in laminate and Al film combination pouches IV: Freeze dried red bell peppers. *Food Tech.* **21**: 212.

Davies BH, Mathews S and Kirt JTO (1970). The nature and biosynthesis of the carotenoids of different colour varieties of *Capsicum annuum. Phytochem.* **9**: 797–805.

De Amitkrishna (1992). The wonders of chilli. *Indian Spices* **29(3)**: 15–18.

Fallik E, Grindberg S and Ziv (1997). Potassium bicarbonate reduces post harvest decay development on bell pepper fruits. *J. Hort. Sci.* **72(1)**: 35–41.

Garg HP and Krishnan A (1974). Solar drying of agricultural products. Drying of chillies in a solar cabinet drier. *Ann. Arid Zone* **13**: 285–92.

Gbolade AA, Omobuwajo OR and Soremekun (1997). Evaluation of the quality of Nigerian chillies for pharmaceutical formulations. *J. Pharma.Biomed.Analysis* **15(4)**: 545–8.

Gomez R, Pardo JE, Costa J, Navarro F and Varon R (1997). Methods for estimating colour in paprika pepper varieties (*Capsicum annuum* L.) *Rivista-di-Scienza- dell Alimemtazione* **26(3–4)**: 91–96.

Gopalakrishna AG and Babylatha R (2000). Effect of roasting of spices on the retention of active principles during storage. *Indian Spices* **37(3)**: 19–21.

Govindarajan VS (1985). Capsicum production, technology, chemistry and quality. Part 1. History, botany, cultivation and primary processing. Critical Reviews Food Science and Nutrition CRC press, USA.

Govindarajan VS, Rajalakshmi D and Naginchand (1986). Capsicum production, technology, chemistry and quality. Part IV—Evaluation of quality. Critical Reviews Food Science and Nutrition, CRC press, USA.

Govindareddy K, Lalithakumari A and Bavaji JN (1999). Variations in the biochemical constituents in chilli varieties. *Indian Spices* **36**: 13–14.

Harold Egar, Kirk RS and Ronald Sawyer (1981). Pearsons chemical analysis of foods, 8th edn., 236–328.

Joshi S, Thakur PC, Verma TS and Verma HC (1993). Selection of spice paprika breeding lines. *Capsicum & Eggplant Newslett.* **12**: 50–52.

Kachru RP and Srivastava PK (1990). Status of chilli processing, *Spice India*, **3**: 13–16.

Kal-Coulibaly S, Coxam V and Barlet JP (1998). Physiological, pharmacological and toxicological interests of chillies and their active principle-capsaicin. *Medecineet Nutrition* **34(6)**: 236–45.

Kanner J, Mendel H and Budowski P (1978). Carotene oxidizing factors in red pepper fruits oleoresin-cellulose solid model. *J. Food Sci.* **43(3)**: 709–12.

Krishnamoorthy MN and Natarajan CP (1973). Colour and its changes in chillies. *Indian Food Packer* **27**: 39–43.

Lakshmanachar MS (1993). Spice oils and oleoresins. *Spice India* **6(8)**: 8–10.

Laszlo L (1970). Chlorophyll and pigments analyses with vegetable and spice paprika varieties. *Zoldsegtermesztes* **4**: 165–72.

Laul MS, Bhalerao SD, Rane VR and Amia BL (1970). Studies on the sun drying of chillies. *Indian Food Packer* **24(2)**: 22–27.

Laxmichand R, Saxena P, Garg GK and Singh BPN (1990). Drying of chilli (*Capsicum*) fruits and effects of drying temperature and storage on capsaicin content. *Crop Res.* **3(2)**: 252.

Lease JG and Lease EJ (1956). Effect of soluble antioxidants on the stability of red colour of pepper. *Food Tech.* **10**: 403.

Lewis YS (1972). The importance of selecting the proper variety of spice for oil and oleoresin extraction. Proc. International Conf. Spices, Tropical Production, London, 183.

Lindsay RC (1985). Food additives, *In.* Food chemistry (Fennema OR, *ed.*) Marcel Decker Inc. New York, USA.

Mangaraj S, Singh A, Samuel DVK and Singhal OP (2001). Comparative performance evaluation of different drying methods for chillies. *J. Food Sci.Tech.* **38(3)**: 296–9.

Mathew AG and Shankaracharya NB (1974). Oleoresin from Indian chillies. *In*: Proc. Symposium on the development and prospects of spice industry in India, AFST, CFTRI, Mysore.

Mathew AG, Lewis YS, Gadget RH, Nambudri ES and Krishnamoorthy N (1971). Oleoresin capsicum. *Flavour India* **2**: 23–26.

Mini C (1997). Oleoresin recovery, quality characterization and storage stability in chilli (*Capsicum* spp) genotypes. Ph. D. Thesis, College of Horticulture, Vellanikkara, Thrissur, Kerala Agricultural University, Kerala.

Miyasaki KT, Genco RJ and Wilson ME (1986). Antimicrobial properties of hydrogen peroxide and sodium bicarbonate individually and in combination against selected oral gram negative facultative bacteria. *J. Dental Res.* **65**: 1142–8.

Mohammed M, Wilson LA and Gomes PI (1991). Effect of post-harvest dips on the storage quality of fruit of two hot pepper (*Capsicum frutescens* L.) cultivars. *Trop. Agric.* **68(1)**: 81–87.

Nagargoji KM and Singh JN (1986). Vegetable processing, opportunities and challenges. *Indian Food Industry*, Vol. 5 April-June: 77–79.

Narayanan CS, Sankarikutty B, Raja Raman K, Bhat AV and Matthew AG (1980). Studies on separation of high pungent oleoresin from Indian chilli. J. *Food Sci. Tech.* **17**: 136–8.

Narayanan KM, Dwarakanath CT, Ramachandrarao TN and Johar TS (1964). Studies on pepper oleoresin from pepper rejections. Part II—Storage studies on oleoresin. *Spices Bulletin* **3(8)**: 117.

Natarajan CP, Krishnamoorthy MN, Padmabai R and Kuppuswamy S (1969). Annual Report, CFTRI Mysore, India.

Nisar Ahmed, Krishnappa KM, Upperi SN and Khot AB (1987). Pungency in chillies as influenced by variety and maturity. *Current Research* (UAS, Bangalore). **16(12)**: 161–75.

Pawar VD, Patil DA, Khedkar DM and Ingle UM (1985). Studies on drying and dehydration of pumpkin. *Indian Food Packer* **34(4)**: 58–66.

Pradeepkumar T (1990). Interspecific hybridization in *Capsicum*. M. Sc. thesis, Kerala Agricultural University, Vellanikkara pp. 102.

Pruthi JS (1993). Major spice of India—Crop management and post harvest technology. ICAR, (2) 221–2.

Pruthi JS, Saxena AK and Afanam JK (1980). Studies on the determination of optimum conditions of preservation of fresh vegetables in acidified sulphited brine for subsequent use in Indian style curries. *Indian Food Packer* **34(6)**: 9.

Puranaik JS, Nagalakshmi M, Balasubramanyam S and Shankaracharya NB (2001). Packaging and storage studies on commercial varieties of Indian chillies (*Capsicum annuum* L.) *J. Food Sci. Tech.* **38(3)**: 227–30.

Purseglove JW, Brown EG, Green CL and Robbins SRJ (1981). Species, (Vol. 1) Tropical Agricultural Series, Longman Group Ltd. UK. pp. 356.

Raina BL and Teotia MS (1986). Evaluation of chillies (*Capsicum annuum* L.) grown in Jammu & Kashmir. *Indian Food Packer* **40(5)**: 6–10.

Rani PU (1996). Evaluation of chilli germplasm for capsanthin and capsaicin contents and effect of storage on ground chilli. *Madras Agric. J.* **83(5)**: 288–91.

Sankaranarayana ML (1987) Research and development in spices, spice products and their end uses. *Indian Spices*: **24(2&3)**: 5–13.

Satyanarayana VV and Sukumaran CR (2002). Post-harvest profile of chillies. Technical bulletin AICRP-PHT, PHT Centre, Bapatla.

Shivhare US, Narain M, Agarwal US, Saxena RP and Singh BPN (1987). Some physico-chemical properties of chillies. J. *Food Sci. Tech.* **24(2)**: 98–99.

Srivastava M, Nagamadir M, Pandey PH and Salokhe VM (1990). Design and development of a low cost chilli dryer. Proc. International agricultural engineering conference and exhibition. Bangkok Thailand 3–6 December 505–15.

Sumathykutty MA and Mathew AG (1987). Chilli processing. *Indian Cacao, Arecanut and Spices J.* **4**: 112–3.

Teotia MS and Raina BL (1987). Evaluation of chillies grown in Punjab. *Indian Food Packer* **41(3)**: 81–86.

Tewari VP (1988). Genetic improvement of capsaicin content in hot pepper (C. *frutescens* L.) *Capsicum News Letter* **7**: 41.

Tewari VP (1990). Development of high capsaicin chillies (*Capsicum annuum*) and their implications for the manufacture of export products. *J. Plantation Crops* **18(10)**: 1–13.

Thain EM, Robbins SRJ and Green CI (1980). World trade in chillies. All India workshop on chillies organised by SEPI, Hyderabad March 1980: pp. 143–57.

Thirupathireddy A and Mahendranathreddy T (1994). Variability in the biochemical constituents of some promising varieties of chilli fruits. *Andhra Agric. J.* **41(1–4)**: 70–71.

Tirumalachar DK (1967). Variability for capsaicin content in chilli. *Curr. Sci.* **36**: 269–70.

Uma Jyothi K and Ravi Shankar C (2006). Effect of post harvest chemical treatments on storage of dried chilli cv. LCA 334, *Veg. Sci.* **33(1)**: 38–44.

Uma Jyothi K, Vijaya Lakshmi and Khalid Ahmed (2003). Screening of chilli cultivars for export. *Andhra Agric. J.* **50**: 475–6.

Uma Jyothi K, Surya Kumari S, Venkata Reddy P and Ravi Shankar C (2008). Study of drying methods and chemical treatments on quality aspects of chilli cv. LCA 334. Souvenir and Abstr. National Symposium on Spices and Aromatic crops (SYMSAC IV) 25–26 November 2007, OUAT Bhubaneshwar, pp. 357.

Uma Jyothi K, Vijaya Lakshmi T, Surya Kumari S, Venkata Reddy P and Siva Reddy KV (2007). Variation in the biochemical constituents of chilli germplasm. UGC Sponsored National Seminar on recent trends in plant sciences 28–29 June, p. 49.

Zeichmeister L and Cholnoky L (1950). Principles and practices of chromatography, Chapman Hall Ltd. London.

# 13

# Chemistry and Quality of Paprika and Paprika Products—Research Needs

JAGDISH SINGH, RAJESH KUMAR AND MATHURA RAI

## 1. INTRODUCTION

Paprika chilli (*Capsicum annuum* L.), is known by different names in different parts of the world. The genus *Capsicum,* which is commonly known as "red chile", "chilli pepper", "hot red pepper", "Tabasco", "paprika", "cayenne", etc., belongs to the family solanaceae. Species of *Capsicum* which are native to South and Central America, Mexico and the West Indies are now widely cultivated throughout temperate, subtropical Europe, the Southern US, tropical Africa, India, East Africa and China. Indian chillies are medium to highly pungent while African chillies have higher pungency. It is believed that about 30 species are distributed throughout the world, and range from the hot variety to the sweet variety. The five major species of cultivated *Capsicum* are: *Capsicum annuum, Capsicum frutescens, Capsicum chinense, Capsicum pendulum and Capsicum pubescens.*

Columbus' discovery of the New World resulted in the discovery of the chillies which was found to be an excellent substitute to black pepper, called the chilli pepper. This chilli, which was once native to the warm temperate and tropical regions of America, found its way to the Indian coast. It had its roots in the West Asian coast, and also in the islands of East Africa. This augmented the use of chillies in the food of various nationalities and societies and has become an integral part of indigenous diets. It is now an important vegetable and spice crop grown in the tropical and subtropical regions of the world. Both, ripe red and green chillies are an important constituent, used to impart pungency, flavour and colour to food. It takes about 30–35 days from fruit set to complete development of green fruits. The fruit starts ripening 80–90 days after fruit set depending on the cultivars. The carotenoid content increases to about 120 per cent at the time of full ripeness as compared to the green stage. Green chillies are

Indian Institute of Vegetable Research, Varanasi.

rich in vitamin A and C and the seed contains traces of starch. The amount of vitamin C varies from 110–210 mg/100 g in green chillies, whereas, the red ripe fruit contains around 113–160 mg/100 g.

## 2. INNOVATIVE USES OF CHILLI PEPPER

Among the spices, hot pepper or chilli (*Capsicum* spp.) has been popular from ancient time. Many members of the genus *Capsicum* are of commercial interest, not only for their taste and colour, but also because of their essential oil and active principles. The *Capsicum* genus represents a diverse plant group, from the well-known fruits of the sweet green bell pepper and bright red paprika to the fiery hot small red chilli fruit. The fruits of the genus *Capsicum* have as many versatile uses as its diversity. It is the most popular and widely used spice all over the world. The red pepper fruit has been used since centuries as a source of pigments to add to or change the colour of foodstuffs. The pungency for the red peppers and the colour value for the paprikas are the most important quality traits. Fruits are consumed in fresh, dried or processed forms, as table vegetable or spice. Fruit carotenoids (colour), capsaicinoids and flavour extracts are used in food, feed, medicine and the cosmetic industries. The essential oil is obtained by steam distillation of the powder. Oleoresins are produced by solvent extraction. These are used by spice industries and have a very high potential. Oleoresin of paprika (colour extracts from non-pungent fruits) is still the single largest selling oleoresin in the world market. Currently, this is manufactured in countries like Spain, Morocco, Ethiopia, Mexico and the United States. The paprika oleoresin is natural colouring agent, therefore it is considered to be among the best substitute of synthetic colour used in food and cosmetic industries. In food processing industries, especially in meat industry, concentrated oleoresin is added to the processed meat to impart attractive colour. In beverages industries, oleoresins are used to improve colour and flavour of its products. In certain regions of the world (*e.g.* Japan and South Korea) oleoresins are mixed with chicken feed in order to impart attractive red colour to chicken skin and red/yellow colour to yolk. Oleoresins are mixed with the feed of flamingoes in zoo and koi in aquariums for improving the feather colour (Bosland, 1996).

Capsaicin, the active hot principle of the *Capsicum* species, has many medicinal properties. With the discovery of a capsaicin-like receptor in the human body in 1997, the demand for capsaicin in the therapeutic field for the treatment of pain and inflammatory diseases is of growing importance. Recently, capsaicin has also been used as an antiobesity agent because of its cholesterol lowering properties. The antimicrobial and antiseptic properties of some *Capsicum* species are also well documented. The pharmaceutical industry uses capsaicin as a counter irritant balm for external application (Carmichael, 1991). Capsaicinoids (mainly capsaicins) are active ingredient in "Heet" and "Sloan's Liniment" used for sore muscles (Boasland, 1996). Pharmaceutical industry uses capsaicinoid extracts to prepare certain drugs (sprays), which are applied externally to stop pains of arthritis (rheumatoid arthritis and osteoarthritis) and to relieve cramps. The capsanthin and capsorubin (exclusively present in pepper fruits) can improve the cytotoxic action of anticancer chemotherapy and is considered to be potential carotenoid for possible resistance modifiers in cancer chemotherapy.

### 2.1. Capsaicinoids in Self-defense Weapons

An interesting application of capsaicinoids is their use in self-defense weapons. Defense sprays have become popular for both police use and personal protection. Most of them contain o-chlorobenzylidene, malononitrile, chloroacetophenone, oleoresin or a combination of these ingredients. When applied topically, capsaicin produces a spontaneous inflammatory reaction in mucous membranes, and contact with eyes results in blepharospasm caused by irritation of the corneal nerves. Additional symptoms of eye contact include extreme burning heat, lacrimation, conjunctival edema and hyperemia (Gonzalez *et al.,* 1993). In the nasal mucosa, capsaicin produces burning pain, sneezing and a dose-dependent serious discharge (Geppeti *et al.,* 1988). Contact with skin produces a burning sensation, erythema without vesiculation, while its inhalation results in transitory bronchoconstriction cough and retrosternal discomfort (Fuller *et al.*, 1985).

## 3. QUALITY EVALUATION OF CHILLI PAPRIKA AND ITS PRODUCTS

Quality evaluation techniques include physical, chemical-microbiological, instrumental and sensory evaluation of chillies (whole and ground) and their processed products, like chilli oleoresin, etc. *Capsicum* fruits contain colouring pigments, pungent principles, resin, protein, cellulose, pentosans, mineral elements and a very little volatile oil, while seeds contain fixed (non-volatile) oil. The fruits of most *Capsicum* species contain significant amounts of vitamins B, C, E and provitamin A (carotene) when in a fresh state. The large type of *C. annuum* is among the richest known sources of vitamin C, which may be present up to 340 mg/100 g in some varieties. Fruits of *Capsicum* species have relatively low volatile oil content, ranging from about 0.1 to 2.6 per cent in paprika. The characteristic aroma and flavour of the fresh fruit is imparted by the volatile oil. The composition of the volatile oil of fresh green bell peppers has been examined by Buttery *et al.* (1969) using Gas Chromatography. Twenty four components in this oil were positively identified. One of the major components, 2-methoxyisobutyl pyrazine, was considered to possess an aroma characteristic of the fresh fruit and to dominate the organoleptic profile of paprika.

### 3.1. Major Colouring Pigments in Paprika

The colour of the paprika powder is the principal criterion for assessing its quality. The pigment content of paprika powder can range from 0.1 to 0.8 per cent. The major colouring pigments in paprika are capsanthin and capsorubin, comprising 60 per cent of the total carotenoids. Other pigments are β-carotene, zeaxanthin, violaxanthin, neoxanthin and lutein (Anu and Peter, 2000). The pigment content increases as the fruit ripens and continues after maturity. The extractable colour of paprika is generally measured by a spectrophotometric procedure and usually expressed in ASTA (American Spice Trade Association) Colour Value or in Colour Unit, which is 40 times the ASTA colour. The colour of different carotenoids has been given in Table 13.1. The value originally intended to represent the highest dilution at which colour is just noticeable under a standard condition. It is now replaced by the spectrophotometric reading at 460 nm. In international trade the colour intensity of paprika pods is generally expressed as ASTA units, unlike oleoresin which is described in colour value. Nearly 40 per cent of paprika oleoresin is used to blend in chicken feed. The red pigment and trans-capsanthin expressed as percentage of total carotenoids estimated by spectrophotometer and HPLC, respectively, are of significance. But a similar factor to know the red pigment is the absorption ratio which is the ratio of absorption at 470 nm over absorption at 455 nm. Higher the value, *i.e.* 1, higher is the red pigment, while a value of 0.96 shows low level of red pigment. The recommended colour method for paprika is the ASTA Official Method.

**Table 13.1.** The colour of individual carotenoid pigments

| Sl. No. | Pigments | Colour |
|---|---|---|
| 1. | Chlorophyll a | Green |
| 2. | Chlorophyll b | |
| 3. | β-carotene | |
| 4. | β-carotene 5, 6 epoxide | |
| 5. | β-carotene 5, 6, 5′ 6′ diepoxide | |
| 6. | β-cryptoxanthin mono and di epoxide | Yellow |
| 7. | Mutotoxanthin | |
| 8. | Antheraxanthin | |
| 9. | Violaxanthin | |
| 10. | Luteoxanthin-a | |
| 11. | Luteoxanthin-b | |

**Table 13.1.** *Contd.*

| Sl. No. | Pigments | Colour |
|---|---|---|
| 12. | Neoxanthin | |
| 13. | β-cryptoxanthin | |
| 14. | Lutein | Orange |
| 15. | Zeaxanthin | |
| 16. | Capsanthin | |
| 17. | Capsanthin epoxide | Red |
| 18. | Capsorubin | |

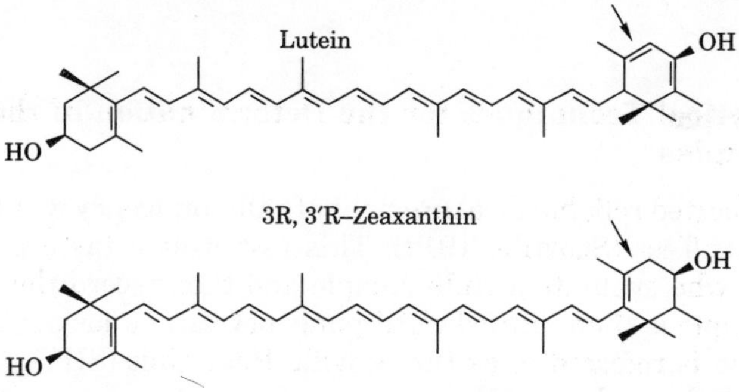

**Fig. 13.1.** Lutein and zeaxanthin also known as macular carotenoids benefit eye health

## 3.2. Pungency Principles in Chillies

The most important pungent principle naturally present in chillies is capsaicin (8-methyl-N-vanillyl-6-enamide), an alkaloid compound characterized by acrid and burning taste, which is located in the internal partitions of the fruit (Kulka, 1967). Capsaicin and the four related compounds-dihydro-capsaicin, nor-dihydrocapsaicin, homocapsaicin and homo-dihydrocapsaicin are all responsible for pungency. Of these, the first two compounds are equally pungent and constitute 80–90 per cent of the total capsaicinoids; the later two are present in minor amounts. The degree of pungency and the character of taste sensation vary markedly with different varieties of chillies. The heat of *Capsicum* powder is measured by Scoville heat units (SHU, Scoville, 1912). One part per million concentration of capsaicinoids is measured as 15 SHU. The nature of pungency has been established as a mixture of seven homologous branched-chain alkyl vanillyl amides, named capsaicinoids. Capsaicin stimulates gastric secretions and, if used in excess, causes inflammation. Its level varies widely in chilli peppers, from less than 0.05 per cent in the mildly pungent types to as high as 4.28 per cent (Mathur *et al.*, 2000) in the hottest chillies.

### 3.3. Biosynthesis of Capsaicinoids

Capsaicinoids are synthesized from phenylpropanoid intermediates through the cinnamic acid pathway. One of the enzymes involved in the biosynthesis is the caffeic acid-3-0-methyltransferase which has been reported to be important in lignification (Inoue *et al.*, 1998). In capsaicinoid synthesis, it catalyzes the conversion of caffeic acid to ferulic acid by methylating the hydroxyl group at the 3′ position on the benzene ring. Ferulic acid is converted to vanillin. Vanillin is first converted to vanillylamine, which later condenses with branched chain fatty acid to form capsaicinoids. At all the stages of growth, the total or individual capsaicinoids based on dry weight is far higher in the placenta than in the pericarp, thus the main site of capsaicinoids synthesis is the placental tissue of the fruits (Fujiwake *et al.*, 1982).

### 3.4. Analytical Techniques for the Determination of the Pungent Principles

The first reported reliable measurement of chilli pungency was the Scoville Organoleptic Test (Scoville, 1912). This test uses a taste panel of five individuals who evaluate a chilli sample and then record the hot flavour level. A sample is then diluted until pungency can no longer be detected. This dilution is referred to as the Scoville Heat Unit (SHU). This test is subjective, and members of the taste panel cannot determine the amount of each of the capsaicinoids present in the sample. There are seven different capsaicinoids, and the hot flavour is generally made up of at least two of these compounds. Thus the SHU test is not accurate since it depends on the individual's taste sensitivity, which changes from person to person and does not measure the actual chemical percentage within the product. SHU test may be an appropriate test for the food spice community; however it cannot serve as the standard. One Scoville Unit corresponds to about fifteen parts per million (ppm) of the capsaicin. This corresponds to fifteen microgram of capsaicin per gram of the dry chilli. Therefore, the Scoville organoleptic test has since been replaced with instrumental methods to scientifically measure the amount of capsaicin. Currently, analysis of capsaicinoids is conducted by using spectrophotometric (Meilgaard *et al.*, 1987), gas chromatographic (Krajewska and Powers, 1987) and high performance liquid chromatography (HPLC) procedures (ASTA, 1985). Recent studies have further shown that the quantification of capsaicinoids using LC-MS-MS was more sensitive and exhibited greater accuracy, even at low analyte concentrations.

Although the choice of an analytical method for capsaicinoid determination usually depends upon the available instrumentation, the following considerations should be taken into account. The sensory, colorimetric, spectrophotometric or thin layer chromatography (TLC) procedures are not convenient for the determination of individual compounds and are, therefore, recommended only for the analysis of total capsaicinoids. Reverse-phase HPLC with a complexation step could be an alternative method. Although GCMS is believed to offer good results, it still requires specific sample preparation and costly instrumentation. For its ability to identify capsaicinoids in dilute or very small samples and also for its high costs, LC-MS-MS should be employed for forensic analysis.

### 3.5. Analytical Techniques for the Determination of the Colouring Principles

Column chromatography has been the traditional method for the separation of carotenoids (Davis, 1965). For example, with paprika carotenoids, SeaSorb 43 is effective, but it does cause some isomerization and oxidation of pigments; hence, a more rapid method is desirable. Paper chromatography based on impregnated papers (Booth, 1962) and papers with suitable fillers (Jensen, 1960) have been suggested. Separations are described based on thin-layer systems on alumina (benzene), silica gel G (methylene chloride-ether, petroleum ether-benzene, undecane-methylene chloride and methylene chloride-ethyl acetate), calcium hydroxide (hydrocarbon mixture-methylene chloride), secondary magnesium phosphate (carbon tetrachloride, benzene and petroleum ether-ether), silica gel G mixed with rice starch (12-hexane-ether), calcium hydroxide (benzene, benzene-methanol and petroleum ether benzene) (Stahl *et al.*, 1963). It has also been reported that not all carotenoid mixtures could be separated with a single solvent and a single absorbent in a thin-layer system. Rosebrook (1971) reported a method for the detection of the extractable colour in paprika and paprika oleoresin. The colour in paprika is extracted with acetone, and absorbance is measured at 460 nm.

### 3.6. Quality Evaluation of Oleoresins of Capsicum, Paprika, red Pepper and Chillies

The term 'oleoresin' has been used for the fractionated components that constitute the colour and flavour (pungency, aroma and related sensory factors) and truly recreate upon dilution of the sensory qualities of the original fresh material. Oleoresins are essentially divided into three types. Oleoresin paprika is used as a food colouring agent in processed meats, dairy products, soups, sauces and snacks. Oleoresin red pepper is a source of both colour and pungency, essentially used in canned meats, sausages, in some snacks and in a dispersed form in some drinks, such as gingerale. Oleoresin capsicum (African) is the most pungent and is used for the counter-irritant property in plasters and pharmaceutical preparations.

There is an apparent confusion in the nomenclature of oleoresins of capsicum, chillies, paprika and red pepper, which has been sorted out by the Essential Oils Association (EOA) of USA (1979) as follows:

**Table 13.2.** Nomenclature of oleoresins of capsicum, paprika, red pepper and chillies

| Type of oleoresin of capsicums | Colour value | Colour description | Pungency of bite intensity |
|---|---|---|---|
| Oleoresin capsicum | 4,000 max | Clear red, Light amber | Very high bite/ pungency |
| Oleoresin red pepper | 240,000 max | Very deep red colour | high bite/pungency |
| Oleoresin paprika | 40,000–100,000 | Deep red colour | Nil or negligible |
| Oleoresin chilli | 4 000–10,000 | Deep red colour | Highly pungent |

*Source*: Essential Oils Association of USA; Eisvale (1981).

Various solvents are used for oleoresins extraction, and the yield and quality of oleoresins depends upon the solvents used. Ethylene dichloride is a commonly used solvent for the extraction of oleoresins.

## 4. QUALITY OF OLEORESIN PAPRIKA

Oleoresin paprika is the product obtained by solvent extraction of the powdered dried ripe red pods of paprika *Capsicum annuum* L., with the subsequent removal of the solvent. Oleoresin paprika is evaluated strictly on a unit colour basis. The bulk of the oleoresin paprika has a colour value of 40,000–100,000. It is deep red coloured, viscous liquid, with a characteristic odour of paprika.

### 4.1. Quality of Chilli Oleoresin

Oleoresin chilli is extracted from chillies (the fruit of red pepper, *Capsicum annuum* L., or *Capsicum frutescens* L.) using approved food grade solvent and subsequent careful removal of the residual solvent by distillation. In addition to the intense pungency caused by the capsaicin and the small quantities of allied alkaloids, the oleoresin chilli has dark red carotenoid pigments. The oleoresin should be free from rancid flavour.

## 5. CAPSAICINOIDS AND TOXICITY

Many studies have been conducted on the most abundant capsaicinoids in the *Capsicum* pepper, capsaicin and dihydrocapsaicin. Capsaicin, the pungent phenolic compound of the *Capsicum* species, has shown a wide range of pharmacological properties, including antigenotoxic, antimutagenic and anticarcinogenic effects. Other studies, however, have shown it to be a tumour promoter and potential mutagen and carcinogen, resulting in capsaicin being termed as "a double edged sword" (Richeux *et al.*, 1999). Many other studies on capsaicinoid toxicity have shown that capsaicin directly inhibits protein synthesis by competition with tyrosine in a cell-free system, as well as in a cell culture system according to the cell type and the extracellular concencratlon (Cochereau *et al.*, 1996). Reports in the literature (Nagabhushan and Bhide, 1985) indicate that capsaicin and its metabolites are able to induce DNA damage and mutagenicity observed in bacteria and *in vivo* micronucleus formation in mice. In the presence of Cu (II) and molecular oxygen, capsaicin was reported to cause strand scission in DNA through an oxidative mechanism (Morri *et al.*, 1995). The green chillies are rich in rutin, which is of immense pharmaceutical importance.

## 6. CONCLUSION AND RESEARCH NEEDS

There is a need for the development of improved high yielding varieties of paprika that are rich in attractive red pigments and which have higher drying yield and disease resistance. There is also a need to develop better, simpler and quicker methods of quality evaluation of chillies and chilli oleoresin in conjunction with better sensory evaluation techniques. There is a great potential for the development and export of paprika and other products. There is an estimated world demand for about 50,000 tonnes, of which Spain alone utilizes 20,000 tonnes annually. Greater emphasis may also be laid on quality production of chilli and paprika. Oleoresin, including tailored oleoresins, to meet the varying demands of the various types of food industries and the pharmaceutical industry. It may also be noted that the paprika oleoresin is one of the most celebrated and most popular natural colourants for different categories of processed foods. Capsicum, which is traditionally consumed as a vegetable or food additive, is now gaining importance also for its capsaicinoids and carotenoids, owing to their nutraceutical and therapeutic value. Production of these constituents can be achieved by genetic engineering and biotechnology; however, adequate efforts have not been made in this direction. The challenge of obtaining a zero capsaicin and high colour value crop will be of great economic value, which is achievable through genetic engineering.

## REFERENCES

American Spice Trade Association (1968). Official Analytical Methods for Spices, 2nd Edition, ASTA, New York, USA.

Anu A and Peter KV (2000). The chemistry of paprika. *Indian Spices* **37(2)**: 15–18.

Association of Official Analytical Chemists (AOAC) (1990). Official and tentative methods of Analysis, AOAC, Arlington, Virgina, USA.

Booth VH (1962). A method for separating lipid components of leaves. *Biochem. J.* **84**: 444.

Bosland PW and Votava EJ (1999). Pepper: vegetable and spice capsicums. pp. 91–95. Wallingford CAB1 publishing.

Buttery RC, Selfort RM, Gudanini DG and Ling LC (1969). Characterization of some volatile constituents of bell peppers. *J. Agric. Food Chem.* **17**: 1322–7.

Carmichael JK (1991). Treatment of Herpes Zoster and post therapeutic neuralgia. *Am. Family Physician* **44**: 203–10.

Cochertau C, Sanchez D, Bourhaoui A and Creppy EE (1996). Capsaicin: a structural analog of tyrosine inhibits the aminoacylation of tRNAtyr. *Toxicol. Appl. Pharmacol.* **14**: 135–7.

Essential Oil Association of America (EOA), USA (1979). EOA Book of Standards and specifications. EOA of USA. No. 60, 42nd Street, NY 10017, USA.

Fujiwake H, Suzuki T and Iwai K (l982). Intracellular distribution of enzymes and intermediates involved in the biosynthesis of capsaicin and its analogs in capsicum fruits. *Agric. Biol. Chem.* **46**: 2685–9.

Fuller K, Dixon C and Barnes P (1985). Bronchocontsrictor response to inhaled capsaicin in humans. J. *Appl. Physiol.* **58(4)**: 1080–1.

Geppeti P, Fusco B and Marabine S (1988). Secretion, pain and sneezing induced the application of capsaicin to the nasal mucosa. *British J. Pharmacol.* **93**: 509–14.

Inoue K, Sewalt VJH, Balance GM, Ni W, Sturzen C and Dixon KA (1998). Development, expression and substrate specificities of α, α-caffeic acid-3-0 methyltransferase and caffeoylcoenzyme-A-3-0-methyltransferase in relation to lignifications. *Plant Physiol.* **17**: 761–70.

Kosuge S and Furuca P (1970). Studies on pungent principles of *Capsicum.* Chemical composition of pungent principles. *Agric. Biol. Chem.* **34:** 268.

Krajewska AM and Powers JJ (1987). Gas chromatographic determination of capsaicinoids in green *Capsicum* fruits. *J. Assoc. Off. Anal. Chem. Int.* **70(5)**: 926–8.

Kulka K (1967). Aspects of functional groups and flavour. *J. Agri. Food Chem.* **15**: 48.

Mathur Ritesh, Dangi RS, Dass SC and Malhotra RC (2000). The hottest chilli variety in India. *Curr. Sci.* **79(3)**: 287–8.

Meilgaard M, Civille GV and Carr BT (1987). Sensory evaluation techniques. CRC Press, Boca Rotan, EL, Vol. 2.

Morre DJ, Chueh PJ and Morre DM (1995). Capsaicin inhibits preferentially the NADH oxidase and growth of transformed cells in culture. *Proc. Natl Acad. Sci.* USA, **92**: 1531–3.

Nagabhushan M and Bhide SV (1985). Mutagenicity of chilli extract and capsaicin in short-term tests. *Environ. Mutagen* **7**: 881–8.

Pruthi JS (1980). Spices and condiments-chemistry, microbiology and technology, Acad. Press Inc., New York, USA, pp. 1–150.

Purseglove JW, Brown EG, Green CL and Robins SKJ (1981). Spices. 1st *edn*, Longman Inc., New York, USA, p. 331, (Chillies/Capsicum).

Richeux F, Cascante M, Ennamany R, Saboureau D and Creppy EE (1999). Cytotxicity and genotoxicity of capsaicin in human neuroblastoma cells SHSY-5Y. *Arch. Toxicol.***73**: 403–109.

Rosebrook DD (1971). Collaborative study of a method for the extractable colour in paprika oleoresin. *J. Assoc. Off. Anal. Chem.* **54**: 37.

Scoville WL (1912). Note on *Capsicum*. *J. Amer. Pharm. Assoc.* **1**: 453.

Stahl WH (1963). Critical reviews of methods of analysis of oleoresins (quality assessment). *J. Assoc. Off. Anal. Chem.* **48**: 515.

# Appendix-I

## Highlights of Brain Storming Session

The Brain Storming Session on 'Advances in Chilli Research' was held on 28$^{th}$ March 2008 at Indian Institute of Vegetable Research, Varanasi. The inaugural session was chaired by Dr. Mathura Rai, Director, IIVR, Varanasi and co-chaired by Dr. M.L Chadha, Director, World Vegetable Center (formerly AVRDC), Regional Center, Hyderabad. It was attended by more than 50 participants involved in chilli research and development. The message of Dr. H. P. Singh, DDG (Hort.) ICAR, New Delhi was read out by Dr. Jagdish Singh, Principal Scientist, IIVR, Varanasi. Inaugurating the session, Dr. Mathura Rai, Director, IIVR, Varanasi highlighted the contribution of IIVR in research and development of vegetable as whole and chilli in particular. He stressed upon the need of putting more indent for the seeds by the seed sectors of state governments in order to increase the seed replacement by new and improved cultivars and thereby the productivity of the crop.

The technical session was chaired by Dr. M. L. Chadha, co-chaired by Dr. M. S. Dhaliwal and the rapporteurs were Dr. K. Madhavi Reddy and Dr. Rajesh Kumar. There were fourteen lead talks during the session on the following topics;

(*i*) Advances in chilli research: an overview

(*ii*) International scenario on research and development in chillies

(*iii*) Development of hybrids in chillies for higher yield and quality

(*iv*) Breeding for virus resistance in chillies

(*v*) Exploitation of male sterility systems in chillies

(*vi*) Indian Paprika: present status and future needs

(*vii*) Status of chillies research in Eastern India

(*viii*) Status of chilli research at NRC Seed Spices

(*ix*) Status of Capsicum research at IARI, Reg. Station, Katrain (*x*) Current trends in production technology of chillies

(*xi*) Pest management in chillies

(*xii*) Disease management in chillies

(*xiii*) Marketing strategies for chillies

(*xiv*) Role of private sectors in chilli improvement in India.

Dr. Mathura Rai, Director IIVR explained an overview on advances in chilli research in India and with particular reference to IIVR. Vast variability of chilli germplasm that is being prevalent in India and also about the land races was discussed. In north-eastern India a natural variability of *Capsicum baccatum*, *C. chinense*, *C. frutescens* and derivatives within *C. annuum* complex is prevalent. The number of germplasm lines of *Capsicum* sp. in different public institutions was highlighted. Priority areas of crop improvement that is being carried out at different institutes mainly about PGR management, development of varieties for different uses, breeding for biotic and abiotic stresses, development of varieties for oleoresin and value addition, development of mapping populations, development of GMS and CGMS lines by different institutes were discussed. He also discussed about the improvement of chilli under AICRP, since 1971 and till 2009 twenty OP varieties and ten hybrids of chilli and capsicum were released at national level. Under Network project on promotion of hybrid research in vegetables, research on chilli hybrid development was emphasized at IIHR, IIVR, IARI, PAU, TNAU, Kanpur since 1996 and the work continued under NATP project up to 2004. Six chilli hybrids were identified for release under this project *viz*. CH 1 and CH 3 (GMS based) from PAU, Ludhiana by Punjab State Varietal Release Committee; CCH 2 (IIVR), KCH 3 (Kanpur), MSH 172 (IIHR) and MSH 149 (IIHR) with CGMS background were identified for release by Central Varietal Release Committee at national level. At IIVR, Varanasi chilli lines *viz*., BS 35, GKC 29 and EC 497636 collected from north-east hill region were showing resistance to chilli leaf curl virus under artificial inoculation conditions. Chilli lines resistance to *Phytophthora* blight (CM 334, GKC 29 & IC 364063) and anthracnose fruit rot were identified. Nine sets of CGMS lines received from AVRDC, Taiwan are being maintained at IIVR. Male sterility derived from of *C. annuum* and *C. chacoense* cross is being transfered in *annuum* backgrounds. Recombinant inbred lines (RILs) were developed using California Wonder and LCA 235 to identify QTLs for pungency. Paprika type and pickle type chilli lines are being maintained at IIVR. Work on aflotoxins in dry fruits of chilli is being done.

Dr. M. L. Chadha, Director, Regional Station for South Asia, The World Vegetable Centre has given international scenario on research and development in chillies. He emphasized the nutritional value of hot pepper. They are rich in Provit A, Vit C, A & E, antioxidant, anticancerous and reduce fat. Key issues discussed were maintenance of genetic diversity, nutritive value, development of varieties resistant to diseases and pests and good agricultural practices. As seed germination is poor, seed priming with PEG, acetyl salicylic, $Na_3PO_4$ is recommended. Planting method technologies need to be widely tested and adopted properly. Adoption of polythene mulch, polyhouse/tunnel/plastic shelter cultivation with training was discussed. Major production constraints in chilli are diseases and pests and average loss due to diseases is 7–43% in India. Important points discussed were on water management, application of form and method of fertilizers, weed management, male sterility, molecular research and post-harvest management. The number of pepper accessions at AVRDC, Taiwan are 7855 including wild species. Resistant sources for major biotic and abiotic stresses are identified and are ready for distribution. A collaborative research program with IIHR, Bangalore on chilli improvement is going on.

Dr. K. Madhavi Reddy, Chilli Breeder, IIHR delivered two talks; (*i*) Development of hybrids in chilli for higher yield and quality, and (*ii*) Breeding for virus resistance in chilli. In the first talk, need for chilli hybrid development, methods of $F_1$ hybrid development in chilli and development of male sterility systems was discussed. Gaps in hybrid seed production of public bred hybrids were also discussed. Strong need for popularization of public hybrids was also emphasized. In the second talk, she discussed about major viruses *viz*., Chilli veinal mottle virus (ChiVMV) belonging to the potyvirus group, Cucumber mosaic virus (CMV) belonging to cucumovirus group, Chilli leaf curl virus (ChLCV) belonging to Begomovirus and Groundnut bud necrosis virus (GBNV) of Tospovirus causing major set back in chilli production in India. International and national status of research work done on breeding for these virus resistances was discussed, emphasizing on marker-assisted selection to speed up virus resistance breeding program. Approaches discussed to fill the gaps in virus resistance breeding program were; (*i*) Isolation and characterization (biological and molecular) of major viruses affecting chilli; development of reliable disease screening methodologies and confirmation of resistance by ELISA, PCR and nucleic acid probes, (*ii*) Being secondary centre of diversity for chilli, excellent biodiversity is prevalent in India. Systematic effort in identification of resistant sources will enable to identify newer sources of resistance from both cultivated and wild species, (*iii*) Cataloguing the isolates of CMV, ChiVMV, ChLCV and GBNV and resistance sources against existing severe isolates affecting chilli in different chilli growing areas in India, (*iv*) Identification of resistant genes against different isolates helps in pooling resistant genes for stable and durable resistance, and (*v*) Introgression of stable virus resistant genes from wild species through marker assisted selection will help in the development of useful pre-breeding lines, therefore broadening the genetic base.

Dr. M. S. Dhaliwal, Professor, PAU delivered talk on exploitation of male sterility systems in chillies. Explained about GMS and CGMS systems that have been commercially exploited for chilli hybrid seed production and the chilli hybrids developed using male sterile systems in public institutes were discussed. The GMS based chilli hybrids, *viz.* CH 1 and CH 3 from PAU, Ludhiana and CGMS based hybrids, *viz.* Arka Meghna, Arka Sweta, Arka Harita from IIHR, Bangalore and Kashi Surkh from IIVR, Varanasi have been developed by public institutes in India using male sterility systems. Dr. Rajesh Kumar, Senior Scientist, IIVR discussed about Indian Paprika: its present status and future needs. The chemical constituents and analytical methods for red colour development and estimation of color value in chilli were reviewed.

Dr. S. K. Samanta, Joint Director Research, BCKV, Kalyani presented status on chilli research in Eastern India. Genetic diversity of *Capsicum* sp. prevalent in eastern India was explained. Different types of chillies grown in eastern India are Assam local collections, Sukhia bullet, Jalapeno types, tomato chilli, Naga Jolokia, golden habanero, Dorsetnaga, Tezpur chilli, ornamental chilli types of *C. chinense* which are highly temperature sensitive and highly susceptible to tospovirus, Anaheim types, chilli types with viviparous germination etc. were discussed. Points emphasized during the talk were maintenance of diversity, collection and registration of valuable germplasm, development of CMS lines from *C. chinense* and *C. frutescens*, organic practices to be adopted for chilli production, low cost production techniques to be adopted, identification to temperature insensitive gene to incorporate into sweet pepper types, identification of thrips and mites resistant genes, etc.

Dr. B. B. Vashishtha, Director, NRC Seed Spices, Ajmer emphasized on local types of chilli being grown in Rajasthan such as Haripur, bell and Mathania types. In Rajasthan more than 70% of the crop is harvested as green chilli and also briefed about different popular food preparations of Rajasthan using local chilli types.

Dr. P. R. Kumar, Scientist, IARI Regional Station, Katrain discussed the status of capsicum (bell pepper) research at the station. Work on paprika and sweet pepper types suitable to higher altitudes is being emphasized at the station. Two bell pepper and three paprika lines tolerant to cold were identified.

Dr. L. B. Naik, Principal Scientist, IIHR, Bangalore talked on current trends in production technology of chillies. To overcome the bottlenecks/ gaps in chilli production advanced technologies for high productivity such as use of high yielding varieties/hybrids, use of quality seed, net/polyhouse production of chilli hybrid seeds, container raised seedling production, INM with use of micronutrients, water soluble fertilizer foliar sprays, use of biofertilizers, VAM, vermicompost, vermiwash, nutrient rich media, organic cultivation, efficient water management like micro-irrigation, fertigation, rainfed production systems, plastic mulching, herbicide use, use of plant growth hormones and chemicals to control flower and fruit drop and IPM practices were discussed.

Dr. A. B. Rai, Principal Scientist, IIVR emphasized on excessive use of pesticides by chilli growing farmers causing resurgence of pests. Excessive use of imidachloprid caused resurgence of mites in chilli crop. Histological anatomy of leaf infected with thrips and mites in comparison with healthy leaves was explained. Dr. T. K. Bag, Senior Scientist, IIVR talked on disease management in chillies.

Dr. Sameer Agrawal, Bejo Sheetal Seeds Pvt. Ltd., Jalna talked on role of private sectors in chilli improvement. Chilli has a very diverse segment as the preference varies with the region. Twenty six percent of chilli area is in Andhra Pradesh with 53% productivity with a record of 8t dry chilli yield/ha. Bottlenecks in chilli production in India are: chilli is traditionally grown crop, mostly local/own seeds are being used, low average yield and unorganized market across the country. Possibilities for good chilli production in future are cultivation of high yielding hybrids tolerant to different diseases, adoption of better agronomic practices to increase average yield and more accessibility to markets thus giving a fair price to the farmers. For international market chilli types/cultivars with high colour, jalapeno types, Korean types, Malaysian types, cayenne types, sweet pepper types besides segments like Habernero types are preferred. Favourable factors for wide cultivation of chilli in India are suitable climate, can be stored for two years and good international demand for Indian chilli.

Dr. Manju Vishwakarma, Ankur seeds, Nagpur also explained role of private sectors in chilli improvement. Chilli is a revenue generating crop for cultivators and traders. Only 5% of total chilli production is getting exported from India as against 30% export by China. Dominating varieties of south are G4 and for north and central India is Pusa Jwala. Bhivapuri (Maharashtra), Nalchatti (Madhya Pradesh), Naga Jolokia (North East), Byadagi (Karnataka) and Saripudi (South) are most popular local varieties of India being cultivated extensively. In 1992–94 the private sector had exploited the heterosis for productivity by introducing $F_1$ hybrid seed of chilli in the market. Due thought was given for plant habit, yield, pungency, colour persistence, storability, drying period, physical damages and transportability. Quality parameters of paprika type chilli were explained. $F_1$ cultivation of chilli increased to 20–25% in India with highest seed replacement ratio of 83.7%. Heterosis is as high as 45% both for quantitative and qualitative traits. Chilli drying is directly related to type of skin and moisture. Compact chillies with 65–75% moisture loss were introduced with excellent storage quality in cool and ambient conditions. Indian chilli have highest demand in European market with 25% share in global market. More number of private sector chilli $F_1$ hybrids are being cultivated in segmentised chilli growing regions in India.

To formulate the road map for advances in chilli research five groups based on the following theme areas were formed:

- Collection and characterization of chilli germplasm
- Crop improvement, molecular breeding including male sterility systems
- Crop production
- Crop protection
- Post-harvest and value addition.

Crop improvement group emphasized the need for disease/pest resistance, with special reference to anthracnose fruit rot, *Phytophthora* blight, *Fusarium* wilt, cucumber mosaic virus, chilli leaf curl virus, tospovirus, thrips and mites resistance. Breeding for quality traits such as high pungency, low pungency and high colour is also to be prioritized. There is need to diversify the male sterile systems by incorporating the disease resistance genes in suitable background. Marker assisted breeding to be emphasized for important traits.

## RECOMMENDATIONS

### Strategies for Chilli Improvement in India

1. Basic research in chilli and bell pepper to be taken up by public sector (ICAR institutes/SAUs)
2. Applied research can be done by both public and private sectors
3. Priority Areas
    - Diseases
        - Viral–PepLCV and CMV
        - Anthracnose
        - *Fusarium* wilt

        - *Phytophthora* spp.
    - Insects
        - Thrips and mites
    - Quality traits
        - Pungency (capsaicin)
        - Paprika type (less pungency & high colour-high oleoresin)
    - Male sterility
        - Identification and transfer of morphological molecular markers to identify GMS lines at seedling stage
        - Identification of molecular markers for *Rf* gene to identfy A and C lines
        - Broadening genetic base of GMS system to get high heterosis
        - Incorporating disease resistance gene in CMS lines
        - Development of stable GMS and CMS systems
        - Transfer of GMS system in different backgrounds
- Molecular breeding
    - Development of mapping population for QTL analysis for important traits like pungency, paprika oleoresin, and plant and fruit morphological traits.

4. Production of quality seeds and its distribution to the farmers
    - Frontline demonstration
    - Organization of training and awareness programmes to the farmers
    - Development of bio-intensive module for pest management in chillies

5. Validation of technologies before dissemination.

# Appendix-II

## Agro-climatic Zones of India

| Agro climatic zones | States |
|---|---|
| I. Humid western Himalayan region | Jammu & Kashmir, Himachal Pradesh and Uttaranchal |
| II. Humid Bengal-Assam basin | West Bengal and Assam |
| III. Humid eastern Himalayan and bay islands | Sikkim, Meghalaya, Manipur, Nagaland, Mizoram, Tripura, Arunanchal Pradesh and Andaman & Nicobar Islands |
| IV. Sub-humid Sutlej Ganga Alluvial plains | Punjab, Uttar Pradesh and Bihar |
| V. Sub-humid eastern and south eastern uplands | Chhatissgarh, Orissa and Andhra Pradesh |
| VI. Arid western plain | Rajasthan, Gujarat, Haryana and Delhi |
| VII. Semi arid plateau and central highlands | Madhya Pradesh and Maharastra |
| VIII. Humid to semi arid western ghats and Karnataka plateau | Karnataka, Tamil Nadu and Kerala |

# Appendix-III

**ACRONYMS**

| | |
|---|---|
| Ac/Ds | Acivator/Dissociator |
| AFLP | Amplified Fragment Length Polymorphism |
| AICRP-VC | All India Coordinated Research Project-Vegetable Crops |
| ANGRAU | Acharya N G Ranga Agricultural University |
| AOA | Antioxidant activity |
| AsA | Ascorbic acid |
| ASTA | American Spice Trade Association |
| AVRDC | Asian Vegetable Research and Development Centre |
| BA | Benzyl adenine |
| BAC | Bacterial Artificial Chromosome |
| BBWV | Broad bean wilt virus |
| CaCV | Capsicum chlorosis virus |
| CAPS | Cleaved Amplified Polymorphic Sequence |
| CCS | Capsanthin-capsorubin synthase |
| cDNA | Complementary Deoxy Ribose Nucleic Acid |
| CdTV | Chino del tomate virus |
| CGMS | Cytoplasmic genetic male sterility |
| ChiVMV | Chilli veinal mottle virus |
| CMMV | Chilli mottle mosaic virus |
| CMS | Cytoplasmic male sterility |
| CMV | Cucumber mosaic virus |

| | |
|---|---|
| CP | Coat protein |
| CSAUA&T | Chandra Shekhar Azad University of Agriculture & Technology |
| CVMV | Chilli veinal mottle virus |
| DAA | Days after anthesis |
| DAT | Days after transplanting |
| DH | Doubled haploid |
| DNA | Deoxy Ribose Nucleic Acid |
| EC | Exotic collection |
| ELISA | Enzyme Linked Immuno Sorbant Assay |
| EOA | Essential Oil Association of America |
| FAO | Food and Agriculture Organization |
| FYM | Farm yard manure |
| GA | Gibberalic acid |
| GAP | Good agricultural practices |
| GAU | Gujarat Agricultural University |
| GBNV | Groundnut bud necrosis virus |
| GBPUA&T | Govind Ballabh Pant University of Agriculture & Technology |
| GCMS | Gas chromatography mass spectrophotometry |
| GMS | Genetic male sterility |
| HDPE | High density polyethylene |
| HPLC | High performance liquid chromatography |
| HPR | Host plant resistance |
| IAHS | Indo-American hybrid seeds |
| IARI | Indian Agricultural Research Institute |
| IARI-RS | Indian Agricultural Research Institute-Regional Station |

| | |
|---|---|
| IBA | Indole butyric acid |
| IC | Indigenous collection |
| ICAR | Indian Council of Agricultural Research |
| ICPM | Integrated crop and pest management |
| ICPN | International Chilli Pepper Nursery |
| IIHR | Indian Institute of Horticultural Research |
| IIVR | Indian Institute of Vegetable Research |
| INSV | Impatiens necrotic spot virus |
| INTHOPE | International hot pepper trial |
| IPM | Integrated pest management |
| IPR | Intellectual Property Rights |
| ISPN | International Sweet Pepper Nursery |
| ISR | Induced systemic resistance |
| ITC | Indian Tobacco Corporation |
| ITS | Internal transcribed spacer |
| JNKVV | Jawahar Lal Nehru Krishi Vishwavidyalaya |
| KAU | Kerala Agricultural University |
| LC-MS-MS | Liquid chromatography-mass spectrophotometry |
| LCV | Leaf curl virus |
| LOD | Log of odds |
| MAHYCO | Maharashtra hybrid seed company |
| MAS | Marker assisted selection |
| MH | Maliec hydrazide |
| MPKV | Mahatma Phule Krishi Vishwavidyalaya |
| MPP | Metallised polyester polyethylene |
| MRL | Maximum residual limit |

| | |
|---|---|
| NAA | Naphthalene acetic acid |
| NAAS | National Academy of Agricultural Sciences |
| NAG | National active germplasm |
| NARS | National Agriculture Research Systems |
| NBPGR | National Bureau of Plant Genetic Resources |
| NCP | Natural cross-pollination |
| NCR | Non-coding region |
| NEH | North eastern hill |
| NGOs | Non-Governmental Organizations |
| NPK | Nitrogen phosphorous potassium |
| NUE | Nitrogen use efficiency |
| OP | Open pollinated |
| OUA&T | Orissa University of Agricultural & Technology |
| PAP-SSR | Plastid-lipid-associated protein-simple sequence repeats |
| PAU | Punjab Agricultural University |
| PBNV | Peanut bud necrosis virus |
| PCR | Polymerase chain reaction |
| PEG | Poly ethylene glycol |
| PepLCV | Pepper leaf curl virus |
| PeSMV | Pepper severe mosaic virus |
| PG | Polygalacturonase |
| PGPR | Plant growth promoting rhizobacteria |
| PI | Plant introduction |
| PMMV | Pepper mild mottle virus |
| ppm | Parts per million |
| PVMV | Pepper veinal mottle virus |

| | |
|---|---|
| PVX | Potato virus X |
| PVY | Potato virus Y |
| QTL | Quantitative trait loci |
| RAPD | Randomly Amplified Polymorphic DNA |
| RARS | Regional Agricultural Research Station |
| RFLP | Restriction Fragment Length Polymorphism |
| RIL | Recombinant inbred line |
| RNA | Ribose nucleic acid |
| RT-PCR | Real time PCR |
| SA | Salicylic acid |
| SCA | Specific combining ability |
| SCAR | Sequence characterized amplified region |
| SGN | Solanaceae genome network |
| SHU | Scoville heat units |
| SKUA&T | Sher-e-Kashmir University of Agriculture & Technology |
| SNP | Single nucleotide polymorphism |
| SSAP | Sequence-specific amplification polymorphism |
| SSR | Simple sequence repeat |
| SST | Starter solution technology |
| TEV | Tobacco etch virus |
| TLC | Thin layer chromatography |
| TLCV | Tobacco leaf curl virus |
| TMV | Tobacco mosaic virus |
| TNAU | Tamil Nadu Agricultural University |
| ToMV | Tomato mosaic virus |
| TSS | Total soluble solids |

| | |
|---|---|
| TSWV | Tomato spotted wilt virus |
| TYLCV | Tomato yellow leaf curl virus |
| UAS | University of Agricultural Sciences |
| UNDP | United Nations Development Programme |
| USDA | United States Department of Agriculture |
| UV | Ultra violet |
| VAM | Vesicular-Arbuscular Mycorrhizae |
| VIGS | Virus-induced gene silencing |
| WAT | Week after transplanting |
| WFT | White fly mediated transfer |
| WSMV | Watermelon silver mottle virus |
| WUE | Water use efficiency |
| WVC | The world Vegetable Centre |
| YAC | Yeast Artificial Chromosome |

# Subject Index

## D

## I



## K



## L



## M

## N



## O



## P

## Q



## R

## S



## T



## U



## V

**W**



**Y**